Environmental Physiology of Marine Animals

"Oh, what is abroad in the marsh and the terminal sea?"
From: The Marshes of Glynn, by Sidney Lanier.

Environmental Physiology of Marine Animals

Winona B. Vernberg
and
F. John Vernberg

Springer-Verlag
New York • Heidelberg • Berlin
1972

Winona B. Vernberg
Research Professor of Biology
Belle W. Baruch
Coastal Research Institute
University of South Carolina

F. John Vernberg
Belle W. Baruch Professor
of Marine Ecology
Belle W. Baruch
Coastal Research Institute
University of South Carolina

© 1972 by Springer-Verlag New York Inc.
Library of Congress Catalog Card Number 70-183485.

Printed in the United States of America.

ISBN 0-387-05721-8 Springer-Verlag New York • Heidelberg • Berlin
ISBN 3-540-05721-8 Springer-Verlag Berlin • Heidelberg • New York

To Nicholas Scott Lewis,
a new generation

Preface

Within recent years man has become increasingly aware of the disastrous environmental changes that he has introduced, and therefore society is now more concerned about understanding the adaptations organisms have evolved in order to survive and flourish in their environment. Because much of the information pertaining to this subject is scattered in various journals or special symposia proceedings, our purpose in writing this book is to bring together in a college- and graduate-student text the principal concepts of the environmental physiology of the animals that inhabit one of the major realms of the earth, the sea.

Our book is not meant to be a definitive treatise on the physiological adaptation of the animals that inhabit the marine environment. Instead, we have tried to highlight some of the physiological mechanisms through which these animals have been able to meet the challenges of their environment. We have not written this book for any one particular scientific discipline; rather, we hope that it will have an interdisciplinary appeal. It is meant to be both a reference text and a text for teaching senior undergraduate and graduate courses in marine biology, physiological ecology of marine animals, and environmental physiology of marine animals.

We are indebted to our students and colleagues present and past, who have helped formulate some of our ideas. We are grateful to the following people for critical reading of various parts of the manuscript: Mr. John Baptist and Drs. Anna Ruth Brummett, Richard Dame, Rezneat Darnell, Charlotte Mangum, Ernest Naylor, A. N. Sastry, Carl Schlieper, S. E. Woods, Ms. Barbara Caldwell, Ms. Cary Clark, and Ms. Susan Ivestor. We are especially indebted to Dr. P. J. DeCoursey for reading most of the manuscript and giving constant encouragement, Ms. Susan Counts for typing the manuscript, Ms. Vicky Macintyre for invaluable editorial assistance, and Mr. Gary Anderson for original art work. We also gratefully acknowledge assistance on the home front from Mr. Eric Vernberg and from the Misses Amy Vernberg, Melissa Holmes, and Bess Roberts.

January 5, 1972

Columbia, South Carolina

Winona B. Vernberg

F. John Vernberg

Contents

The fiddler crab, *Uca pugilator,* one of the characteristic inhabitants in coastal marine environments. (Courtesy of Coates Crewe).

CHAPTER I

THE ORGANISM

Man has always had a vital and continuing interest in the sea. Coastal waters were an early source of food and means of travel, but the extent of the oceans was not recognized until the explorations of the 14th and 15th centuries. Another five centuries elapsed before man had the technology to truly begin underwater exploration.

Modern interest in the marine environment is based mainly on its scientific and economic value. Competition for the natural resources of the sea is already great in some parts of the world. Estuaries, for example, are multipurpose regions, serving as centers for living, industrial development, shipping, fishing, and recreation. Full utilization of the resources of the open ocean has not yet been realized, but the demand and competition for these resources are destined to increase rapidly within the next few years.

Until recently, the resources of the sea were assumed to be almost limitless. Now, of course, we know this assumption is false. In some areas of the world, estuaries already have been exploited to such an extent that they are biologically dead; hence, they are of limited use to man. Effective utilization of the great natural marine resources depends on proper management and manipulation based on scientific fact. The outcome of environmental manipulation is difficult to predict, mainly because little is known of the functional responses of marine animals to an altered environment. The principal goal of this book is to emphasize this one important aspect of the broad subject of marine science.

Rather than present a compendium of published papers on the physiology of marine animals, we propose to review the existing literature and present principles that may serve as a framework for present-day students. We hope that future generations of environmentalists will both alter and add to this framework. Certain physiological phases of marine organisms have been intensively studied, others scarcely touched. Many basic physiological processes common to all animals have been best studied using marine organisms, but these investigations have not been strongly oriented to answering environmental questions; hence, they are only of tangential interest to the theme of this book. Our theme is to emphasize the functional interrelationships of the component parts of an organism in meeting the stress of environmental fluctuation. To us the intact organism is the basic unit of biological organization, but at the same time we recognize that the various component physiological systems play a vital role in maintaining the integrity of the organism.

The organism is a discrete, easily recognized unit. It is a dynamic and energetic system, consisting of many functional subunits finely tuned to interact harmoniously and thus insure survival of the total system. To the organism in a harsh environmental complex, mere survival is the minimal level of performance; but for the perpetuation of the species, more highly integrated processes leading to reproduction must be operational. Irrespective of the levels of performance demanded of an organism, a constant interchange of energy takes place between the organism and its ambient environment. The sum total of the energetics of the organism is called *metabolism*. Various functional subunits of metabolism have been recognized, and these play an important role in the integrative functioning of the intact organism. The operation of the subunits requires a specific form of energy, each of which may in turn release energy in a different form. Specialization of cells into tissues and organ systems occurred as organisms evolved from unicellular to multicellular units. Although the degree of complexity in these organ systems is exceedingly diverse throughout the animal kingdom, certain principles are fundamental to all of these variant forms. In this chapter, physiological similarities will be stressed; later sections will describe modifications that relate organismic adaptation to a given set of environmental conditions characteristic of the organism's habitat.

To be functional, the organism must have an energy input followed by effective utilization of this energy. Over an extended period of time, energy input must equal energy utilization if an organism is to maintain itself. For brief periods of time, an organism can rely on stored energy reserves, which results in less energy input than energy utilization. This ability to operate at a negative level varies from species to species. Without food some animals will die quickly, but others can survive for years. When excess energy is available, the animal may either undergo a rapid period of growth and/or store the excess energy for future use.

Maintenance of a relatively balanced energy budget demands the integration of a number of processes. Energy input into a system requires that the organism be able to perceive the environment in terms of seeking and capturing food. Once food has been located, the animal must be able to go through the process of ingestion, digestion, and assimilation, circulation of the assimilated food to the body cells, and egestion of unusable portions. The energy that has been gained then must be released by the cells for utilization in active processes or stockpiled for later use. Most animals require oxygen for the release of this energy. During these metabolic processes, wastes are produced which must be removed, processed, and eliminated into the external environment. Other processes, including locomotion, mating activities, and production of gametes, also make demands on the available energy. In addition, energy is required for the chemical and nervous coordination of all of these activities necessary to sustain life.

A. METABOLIC SUBUNITS

1. Perception of the Environment

Although all cells are capable of performing basic functional processes consistent with their existence, mechanisms have evolved which permit all parts of multicellular organisms to function as an integrated unit. The degree of organismic integration varies from a more or less loose clump of cells to multicellular organisms consisting of tissues which have become specialized to perform certain functions. In general, both nervous and chemical controls are involved in the coordination of multicellular marine animals.

a. Nervous Coordination—Sensory Modalities

A highly coordinated organism has four main components: information-gathering modalities; methods of transmitting this information either to integrative centers or directly to effector organs; coordination centers; and effector structures.

Organisms must not only be sensitive to changes in their external environment in order to survive, but they must also have operative internal information-gathering mechanisms. Various forms of *exteroceptive organs* have evolved that permit the organism to perceive the external environment, and the relative role of each in the success of an organism is correlated in part with the organism's ecology. Laverack (1968) has extensively reviewed the sensory organs of marine animals. Internal receptors which perceive changes within an organism also show diversity in form, but relatively little is known about them in marine animals.

(1) Vision

In the marine environment, the most commonly used sensory modalities are those for light reception, phonoreception, chemoreception, and those largely undefined organs for the perception of gravity and pressure. Light receptors are common to almost all animals. These receptors may be dermal light receptors or well-defined organs, such as the vertebrate eye, the compound eyes of crustacea and insects, the simple eyes or ocelli of other arthropods and invertebrates. Only a small fraction of the total spectrum of radiation is perceived as light, and the visible spectrum of all light is limited to wavelengths between about 300 and 700 mμ. Photosensitivity is diffuse with dermal light receptors, whereas the more complex optic receptors can perceive qualitative, temporal, and spatial qualities of light. Sensitivity is several thousand times less in the dermal light receptors than in the more well-developed eyes, but it is thought that the photochemical systems responsible for all light reactions are similar.

Dermal light receptors are widespread throughout the animal kingdom, but they are most common in aquatic invertebrates having a simple epithelium or superficial nervous system. After reviewing the work on dermal light receptors, Steven (1963) concluded that the light response probably results from accumulation of small amounts of photosensitive substances in the cytoplasm.

There are three major types of responses in animals caused by the dermal light sense: orientation by bending the body; the shadow reflex or withdrawal of exposed parts of the body in response to a sudden change in illumination; and environment selection. In animals with well-developed visual organs, reception of light is used in environment selection, predator-prey relationships, and close-range communication.

(2) Audition

Animals are able to detect vibrations in the surrounding medium and the substrates by means of many different structures. The perception of vibrations may enable an organism to distinguish between frequencies, to measure relative intensity, and to determine the direction of sound. The term *phonoreception* has been suggested to designate reception of sound, irrespective of the type of receptor that is stimulated or the medium through which the sound reaches the body—be it solid, liquid, or air (Frings, 1964; Frings and Frings, 1967).

Sound receptor systems may be one of two types. One type, which is sensitive to molecular displacement, consists of a membrane or a hair suspended so that it responds to slight movements of the medium. Sound waves must have free access to it from all directions if it is a membrane. The other type of receptor is based on a pressure-sensitive system. In this type of system a heavy membrane is shielded on its inner surface from sound waves, and the inner chamber is kept at some stable reference pressure that is usually equal to the atmosphere. The stimulus is registered directly by displacement receptors; pressure receptors register only indirectly.

Many invertebrates will respond to substrate vibrations. A number of potential sound receptors in invertebrates have been described, including innervated projections from the body wall, sensillae in the body wall itself, proprioceptors located deep in the muscles or external organs, and specialized sensillae (Frings, 1964). Only some arthropods and vertebrates have an ability to detect airborne sound as well. In vertebrates, the ear functions as a pressure receptor and is the basic organ for sound perception. In aquatic mammals, the ear responds to pressure change in the surrounding medium. Fish that have a gas-filled swim bladder connected to the inner ear also respond in this manner. The lateral line organ of fish is a displacement receptor and enables the fish to perceive low-pitched vibrations. This organ is used as a "distant" touch receptor to detect and locate moving animals. Such a receptor can be very important for the orientation of fish in space, especially for those in murky waters. Animals respond to a very wide range of frequencies. Some fish perceive frequencies from below 100 Hz up to 3,000 Hz. In contrast the range of frequencies in man is from 20–20,000 Hz. Phonoreception functions in exploration of the environment, in finding food, in avoiding danger, in reproductive behavior, and in communication.

(3) Chemoreception

All living animals are able to perceive the chemical characteristics of the environment. There are three main types of chemoreceptors, classified according to relative sensitivity and the location of receptors on the body surface. The "general chemical sense," which may extend over large parts of the body, is the least sensitive of the receptors. Mediated by free nerve endings and by rarely identified, unspecialized receptors it is characteristic of both invertebrates and vertebrates. Usually the response to this sense is a simple avoidance reaction. The other types of chemoreception, taste and smell, are more varied responses. Taste requires contact between relatively large quantities of water-soluble substances and specialized taste receptors. *Olfaction*, or the sense of smell, responds to low chemical concentrations and is the most sensitive of the chemoreceptors. The chemical senses are used in selecting an environment, finding food, detecting enemies, and locating a mate.

(4) Perception of gravity

Gravity, one of the most constant mechanical forces in the environment, does not vary in either the direction of pull or the intensity in any given location. Gravity is used by animals as the basic plane of reference enabling them to maintain a definite attitude in space, and they may also have an orientation response to or from the gravitational force. The sensory equipment that allows perception of gravity basically consists of some free or attached structure that responds to gravity and, by a shearing force or pressure, acts upon sensory cells that provide information to the animals. Statocysts, which are present in animals from coelenterates to man, are one of the most common gravity detection devices. Usually these statocysts consist of a fluid-filled vessel lined with sensory hairs and containing a statolith or statoliths, which may be either grains of sand or secreted substances. Gravity is detected through the vestibular labyrinth of the inner ear in most vertebrates.

(5) Perception of pressure

The nature of the receptor organs in aquatic animals for perceiving pressure changes is not known, but many planktonic animals are sensitive to small changes in hydrostatic pressures—often to changes equivalent to less than a meter of water. Some organisms are restricted to a limited range of depth, while others can move relatively freely from one level to another. Pressure effects and responses in marine animals have been reviewed by Morgan and Knight-Jones (1966).

The conduction of nerve impulses to effector organs and/or to integrative centers involves nervous transmission, ranging in complexity from simple cell to cell (non-nervous) conduction to complex nervous systems. Although this extremely important aspect of coordination is influenced by the environment, the reader is referred to other references for detailed accounts (Bullock and Horridge, 1965). Coordination of the various kinds of information presented to an organism permits it to be more than a passive responder to every stimulus. Marine organisms with highly developed central nervous systems generally have behavior patterns which are complex, varied, and capable of being modified by past experiences.

Effector organs bring about responses to stimuli which reach them from coordination centers and/or a receptor organ. Thus, these organs respond to information from both the external and the internal environment. The various effector organs range in complexity from a ciliated cell to light-generating organs. The degree of complexity and the type of effector organs found in an organism appear to be an adaptation to a particular ecological requirement.

b. Chemical Coordination

Although all chemical substances must potentially exert some regulative effect on the organism, many species have evolved special glands (endocrine glands) which produce hormones effective in coordination of physiological processes. In addition, specialized cells (neurosecretory cells) of the central nervous system produce hormones which travel along the axons to specialized regions for storage until they are subsequently released. For a discussion of hormone action, consult a text in endocrinology.

2. Feeding

The bioenergetics of an animal is completely dependent on an input of energy and chemical necessities in the form of food. The "taking-in" of food involves separate processes: (1) perception and (2) capture and ingestion of food. Perception of food depends upon the ability of the organism to be stimulated by some external environmental cue. This cue may be a chemical entity, or it may have a physical basis, such as a visual or tactile stimulus. The relative dependence of feeding on specific physical and/or chemical mechanisms is associated with the ecology and mode of existence of the species. Animals living in clear and open areas (for example, mobile species in the intertidal zone) may use vision as the primary means for locating food. In poorly lighted habitats, audition, chemoreception, or thermal sense may be more useful. In many environments the animal is apt to rely on chemoreception if food must be located over long distances. In some animals the process of feeding requires more than an external cue. For example, food may be continuously present, but the animal feeds only intermittently; thus, in these animals feeding is dependent on internal factors, such as different thresholds of hunger.

Mechanisms for the capture and ingestion of food can be correlated with the niche an animal occupies. This relationship was clearly shown in the earlier classification of the feeding mechanism proposed by Yonge (1928), which was not based on phylogenetic relationships, but instead reflected the ecology of the organism.

Three main categories of feeding were proposed based on particle size and type of food: small particles; large particles or masses; and soft tissues or fluids. Detailed descriptions of these categories may be found in various textbooks and references (Jennings, 1970; Jørgensen, 1966; Yonge, 1928).

To be of energetic value to the organism and to provide the basic molecules needed for synthesis and growth, ingested food must be broken into smaller chemical units by digestive enzymes. Three general types of digestive enzymes are recognized: lipases, proteases, and carbohydrases, each acting on one of the three main types of food. The composition and relative activity of these types of enzymes can be correlated with the ecology and food requirements of animals. In general, herbivores have more powerful carbohydrases and weak proteases. The reverse is found in carnivores. In omnivores all three types of enzymes are secreted, and all are approximately of equal importance. The composition and activity of digestive enzymes are also influenced by external environmental factors, such as temperature and type of food. The size and shape of the digestive tract can also be correlated with an animal's environmental demands (Yonge, 1937).

3. Respiration

Energy taken into the body in food molecules must be released within the organism as free energy if it is to be functionally important to an organism for cellular activities. The release of free energy occurs through oxidation-reduction reactions. Classically, chemists thought oxygen was needed for oxidation, but since oxidation is defined as the transfer of an electron from one molecule to another, and reduction is the electron acceptance by a molecule, oxygen is not necessary for this process.

a. Metabolic pathways

Within a cell the total free energy available from a food molecule, such as glycogen, is not released all at once. Instead, the complete breakdown of food molecules is a step-like series of reactions; specific catabolic pathways have been described for various compounds. The relatively simple components that result may be used for various synthetic processes. Again, specific anabolic pathways exist for the synthesis of specific complex cellular constituents.

Detailed analyses of various metabolic pathways are presented in various biochemical textbooks and are beyond the scope of this book. However, exposure to certain general concepts is necessary for understanding pertinent biochemical studies involving marine organisms. Catabolic pathways begin with well-defined complex food substances which are altered into various intermediates, while anabolic pathways begin with these various intermediates

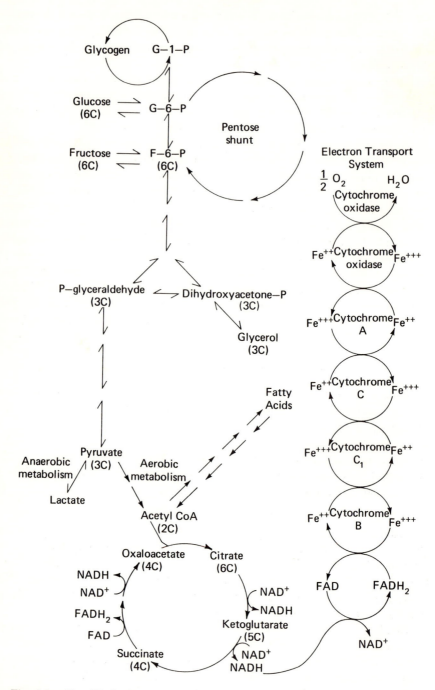

Fig. 1-1. Simplified scheme of one of the principal metabolic pathways involving carbohydrates.

which ultimately are synthesized into distinct end products. One of the principal pathways involving carbohydrates is presented in a simplified scheme in Fig. 1.1. At each step a small amount of energy is released that may be stored in high energy phosphate bond compounds, such as ATP. Whenever oxygen is present, it serves as the terminal hydrogen and electron acceptor.

b. Metabolic diversity

Oxygen utilization, CO_2 production, and utilization of foodstuffs are measurements of the energy metabolism of cells and the intact organism. Respiration is that phase of the metabolic process involved in exchanges of oxygen and carbon dioxide between an organism (or isolated cells) and its external environment. Primitive acellular organisms carry on gaseous exchange across body membranes. However, with increased size, oxygen needs are not adequately supplied by this process. It has been estimated that diffusion is an unsatisfactory mode of supplying oxygen to an active organism if the distance from body surface to a cell is greater than 0.5 mm. Basically the exchange of respiratory gases requires two conditions: a thin, moist membrane or body covering through which respiratory gases can diffuse; and a concentration gradient of gases whereby the ambient oxygen content is higher, and the CO_2 is lower, than that within the organism. Diversity in gaseous exchange mechanisms among marine organisms has proceeded along a number of major lines: (1) Specialized regions of the body modified for gaseous exchange. Crustacea and fish, for example, have external or internal gills, and many polychaetes have parapodia modified for gaseous exchange. In some animals, these regions are retractable or internal and thus are well protected. (2) Maintenance of an oxygen gradient across a diffusible membrane. This can be accomplished by the animal's swimming to new, relatively oxygen-rich regions, by moving the water at the surface of the respiratory membrane, by ciliary action, or by moving a respiratory structure, such as the waving of parapodia or tentacles. (3) Internal structural changes. Increased vascularization in regions of a respiratory membrane, or an improved circulatory system to remove oxygen dissolved in the body fluid from the body surface to the cells, including development of better circulation pathways and a pulsating organ, the heart. (4) Oxygen-carrying compounds. A number of chemical substances which have an affinity for oxygen have evolved. These substances, respiratory pigments, exhibit qualitative and quantitative differences that can be correlated with habitat preferences. (5) Behavior patterns, including swimming to areas of sufficient oxygen content, decreasing locomotor activity, or closing of a shell or protective plates during low oxygen stress. (6) Physiological adaptations, including shifts in metabolic pathways and changes in rate of heart beat or pumping rates.

c. Factors influencing respiration rate

The rate at which organisms consume oxygen is modified by many factors, some environmental, others intra-organismic, and not all animals are influenced in the same manner. Therefore, it is difficult to generalize, and the effects of these factors must be determined for each species. In this chapter, only general terms will be introduced; specific examples will be given later.

(1) Extrinsic factors

(a) *Oxygen tension:* On the basis of two generalized responses to external oxygen tension, organisms may be classified either as: conformers, which have oxygen consumption rates that are dependent on oxygen tension; or regulators, which have oxygen consumption rates that are constant over a wide range of oxygen tension (Fig. 1.2). Typically, at a critical oxygen tension (Pc),

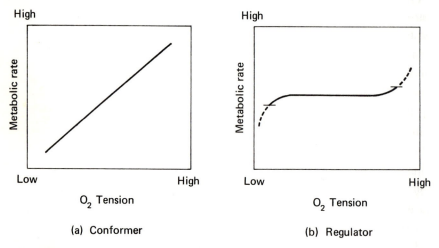

Fig. 1-2. Two generalized metabolic responses to variable external oxygen tension: (a) conformer, the metabolic rate varies directly with oxygen tension; (b) regulator, the metabolic rate is constant over a wide range of oxygen tension.

the oxygen independent rate shifts to become oxygen dependent. The Pc value may vary with both intra-organismic and environmental factors.

During periods of exposure to low oxygen tensions, the organism may be insured of adequate oxygen uptake by pumping more of the ambient medium over its respiratory surface. This is done by increasing the frequency of respiratory movements and/or increasing the volume of water (or air) pumped. Some of the overt expressions of respiratory movements include opercular movements in fish, pumping movements of worms which live in burrows, pleopod beating in many crustacea, and tentacular movements in certain worms. In some organisms increased CO_2 levels are more stimulatory in eliciting these responses than decreased O_2 content.

(b) *Temperature:* As in chemical reactions, an increase in temperature typically is accompanied by an increase in metabolic rate in poikilotherms. Two principal methods exist for expressing the influence of temperature on reactions as a coefficient. One method is termed Van't Hoff's equation, and the temperature coefficient is called the Q_{10}. Typically, for every $10°C$ increase in temperature the rate of a chemical reaction will increase by a factor of 2 to 3 ($Q_{10} = 2$ to 3). A Q_{10} value of less than 2, or more than 3, indicates some process other than a chemical one is involved, e.g., a change in permeability. A Q_{10} of 1 indicates temperature insensitivity. The other principal method of expressing the influence of temperature on reactions is the Arrenhius equation, and the temperature coefficient is designated as μ. The μ value of the Arrenhius equation reflects the amount of energy needed to activate a specific thermochemical reaction. A change in a μ value at a given temperature indicates the activation of a different biochemical reaction. A change in this value over a thermal range indicates a shift in the basic control mechanism(s).

(c) *Other factors:* Factors, such as salinity, light, food, CO_2 and cyclic changes in environmental parameters acting independently or in various combinations, also may influence the rate of oxygen utilization.

(2) *Intrinsic factors*

In general, small organisms consume more oxygen per unit time and weight than large ones (Hemmingsen, 1960). This trend has been reported both intra- and interspecifically (Vernberg, 1959). The relationship of size to metabolic rate is commonly expressed as a power function of body size. Results are written in two general ways. For total metabolism (amount of oxygen consumed per unit time) the equation is:

$O_2 = aW^b$ or $\log O_2 = \log a + b \log W$. For weight-specific metabolism (amount of oxygen consumed per unit time and unit weight), the formula is:

$$\frac{O_2}{W} = aW^{(b-1)} \text{ or } \log \frac{O_2}{W} = \log a + (b\text{-}1) \log W.$$

In these equations, O_2 is the oxygen consumed per unit time; W is the weight of animal, which may be expressed as wet weight, dry weight, nitrogen, etc.; and a and b are coefficients, *a* representing the intercept of the y-axis and *b*, the slope of the function in the logarithmic plot. Respiratory rates are also influenced by other factors, such as locomotor activity, starvation, sex, stage of life cycle, reproductive stage, and molting stage.

Oxygen consumption values are determined under different experimental conditions, and specific terminology is associated with these conditions. *Standard metabolism* is the rate of oxygen uptake when an animal is maintained under a defined state of activity. During periods of controlled, heightened activity of an organism (such as swimming or flying), the oxygen consumption rate is termed *activity metabolism*. *Basal metabolism* refers to the rate of oxygen utilization in an inactive organism and reflects the minimal metabolic rate consistent with life.

4. Circulation

a. Diversity in circulation

In unicellular and simple multicellular organisms, diffusion can supply to the cells the raw materials necessary for them to sustain metabolism and remove metabolic wastes. In more complex multicellular organisms, however, certain functional strategies have evolved which permit efficient circulation of body fluids to the inner cells. An inspection of existing animal groups reveals tremendous diversity in the types of body fluids and in the complexity of circulatory systems. Functionally, there are certain similarities in these groups that indicate the basic role of circulating fluids; dissimilarities may be correlated with unique habitat requirements and specialized functions.

The basic function of circulating fluids is twofold: the transport of substances throughout the body, including nutrients, oxygen, and various regulatory substances; and the removal from the cells of CO_2, metabolic wastes, and cell products. The circulating fluids can be very complex in composition; in addition to supplying cellular needs, these fluids may carry clotting substances, antigens, antibodies, and blood cells. In some species the total osmotic concentration and the specific ionic composition of circulating fluids are carefully controlled. The body fluid tends to give shape to some organisms and also aids in body movements, especially among the soft-bodied invertebrates.

Intracellular circulation of material is described in various cellular textbooks; in this book we will be concerned with the transport of substances to and from cells. Sea water is in close contact with various parts of the body in simpler marine metazoans—for example, sponges move sea water through a system of channels and chambers by flagellar action. Cnidarians offer another variation on this theme in that their cells are in close contact with sea water either through external body layers or through the lining of a convoluted gastrovascular cavity. There is no apparent vascular system that reaches all parts of the body, and diffusion seems to be the principal mechanism of intercellular transport. Generally, metabolic activity is low in animals that lack any type of vascular system.

There are two general types of circulatory systems, open and closed. The closed circulatory system consists of a network of interconnected blood vessels through which blood is pumped by a muscular structure or structures. Usually three types of vessels are recognized: arteries that carry blood away from the heart; veins that carry blood to the heart; and capillaries that are found between arteries and veins. Although the diameter of capillaries is small and their walls thin, it is here that exchange between the vascular fluid and the tissue fluid takes place. Generally, many capillaries are found together, forming a capillary bed. Since the capillary wall is semipermeable, the composition of blood and the tissue fluid differ. In contrast, the open circulatory system does not have a capillary connection between large blood vessels, and the vascular fluid comes into direct contact with tissue cells.

The successful invasion of organisms into diverse marine habitats often is accompanied by cardiovascular adaptations, e.g., rate of beating of the heart or some other muscular structure, volume and composition of the circulatory fluid, blood pressure, or use of vascular and other fluids as a hydraulic means for locomotion.

b. Respiratory pigments

Many organisms have respiratory pigments which aid in providing oxygen for cellular metabolism. Distributed in various parts of an organism, these pigments function to combine reversibly with oxygen. As the pigments pass near a respiratory surface where the external oxygen concentration is higher than that of the body fluids, they form a loose chemical combination with oxygen. Subsequently the oxygen may be released when the pigments reach regions of low tension, such as in tissues and cells. Respiratory pigments may also serve in oxygen storage or as a buffer to control any harmful effects of oxygen.

Several types of respiratory pigments have been described, and excellent reviews of studies on the chemical and physical nature of respiratory pigments are found in Ghiretti (1968) and Manwell (1960). We will be more concerned in this book with the functional significance of these substances in relation to the environmental conditions in which an organism lives.

Various criteria have been used to indicate the physiological significance of respiratory pigments, including: (1) determining whether the amount of oxygen carried by a body fluid (the oxygen capacity) is significantly higher when the respiratory pigment is present, and if so, whether it is sufficient to meet the oxygen demands of the animal; (2) estimating the quantity of respiratory pigment per unit body fluid; (3) ascertaining if the oxygen equilibrium curve is operative within the environmental conditions the organism normally encounters; and (4) observing the effect on the animal's oxygen consumption rate when the pigment has been inactivated by various agents, including CO.

The oxygen capacity, expressed as the number of volumes of oxygen per 100 ml of fluid, for air-saturated sea water at 20°C is 0.538. This value for invertebrates varies from 0.5 to about 10 vol%, while the range of values for cold-blooded vertebrates is 5–12 vol%. The amount of respiratory pigment in the blood and blood cells is proportional to oxygen capacity.

The shape and the position of the oxygen equilibrium curve are determined by measuring the percent saturation of a solution containing a respiratory pigment at different pressures (Fig. 1.3). Two values are particularly important in assessing the physiological-ecological role of respiratory pigments: (1) P_{sat} the pressure required for 95 percent saturation; and (2) P_{50} the pressure required for 50 percent saturation. Obviously if P_{sat} and P_{50} values are high, the fluid is not oxygen saturated until the ambient oxygen tension is high, and the respiratory pigment is said to have low oxygen affinity. This response is usually characteristic of an organism living in regions of high oxygen con-

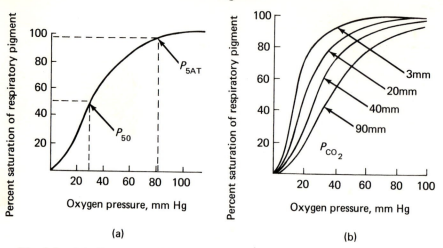

Fig. 1-3. (a) Oxygen equilibrium curve; P_{50}, the oxygen pressure at which 50 percent of the respiratory pigment is saturated, P_{5at}, the oxygen pressure at which 95 percent of the respiratory pigment is saturated; (b) influence of pCO_2 on equilibrium curve.

tent, whereas low values, an indication of high oxygen affinity, are to be expected for animals living in oxygen-poor habitats. Not only oxygen tension but also CO_2 content, temperature, and size of the animal influence the oxygen equilibrium curve. Increased CO_2 content or lowered pH may shift the curve to the right (Bohr effect), to the left (reverse Bohr effect), or it may remain unchanged (no Bohr effect).

If effective inhibition of oxygen uptake results when the respiratory pigments are inactivated, then obviously the respiratory pigments are of functional significance. CO is frequently used as a potentially inhibitory substance, but the concentration of CO must not be too high or the action of cellular hemoproteins or cytochromes will be blocked.

5. Chemical Regulation and Excretion

a. Osmo- and ionic regulation

The maintenance of an internal chemical milieu which will permit the various physiological processes to function is imperative for an organism to survive. Therefore, the amount of water, total salts, ions, and specific chemical compounds must be regulated, and waste products must be excreted. One of the dominant evolutionary trends has been the development of mechanisms to maintain a relatively constant internal environment (*homeostasis*). Simpler animals regulate these functions by using the cell membranes or body covering, but, with the increase in organismic complexity, special regulatory organs and organ systems developed. Many marine organisms are isotonic with sea water, and there is no significant exchange of water between the organism and the sea. In others, the ionic composition of the body fluid and/or cells may be different from sea water, which indicates chemical regulation. The degree of regulation varies not only with the species but also with one individual under different environmental conditions. Specific examples will be presented later. However, there are certain terms and concepts basic to the interpretation and understanding of these special examples which represent evolutionary variations on a fundamental biological theme.

Fluids in the multicellular organism are generally of two types, *extracellular* and *intracellular*. Extracellular fluids include the vascular system, the coelom, and the interstitial spaces; intracellular fluid is found within the cells. In some animals, but not all, the chemical composition of the extracellular fluid is highly regulated and distinctly different from the surrounding media. The intracellular fluid may be very similar to or differ significantly from extracellular fluid, even in those organisms that regulate the chemical composition of this latter fluid. That the chemical composition of the intracellular fluid may be different in various tissues of one animal emphasizes further the interdependence of intracellular and extracellular fluid.

If the chemical composition of the body fluids changes as the external environment changes, the organism is called a *conformer;* but in contrast, if the composition changes only slightly when the external environment fluctuates widely, the organism is a *regulator*. Osmotic characteristics are denoted in the terms *osmoregulators* or *osmoconformers*. However, if ionic composition is being considered, the terms *ionic-regulators* or *ionic-conformers* are used.

Organisms capable of withstanding great variations in salinity are called *euryhalinic,* whereas those restricted to a narrow part of a salinity gradient are *stenohalinic.* It should be noted that an organism could be euryhalinic and an osmoconformer.

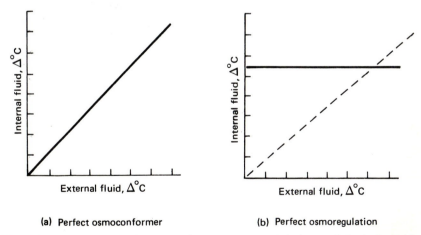

(a) Perfect osmoconformer   (b) Perfect osmoregulation

Fig. 1-4. Two generalized osmoregulatory responses: (a) the $\triangle$ of the body fluids conforms with that of external fluid. The dotted line is the line of iso-osmoticity; (b) perfect osmoregulation, the freezing point depression ($\triangle$) of the body fluids is unchanged when the $\triangle$ of the external fluid changes.

A standard method of estimating osmotic concentration is to determine the freezing point depression ($\triangle$) of a fluid and compare this value with that of the external environment (Fig. 1.4). The freezing point depression of sea water with a salinity of 34.3 0/00 is about 1.87°C. If the osmotic concentration of the body fluid is greater than the medium surrounding the organism, *hyperosmotic regulation* occurs; if concentrations of body fluid are less, the result is *hyposmotic regulation*. If there is no significant difference between the osmotic properties of internal and external fluids, they are *isosmotic* with one another. Freezing point depression determinations are a quantitative estimate of the total number of particles contributing to osmotic concentration and do not qualitatively measure the kinds of particles present. Osmoregulation may occur by regulating the concentration of specific ions or small-sized molecules, such as amino acids, urea, or trimethylamine oxide. Detailed analyses of the chemical composition of fluids have been reported either in tabular form or graphically as an ionogram (Fig. 1.5).

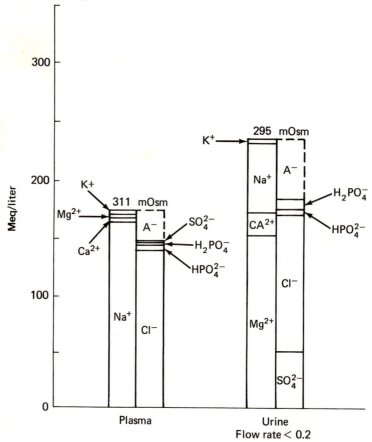

Fig. 1-5. Ionograms of plasma and urine from the southern flounder (From Hickman, 1968). Ionograms are a graphic representation of ionic composition. (National Research Council of Canada.)

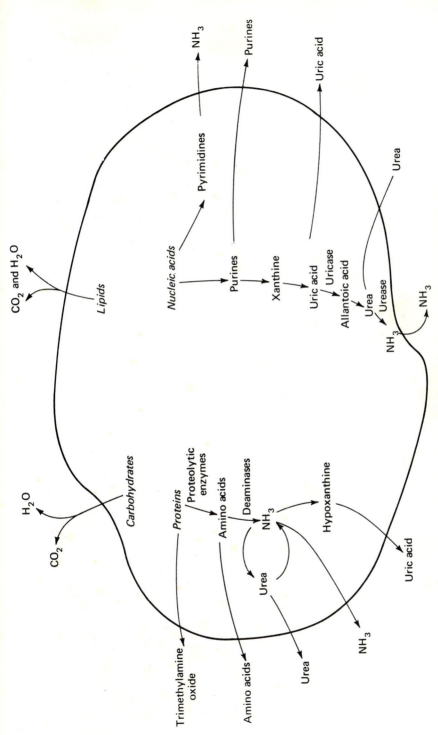

Fig. 1-6. Simplified scheme of the origin of principal excretory products.

b. Excretion

Organisms excrete products to the external environment which are expendable at the moment, including nitrogenous wastes, water, and ions. Various excretory sites have been described in marine animals, and their relative importance differs between species and within an organism, depending upon such factors as the stage of development and the type of excretory products involved. Excretion may take place across the body covering, especially in soft-bodied species, or it may involve special structures, excretory organs, which are found in the more highly evolved animals. These renal organs are called by different names, such as kidneys, antennal glands, or nephridia. These structures are modified to receive the circulatory body fluid, extract those products to be excreted, and deliver these wastes by means of an excretory duct through a body opening to the external milieu. There are also extrarenal routes of excretion, such as the gills, salt-secreting glands (nasal glands or rectal glands), and the gut, which are specialized for specific excretory functions.

The types of excretory products vary among organisms and within an organism depending in part on environmental factors. Fats and carbohydrates are ultimately broken down to water and carbon dioxide, which are easily eliminated. Protein metabolism, however, yields a greater variety of nitrogenous end products, including ammonia, urea, and uric acids (Fig. 1.6). About 90% of the excreted nitrogen comes from the deamination of amino acids, while only approximately 5% to 10% comes from nucleic acids. Ammonia is the principal nitrogenous waste of aquatic animals with the exception of many vertebrates. This substance is toxic and must either be excreted rapidly or converted to a less injurious compound. Rapid excretion is no problem in an aquatic environment, since ammonia is extremely water soluble and diffuses rapidly. Other organisms may excrete urea or uric acid as the principal waste product, although various amounts of a number of nitrogenous wastes may be excreted by one organism. Typically, uric acid is produced by animals living in areas where water is limited: uric acid is less toxic than ammonia and relatively insoluble in water. Urea is also less toxic than ammonia but slightly more water soluble than ammonia. If urea is the chief waste product, the organism is *ureotelic;* if ammonia, *ammonotelic;* and if uric acid, *uricotelic.*

6. Reproduction

In many species reproduction is a major physiological phenomenon dominating all other processes, for, although reproduction is not necessary for an individual to survive, the continuity of the species depends on the ability to reproduce itself. Both sexual and asexual modes of reproduction have evolved in the animal kingdom, but exclusively asexual reproduction is not widespread. Animals that reproduce asexually do not have the opportunity for genetic recombination found in sexually reproducing animals, which is useful for meeting the stresses of a changing environment; consequently, on the evolutionary time scale, some of the most successful asexually reproducing animals are those living in very slowly changing environments (Sonneborn, 1957).

The reproductive process of sexually reproducing animals can be divided into three phases, all of which must function to ensure continuity: gonadal development, spawning and fertilization, and development and growth. Most of the higher animals reproduce on a predictable annual basis during a restricted period of the year which may be limited to a short period of time or may extend over several months. Generally, reproductive cycles consist of a vegetative or resting stage with little gonadal activity and a reproductive stage initiated by the beginning of gametogenesis followed by the maturation and release of gametes.

One common method of assessing the influence of environment on reproduction is to determine the ratio of gonad size to body weight; this ratio is termed the gonad index (GI). The gonad index increases as the gonads increase in size. A representation of the relationship between the GI and various phases of the reproductive cycle is given in Fig. 1.7. The stage of gametogenesis

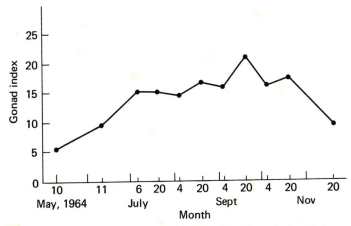

Fig. 1-7. Changes in the mean gonad index of scallops during their reproductive period (From Sastry, 1966.)

can be classified by obtaining quantitative and qualitative data from examination of pieces of gonadal tissue. An indication of the duration of the reproductive cycle is given by the percentage of animals in each of these stages at a different time of the year. Information about stages of the reproductive cycle can also be provided through observations on breeding and spawning periods.

In many marine invertebrates the relationship between the digestive gland and the gonad changes with the phase of the reproductive cycle. Typically, the digestive gland index is high in relation to the gonad index during the vegetative phase of the reproductive cycle, but the relationship is reversed during the reproductive phase. The chemical composition of gonad tissue can also change with the phase of the reproductive cycle. The gravid gonads of limpets, for example, have a high protein and lipid content, while nongravid gonads are high in carbohydrates, but low in proteins and lipids (Giese and Hart, 1967). Many interacting factors, such as food supply, salinity, temperature, and oxygen supply, influence the reproductive cycle in marine animals.

B. PHYSIOLOGICAL ADAPTATION

An organism is exposed to a range of intensities of a given environmental gradient. The *zone of lethality* or the *zone of resistance adaptation* is the extreme of the gradient which results in the organism's death. The intermediate range of an environmental gradient where the organism can survive is called the *zone of compatibility,* the *biokinetic zone,* or the *zone of capacity adaptation.* Interspecific differences in the position of these zones along an environmental gradient may be correlated with different ecological requirements of each individual species. Furthermore, the position of these zones may change during the life cycle of a single species and may reflect ontogenetic differences in the environmental requirements of an organism. One of the fundamental problems in environmental physiology is to understand the physiological mechanisms which have evolved that permit species to adapt to different environmental complexes. Obviously, genetic composition sets the ultimate limits of physiological performance of an organism, but responses

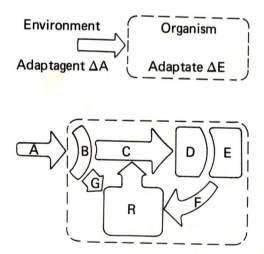

Fig. 1-8. Adaptational response of an organism to environmental alteration. A, affector; B, receptor; C, transmitter; D, effector; E, output; F, feedback transmitter; G, forward transmitter; R, regulator. (After Adolph, 1964. American Physiological Society.)

within these limits depend in part on environmental conditions. Each response represents a different phenotypic expression, sometimes referred to as an environmentally induced variation.

If an organism is to successfully survive in a given environmental complex, its various components must function in a coordinated manner. The failure of one vital process would result in extermination. Thus, although comprehensive studies that interrelate the responses of various physiological systems to the environment are lacking and at present most data represent investigations of isolated physiological systems, the student should not lose sight of the "wholistic" view that an organism is a composite of various functional parts acting together in a complex integrated manner.

When experiencing an environmental alteration, the adaptational response of an organism involves a number of steps (Fig. 1.8). Initially, the strength and duration of the affector must be adequate to induce a response. For example, a short duration of exposure produces an adaptive response, whereas a prolonged exposure to the same intensity results in death. Hence, both time and quantity of the affector must be studied to assess its environmental role. The affector stimulates a receptor, which in turn transmits the information to an effector and/or to a regulator by means of a forward transmitter. As a re-

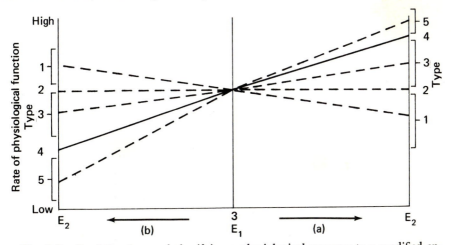

Fig. 1-9. Precht's scheme of classifying a physiological response to a modified environment. Although originally proposed for temperature changes, it is applicable to other environmental parameters. E_1, environment to which an organism is acclimated; E_2, new environment to which an organism is exposed; a, if E_2 is of higher intensity or concentration than E_1; b, if E_2 is of lower intensity or concentration than E_1. The solid line represents the initial response in the new environment. Type 4 (no acclimation), the rate of physiological function varies directly with environmental factor. Type 2 (complete acclimation, the rate is unaltered by environmental change. Type 3 (partial acclimation), the percent of acclimation can be estimated by assuming Type 4 to be 0 percent and Type 2 to be 100 percent acclimation. Type 1 (supra-optimal acclimation) and Type 5 (inverse acclimation). (After Precht, 1958. American Physiological Society.)

sult of the action of the effector, there is some output that may stimulate a
feed-back transmitter, which in turn interacts with the regulator. If numerous
environmental factors are altered simultaneously, highly integrative mechanisms must evolve to permit organisms to cope successfully and completely
with this environment. Obviously, the capability of the experimenter to measure the concomitant changes must also evolve if we are to understand physiological adaptation.

In assessing the ecological and evolutionary influence of a changed environment on an organism, Prosser (1958) has suggested that the duration of
exposure to a stressful environment is important (Table 1.1). Initially, a

TABLE 1.1

*Time Course of Adaptations to Environmental Change**

Reactions	Relative Times (For Metazoans)
Early responses	
Immediate	(minutes)
Sense organ stimulation and direct responses	
Shock-type reactions and overshoots of metabolic rates	
Stabilized state	(hours)
Altered rates of reactions in conformers	
(dependent organisms)	
Rate regulation in independent organisms	
Developmental response patterns which persist in adults	
Compensation for environmental change	(days or weeks)
Acclimation—new equilibrium of rate functions	
Metabolic adaptations in compensation for stress	
(particularly in conforming animals)	
Enzyme induction	
Morphological adaptations	
Non-genetic transmission	(few life cycles)
Cytoplasmic inheritance	
Behavior transmission (social evolution)	
Selection of genetically adapted types	(many generations)
Selection of mutants	
Races, genetic strains, ecotypes	
Reproductive isolation	
Species formation	

American Physiological Society.
* From Prosser, 1958

shock reaction, lasting seconds to minutes, is observed. Following this initial
response, a period of stabilized response lasts for minutes or hours. After a
period of days or weeks, acclimation may occur. We use the term *"acclimation"*

to describe the steady-state compensatory response of an organism to the alteration in one environmental factor, while in contrast, *"acclimatization"* refers to response to fluctuation in multiple factors. The environment may influence the future generations of a species by nongenetic processes or by genetic selection.

Animals may vary tremendously in their compensatory response to environmental parameters. To systematize these changes, investigators have widely used two methods of classifying patterns of response developed by Precht (1958) and Prosser (1958). Precht's scheme (Fig. 1.9) compares responses at only two environmental levels (for example, at two temperatures), while Prosser's system (Fig. 1.10) compares responses over a wide portion of an

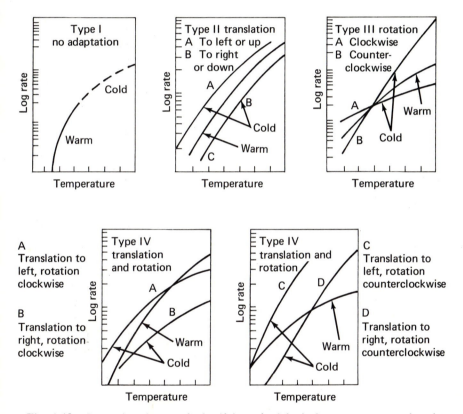

Fig. 1-10. Prosser's scheme of classifying physiological response to a changing temperature. The metabolic-temperature curves of warm- and cold-adapted animals are compared. (After Prosser, 1958. American Physiological Society.)

environmental gradient. Using either of these schemes, experimenters will have a common basis for comparing the acclimation patterns of different individuals to environmental factors.

C. SUMMARY

The sea, long of social-economic-scientific interest to man, is extremely diverse both in habitats and biotic inhabitants. Through unknown periods of time, marine organisms have become adapted to the different environmental complexes found in these diverse habitats. The functional basis of adaptation is one of the basic questions in environmental physiology, and the answers to this question have significance for understanding the nature of the organism occupying the largest ecological unit in our biosphere, the sea. This body of information is relevant not only to a basic understanding of biological phenomena, but also to predicting the effects of environmental alteration on marine animals.

The organism represents a discrete, dynamic, energetic unit of biological organization consisting of many functional subunits. To survive and perpetuate the species, these subunits must interact in a highly integrative manner during the constant exposure of an organism to a fluctuating environmental complex.

REFERENCES

Adolph, E. F. (1964). Perspectives of adaptation: some general properties. *In* Handbook of Physiology. D. B. Dill, ed. Section 4, Adaptation to the Environment. Am. Physiol. Soc., Washington, D.C. Pp. 27–35.

Bullock, T. H., and G. A. Horridge (1965). Structures and Function in the Nervous Systems of Invertebrates. W. H. Freeman & Co., San Francisco and London.

Frings, H. (1964). Sound production and reception. *In* Marine Bio-Acoustics. W. N. Tavolga, ed. Pergamon Press, New York. **1**:155–173.

——, and M. Frings (1967). Underwater sound fields and behavior of marine invertebrates. *In* Marine Bio-Acoustics. W. N. Tavolga, ed. Pergamon Press, New York. **2**:261–281.

Ghiretti, F., ed. (1968). Biochemistry and Physiology of Haemocyanins. Academic Press, New York.

Giese, G. C., and M. A. Hart (1967). Seasonal changes in component indices and chemical composition in *Katharina tunicata*. *J. Exptl. Mar. Biol. Ecol.*, **1**:34–36.

Hemmingsen, A. M. (1960). Energy metabolism as related to body size and respiratory surfaces, and its evolution. Rept. Steno Memorial Hospital, Nordisk Insulinlaboratorium, Copenhagen. **9**:part II.

Hickman, C. P. (1968). Glomerular filtration and urine flow in the euryhaline southern flounder, *Paralichthys lethostigma*, in sea water. *Can. J. Zool.*, **46**:427–437.

Jennings, J. B. (1970). Feeding, Digestion and Assimilation in Animals. Pergamon Press, New York. 2nd ed.

Jørgensen, C. B. (1966). Biology of Suspension Feeding. Pergamon Press, New York.

Laverack, M. S. (1968). On the receptors of marine invertebrates. *Oceanogr. Mar. Biol. Ann. Rev.*, **6**:249–324.

Manwell, C. (1960). Comparative physiology, blood pigments. *Ann. Rev. Physiol.*, **22**:191–244.

Morgan, E., and E. W. Knight-Jones (1966). Responses of marine animals to changes in hydrostatic pressure. *Oceanogr. Mar. Biol. Ann. Rev.,* **4**:267–299.

Precht, H. (1958). Concepts of the temperature adaptation of unchanging reaction systems of cold-blooded animals. *In Physiological Adaptation.* C. L. Prosser, ed. Am. Physiol. Soc., Washington, D.C. Pp. 50–77.

Prosser, C. L. (1958). The nature of physiological adaptation. *In Physiological Adaptation.* C. L. Prosser, ed. Amer. Physiol. Soc., Washington, D.C. Pp. 167–180.

Sastry, A. N. (1966). Temperature effects in reproduction of the bay scallop, *Aequipecten irradians* Lamarck. *Biol. Bull.,* **130**:118–134.

Sonneborn, T. M. (1957). Breeding systems, reproductive methods, and species problems in Protozoa. *In The Species Problem.* E. Mayr, ed. AAAS Symposium, Washington, D.C. Pp. 155–324.

Steven, D. M. (1963). The dermal light sense. *Biol. Rev.,* **38**:204–240.

Vernberg, F. J. (1959). Studies on the physiological variation between tropical and temperate zone fiddler crabs of the genus *Uca.* II Oxygen consumption of whole organisms. *Biol. Bull.,* **117**:163–184.

Yonge, C. M. (1928). Feeding mechanisms in invertebrates. *Biol. Rev.,* **3**:21–76.

——— (1937). Evolution and adaptation in the digestive system of the Metazoa. *Biol. Rev.,* **12**:87–115.

The endless sea (Courtesy of Coates Crewe).

CHAPTER II

THE SEA

A. GENERAL CHARACTERISTICS

Flowing incessantly over 71 percent of the earth's surface is a watery mass known as the sea. This milieu, which has a volume of approximately 315 million cubic miles, offers a great number of distinctive habitats to a rich diversity of biotic species. Organisms must be functionally adapted to withstand the particular set of environmental parameters which they experience in these diverse habitats, or they will perish. To set the stage for understanding physiological adaptation of marine organisms, this chapter will provide a general account of the dominant abiotic factors associated with the sea. Specific environmental characteristics of different habitat-types will be discussed in detail in later chapters.

The entire seawater mass is continuous and, potentially, a drop of water could make its way to any part of the total sea. Although present-day oceanographers recognize many subunits in this great water mass, indiscriminate use of terminology has obscured the physiographic meaning of terms such as oceans, seas, bays, and gulfs. Part of the difficulty in establishing a precise definition is that boundaries of seas and oceans may be defined in several ways: as distinct geographic areas, in terms of well-known oceanographic phenomena, or by given latitude and/or longitude. For example, the Baltic Sea is enclosed by land masses, except where it makes contact with the open sea through a complex series of shallow narrow straits that separate Denmark from Sweden; in contrast, no distinct land masses or constrictive straits separate other oceans or seas. The northern boundary of the Antarctic Ocean, however, is a distinctive oceanographic phenomenon, the Subtropical Convergence, a region of sharp change in surface water temperature, while the boundary between the Pacific and the Indian Oceans south of Australia is the meridian of 147°E to

The Sea

Antarctica (Sverdrup et al., 1942). The principal oceans and seas of the world are listed in Table 2.1.

Increased oceanographic research in the twentieth century has led to many exciting discoveries about the ocean floor, not the least of which is its topographic diversity. Vastly improved bathymetric charts today record impressive features previously unknown: sea mountains, the peaks of which break the ocean's surface to form small islands; tremendously deep trenches, such as the 35,800-foot Mindano Trench, which cleave the ocean floor; or the submerged

TABLE 2.1

The Area, Volume and Mean Depth of Certain Oceans

Name	Area $10^3 \times m^2$	Volume $10^3 \times m^3$	Mean Depth m
Atlantic Ocean†	106,463	354,679	3,332
Pacific Ocean† *including adjacent seas*	179,679	723,699	4,028
Indian Ocean† *including adjacent seas*	74,917	291,945	3,897
Baltic Sea* *including adjacent seas*	386	33	86
Bering Sea*	2,304	3,683	1,598
Sea of Okhotsk*	1,590	1,365	859
Sea of Japan*	978	1,713	1,752
Barents Sea*	1,405	322	229
All oceans and seas†	361,059	1,370,323	3,795

† Based on Kossinna (1921) cited by Sverdrup *et al.* (1942) Prentice-Hall, Inc.
* Zenkevitch, 1963

mountain range (mid-Atlantic Ridge) that extends for 10,000 miles in the North Atlantic Ocean. Marine investigations of the ocean floor are significant to marine biologists since new habitats with distinctive biotic assemblages are also being discovered.

Once thought to be relatively stable, the ocean floor is now considered to be a dynamic system owing to the concepts of continental drift, sea-floor spreading, and plate tectonics (Menard, 1964). These changes have created the diverse benthic habitats that exist at this point in time. On a geological time scale, however, a habitat at one specific geographical location is seen to be changing constantly, and therefore populations of organisms living there must adapt to new environmental stresses in an evolutionary sense. To understand the environmental physiology of these organisms is a challenge. Knowledge gained from accepting this challenge not only will broaden our understanding of the physiological mechanisms of adaptations of species from unique environments, such as the deep sea, but also will help to explain fundamental responses common to all organisms.

Although all marine scientists require some knowledge of oceanography, our specific aim in this book is to emphasize the physiological adaptations of marine organisms to environmental stresses. Thus, the reader is referred to other sources for detailed discussion of the dynamics of physical, geological, and geographical oceanography (Kuenen, 1950; von Arx, 1962; Menard, 1964; Pickard, 1968).

In addition to geographical subunits, horizontal and vertical zones have been recognized in the world sea; these range from the interface between land and sea to the open sea, and from the surface to oceanic bottoms. Although numerous classifications of marine environments have appeared, the one which has been widely accepted, and which is graphically represented in Fig. 2.1, is that proposed by the Committee on Marine Ecology and Paleoecology (Hedgpeth, 1957). The two major divisions in this classification of marine environments are *benthic* (bottom) and *pelagic* (water), which are further subdivided. A third division, the *phytal,* was proposed by Remane (1933) to include both large areas of prominent plants and growths of sessile animals such as bryozoans, corals, or hydroids. Each assemblage contains a characteristic fauna.

There are various benthic environments extending from high ground to the ocean depths: (1) supralittoral, a zone immediately above the high-water mark; (2) littoral or intertidal, the region between the high-water and low-water mark; (3) sublittoral, all bottom regions between the low-water mark and the edge of the continental shelf to a depth of about 100 fathoms; (4) bathyal, the continental slope beginning at the edge of the shelf down to a depth of about 2,000 fathoms; (5) abyssal, the zone from 2,000 fathoms to about 2,800 fathoms; and (6) hadal, all benthic regions below depths of about 2,800 fathoms. The regions of the deep sea, however, are not well defined. Although some oceanographers classify the abyssal zone as that region in which the temperature remains below $4°C$, no definition based on temperature and/or depth has been universally accepted, since the two factors may not be uniform throughout this zone. For example, below 2,000 meters in the Antarctic and Arctic oceans the temperature is less than $-1.2°C$; in the Mediterranean the temperature at these depths is $14°C$; while in the Caribbean it varies from $4°$ to $6°C$. Although its upper boundary is ill-defined, this zone is the largest benthic habitat type, occupying approximately 90 percent of the ocean floor.

The pelagic region has been divided principally into the *neritic zone,* which comprises the water mass over the continental shelf, and the *oceanic zone,* which includes the main mass of sea water. The subunits of the oceanic region are: (1) *epipelagic,* the uppermost layer of the sea including the photic zone and extending to a depth of about 100 fathoms; (2) *mesopelagic,* the layer of

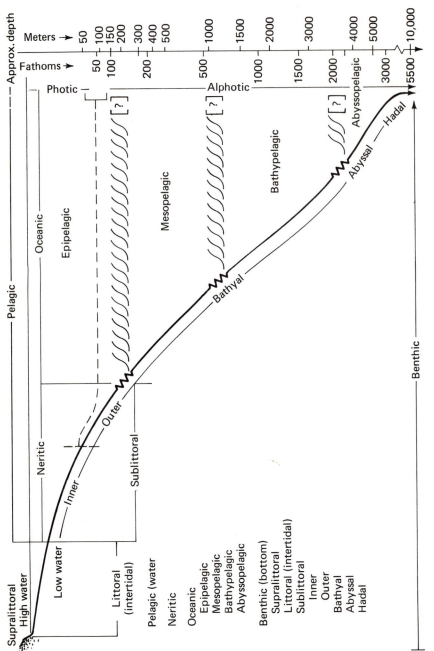

Fig. 2-1. Classification of marine environments (From Hedgpeth, 1957. Geological Society of America).

water between 100 fathoms and about 600 fathoms; (3) *bathypelagic,* the third major water layer which extends from the mesopelagic layer to a depth of about 2,000 fathoms; and (4) *abyssal pelagic,* all water below the preceding layer.

As these subunits are studied in more detail, each is further subdivided into distinctive habitats and microhabitats. Specific examples of detailed zonation patterns of these large subunits will be presented in later sections of this book.

Ecologically significant terms, such as those referring to light, have also been used to describe marine environments. Although light steadily decreases in intensity with depth, the terms *aphotic* and *euphotic* generally denote regions of darkness and light. Some marine scientists specify that in the photic zone sufficient light must be present for photosynthetic activity of plants (assimilation) to exceed their respiration rate (dissimilation), with a resultant increase in organic matter. The point at which these two physiological functions are equal has been called the *compensation point.* Depth of this point varies, depending on numerous factors such as latitude, turbidity of the water, and season. However, light extends below the compensation point to a variable depth, gradually diminishing until vision is no longer possible. This transitional region has been called the *mesophotic* zone; beneath it is the aphotic zone.

Various biogeographic units based on ecological factors have been proposed to describe marine environments, and frequently these refer to given taxonomic groups or to a general region of the sea, such as the divisions of shore and shallow-water seas proposed by Ekman (1953) or the bottom molluscan communities proposed by Thorson (1957). Knowledge of the physiological capability of organisms to meet the environmental stresses encountered in different marine habitats is fundamental to understanding and synthesizing biogeographical principles.

The classification of marine organisms is based largely on habitat type, i.e., the region of the sea they inhabit. Benthic organisms, for example, are associated with the sea bottom; those species associated with the surface of the bottom are the *epifauna,* whereas those that dig and are buried in the substratum represent the *infauna* (Fig. 2.2). In addition, benthic organisms are divided into three groups based on size: *macrobenthos,* organisms too large to pass through a 1 mm mesh sieve; *meiobenthos,* organisms smaller than the macrobenthos but which are retained by a 0.1 mm mesh sieve; and *microbenthos,* organisms smaller than meiobenthos, i.e., they pass freely through a 0.1 mm mesh sieve (Mare, 1942).

Free-moving organisms inhabiting the water column are called *pelagic* species. The ones which can control direction and speed of locomotion are called the *nekton;* those primarily dependent on water movements for locomotion are the *plankton.* Two major categories of plankton are recognized: *phytoplankton,* the plant species; and *zooplankton,* the animal species. The complex terminology resulting from various classifications of plankton includes some terms that are relevant to environmental physiology: *holoplankton,* species in which all developmental stages are part of the plankton; *meroplankton,* species which have

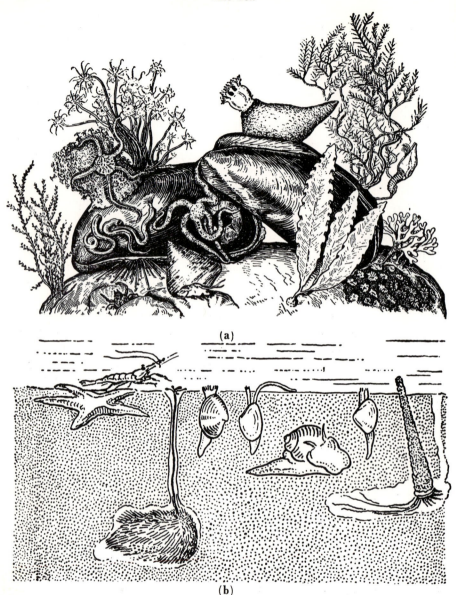

(a)

(b)

Fig. 2-2. (a) Examples of epifauna. One the valves of the left *Modiola modilus* are the hydroid, *Tubularia;* the octocoral *Alcyonium* with the ophiuran *Ophiothrix;* the polychaete *Pomatoceros;* the ophiuran *Ophiopholis* and a *Balanus.* On the valves of the right *Modiola* are the holothurian, *Psolus,* and the hydroid *Hydrallmannia* with the cirriped *Scalpellun.* On the stones in front, left to right, are the hydroid *Ahietinaria;* the crab, *Eurynome,* the prosobranch, *Calliostoma;* the red algae, *Delesseria* and *Rhodymenia,* and the compound ascidian, *Botryllus.* (b) Example of infauna. Starting from the left the animals are the starfish, *Astropecten;* the shrimp, *Leander;* the sea urchin, *Echinocardium;* the lamellibranchs, *Venus, Tellina* and *Spisula;* the prosobranch *Natica* (= Polynices), and the polychaete *Pectinaria.* (From Thorson, 1957. Geological Society of America.)

certain life history stages associated with the benthic environment; *nannoplankton,* small-sized plankton which range from 5 to 60 μ; and *ultraplankton,* which are organisms smaller than 5 μ.

Some organisms regularly occur in the benthic and pelagic regions. If they are *nektonic* species, they are described as *demersal;* if *planktonic,* they are called *meroplanktonic.*

B. CHEMISTRY OF SEA WATER

The chemistry of sea water has received varying degrees of attention from the scientific community since the time of the early Greeks. Knowledge of the chemical properties of sea water is useful not only to physical scientists, but also to biologists, particularly environmental physiologists who are concerned with the active interaction between the external environment and the organism's internal environment. Numerous published tables giving the chemical composition of sea water show variations in values, depending on such differences as collecting techniques and sample analyses. Frequently, only a few water samples are analyzed, and these may have been taken only at the surface, or both the dissolved and/or particulate forms of an element may not be analyzed. Furthermore, since water masses differ greatly in the biochemical cycling of certain elements, the concentrations of a given substance could easily differ by a factor of 10^3. These problems are compounded by the paucity of data concerning chemical reactions and concentrations at the air-sea interface, or at the sea-sediment phase discontinuity.

Marine chemical studies have been arbitrarily grouped into the following categories by Goldberg (1963): (1) qualitative and quantitative composition of elements in sea water; (2) relative reactivities of elements; (3) chemical reactions in the sea and composition of end products; and (4) distribution in time and space of reactants and their end products.

The composition of sea water is commonly expressed in terms of two chemical groups: the major and minor elements. The major constituents, listed in Table 2.2, are those which contribute significantly to the salinity level. A dis-

TABLE 2.2 *Composition of Sea Water**

Constituent	gm/kg of water of salinity 35 0/00	Constituent	gm/kg of water of salinity 35 0/00
Chloride	19.353	Bicarbonate	0.142
Sodium	10.760	Bromide	0.067
Sulphate	2.712	Strontium	0.008
Magnesium	1.294	Boron	0.004
Calcium	0.413	Fluoride	0.001
Potassium	0.387		

* From Culkin, 1965.
Academic Press, Inc.

TABLE 2.3 *Abundances of the Elements in Sea Water**

Element	Concentration in mg/l	Element	Concentration in mg/l
H	108,000	Ag	0.00004
He	0.000005	Cd	0.00011
Li	0.17	In	<0.02
Be	0.0000006	Sn	0.0008
B	4.6	Sb	0.0005
C	28	Te	————
N	0.5	I	0.06
O	857,000	Xe	0.0001
F	1.3	Cs	0.0005
Ne	0.0001	Ba	0.03
Na	10,500	La	1.2×10^{-5}
Mg	1350	Ce	5.2×10^{-6}
Al	0.01	Pr	2.6×10^{-6}
Si	3.0	Nd	9.2×10^{-6}
P	0.07	Pm	————
S	885	Sm	1.7×10^{-6}
Cl	19,000	Eu	4.6×10^{-7}
A	0.6	Gd	2.4×10^{-6}
K	380	Tb	————
Ca	400	Dy	2.9×10^{-6}
Sc	0.00004	Ho	8.8×10^{-7}
Ti	0.001	Er	2.4×10^{-6}
V	0.002	Tm	5.2×10^{-7}
Cr	0.00005	Yb	2.0×10^{-6}
Mn	0.002	Lu	4.8×10^{-7}
Fe	0.01	Hf	————
Co	0.0001	Ta	————
Ni	0.002	W	0.0001
Cu	0.003	Re	————
Zn	0.01	Os	————
Ga	0.00003	Ir	————
Ge	0.00006	Pt	————
As	0.003	Au	0.000004
Se	0.0004	Hg	0.00003
Br	65	Tl	<0.00001
Kr	0.0003	Pb	0.00003
Rb	0.12	Bi	0.00002
Sr	8.0	Po	————
Y	0.0003	At	————
Zr	————	Rn	0.6×10^{-15}
Nb	0.00001	Fr	————
Mo	0.01	Ra	1.0×10^{-10}
Tc	————	Ac	————
Ru	————	Th	0.00005
Rh	————	Pa	2.0×10^{-9}
Pd	————	U	0.003

* From Goldberg, 1965.
Academic Press, Inc.

cussion of salinity is presented later in this chapter; the physiological significance of other constituents will be discussed later in this book.

Minor elements are those present in concentrations less than one part per
million by weight. Although they may exist in minute amounts, some are still
vital to the chemical dynamics of the sea. Table 2.3 includes an estimate of
various elements in sea water according to Goldberg (1965). The major constituents of sea water may be determined by physical-geochemical techniques,
but biological activity must be considered in the analysis of fluctuation in the
composition of minor elements.

Elements are introduced into the ocean as fall-out from the atmosphere or
as run-off from a land mass. Low concentration may occur if the element is
highly reactive and rapidly removed to the sediment, or if the initial concentration is extremely low. The length of time elements remain in sea water before
being removed by some precipitation process is known as the *residence time,*
and it varies greatly for different elements (Table 2.4).

TABLE 2.4 *The Residence Times of Elements in Sea Water as Calculated by
River Input and Sedimentation**

Element	Ocean Content ($\times 10^{20}$ g)	Residence time (millions of years) River Input	Sedimentation
Na	147.8	210	260
Mg	17.8	22	45
Ca	5.6	1	8
K	5.3	10	11
Sr	0.11	10	19
Si	0.052	0.035	0.01
Li	0.0023	12	19
Rb	0.00165	6.1	0.27
Ba	0.00041	0.05	0.084
Al	0.00014	0.0031	0.0001
Mo	0.00014	2.15	0.5
Cu	0.000041	0.043	0.05
Ni	0.000027	0.015	0.018
Ag	0.0000041	0.25	2.1
Pb	0.00000041	0.00056	0.002

* From Goldberg, 1965.
Academic Press, Inc.

The concentration of minor elements may vary with water masses or with
depth. For example, barium concentration increases with depth (Chow and
Goldberg, 1960) while thorium level is different in various oceans, probably
because of its short residence time (Goldberg, 1965). Interestingly, as an apparent result of man's increased use of the internal combustion engine and the
burning of leaded gasolines, a significant increase in the concentration of lead

has been observed in the sea (Tatsumoto and Patterson, 1963). It has been estimated that in recent years 27 times more lead has been introduced into the seas than during the Pleistocene Epoch, and this massive introduction is well illustrated in surface waters off California, where the lead concentration averages 10 times more in surface waters than in deeper waters. Through the use of nuclear energy, man has further altered the marine environment, adding significant amounts of radioactive isotopes to sea water directly or indirectly by fall-out. The continuing introduction into the sea of inorganic as well as organic compounds is particularly significant to the study of organismic-environmental reactions, since organisms must adapt and evolve mechanisms to cope with man-induced environmental changes if they are to survive.

Although many chemical elements of the sea are found in minute amounts, some marine organisms have evolved mechanisms to concentrate these elements by a factor of millions. It has been suggested that for any given chemical element at least one planktonic species exists that can concentrate this element in sea water (Nicholls et al., 1959). However, this ability to concentrate elements is not limited to the plankton. One of the best-known examples is the concentration of vanadium by certain ascidians, which take up most of the vanadium through the alimentary tract. Other examples include fish that concentrate zinc; seaweeds that concentrate nickel; and sponges that concentrate copper. The functional significance of the differential concentration of elements by organisms needs to be explained, for not only do interspecific differences exist, but also tissues concentrate elements at different levels within an organism.

Besides elements, dissolved organic matter in sea water appears to be physiologically significant, although this point has been disputed since the beginning of this century. As demonstrated most convincingly under laboratory conditions, however, the presence and/or absence of organic substances govern the growth and mortality of marine organisms, especially microorganisms. Similar relationships are expected in the sea; for example, some dinoflagellates produce a water-soluble substance which causes mass mortality in fish. Organic substances such as pheromones have been shown to influence interactions between animals, although little is known of the mechanisms involved. Not only do "naturally" occurring organic compounds influence the physiology of the marine biota, but man also introduces a number of "artificial" compounds whose effects are not always known or predictable. Qualitative and quantitative studies of dissolved organic compounds have been hampered by problems of chemical techniques, since many compounds occur in very low concentration, and since various intermediate decomposition products may form during chemical isolation and purification. Despite obstacles to progress in the field of marine chemistry, its singular importance has attracted energetic scientists. In an excellent review of this topic, Duursma (1965) presented an extensive listing of 95 general and specific dissolved organic compounds (Table 2.5), and since then a growing number of papers have dealt with specific chemical problems.

Several generalizations about the chemistry of sea water are pertinent to the physiological theme of this book. First of all, most of the dissolved organic matter originates from dead phytoplankton and detritus (Duursma, 1961). Second, sugar-like compounds are more highly concentrated in the surface waters, where the rate of photosynthesis is high. These compounds appear to be extracellular excretion products resulting from phytoplankton blooms (Guillard and Wangersky, 1958). Third, the concentration of combined amino acids decreases sharply with increasing depth down to the aphotic zone, where it remains relatively constant. In contrast, the free amino acids are more uniformly distributed. Fourth, seasonal and yearly differences in water from one geographical site, as well as in the sea water from different geographic locations, may differentially affect growth of marine organisms. This generalization is based on organic content of seawater samples and on phytoplankton growth as a bioassay of the amount of organic material present.

Of the various substances dissolved in sea water, special attention must be given to the dissolved gases, because their presence or absence may profoundly influence many physiological processes. All the atmospheric gases are found dissolved in sea water. Chemical oceanographers generally assume that regardless of location, every water particle was once at the sea surface and became equilibrated with atmospheric gases. The concentration of a gas in sea water is influenced by its solubility at a given temperature and salinity; solubility is highest at low temperature, and decreases with increasing temperature and salinity. Tables of solubility factors may be consulted in Weiss, 1970. Any change in the concentration of dissolved gases under constant temperatures and salinity conditions may be due to biological activity. Nitrogen levels in marine water appear to be fairly constant, even though nitrogen fixation by algae is known to occur. Oxygen and CO_2 concentration changes, however, have been widely observed; whenever organic material accumulates, oxygen is consumed, and if the amount of organic matter is sufficiently large, an anaerobic environment will eventually result. Certain marine regions typically are anaerobic, such as mud-flats and the deeper regions of the Black Sea; other areas may occasionally experience variable periods of anaerobiosis as a result of pollution.

Oxygen concentration may vary with depth in the water column, and in turn this relationship may vary at one geographical point either seasonally or climatically. Differences between geographical regions have also been reported, as in regions of upwelling, where oxygen deficits typically occur near the surface. Photosynthesis, respiration, vertical mixing, and shifting of the thermocline also influence the distribution of oxygen with depth. At intermediate depths in most oceans there is a layer of water which has a dramatically lower oxygen concentration than that of the layers higher or lower in the water column. This layer, called the *oxygen minimum zone,* may have concentrations of oxygen less than 0.5 ml/liter, whereas the concentration of oxygen in adjacent layers may be in excess of 6 ml/liter. The depth of the oxygen minimum zone

TABLE 2.5 *Specific Dissolved Organic Compounds Identified in Sea Water**

Name of Compound and Chemical Formula	Concentration				Locality

I. CARBOHYDRATES

Name of Compound and Chemical Formula	mg/l	μg/l			Locality
Pentoses $C_5H_{10}O_5$	0–8				Gulf of Mexico
Pentoses $C_5H_{10}O_5$		0.5			Pacific off California
Hexoses		14–36			Pacific off California
Rhamnosides $C_6H_{12}O_5$	0.1–0.4				Pacific Ocean Coast U.S.A.
Rhamnosides					
Dehydroascorbic acid $COCOCOCHCH(OH)CH_2OH$ (O)	0.1				Gulf of Mexico inshore water

II. PROTEINS AND THEIR DERIVATIVES

Name of Compound and Chemical Formula	(a) μg/l	(b) μg/l	(c) μg/l	(d) μg/l	Locality
Peptides C:N ratio = 13.8:1					Gulf of Mexico
Polypeptides and polycondensates of:					
Glutamic acid $COOH(CH_2)_2CH(NH_2)COOH$		8–13	8–13	0.1–1.8	Gulf of Mexico
Lysine $NH_2(CH_2)_4CH(NH_2)COOH$	<1	?	trace–3	0.1–0.9	Gulf of Mexico
Glycine NH_2CH_2COOH		—	trace–3	1.2–3.7	Gulf of Mexico
Aspartic acid $COOHCH_2CH(NH_2)COOH$		3–8	trace–3	0.1–1.0	Pacific off California
Serine $CH_2OHCH(NH_2)COOH$		?	trace–3	1.8–5.6	

Amino acid	Gulf of Mexico	Gulf of Mexico	Gulf of Mexico	Pacific off California
Alanine $CH_3CH(NH_2)COOH$		3–8	trace–3	0.7–3.1
Leucine $(CH_3)_2CHCH_2CH(NH_2)COOH$	0.5–1	8–13	trace–3	0.9–3.8
Valine $(CH_3)_2CHCH(NH_2)COOH$		trace–3	trace–3	0.1–1.7
Cystine $[SCH_2CH(NH_2)COOH]_2$		trace–3	—	0.0–3.8
Iso-leucine $CH_3CH_2CH(CH_3)CH(NH_2)COOH$		8–13	trace–3	—
Leucine $CH_3CH_2CH_2CH_2CH(NH_2)COOH$		—	—	0.9–3.8
Ornithine $NH_2(CH_2)_3CH(NH_2)COOH$		—	trace–3	0.2–2.4
Methioine sulphoxide $CH_3S(:O)CH_2CH_2CH(NH_2)COOH$				—
Threonine $CH_3CHOHCH(NH_2)COOH$		—	3–8	0.3–1.3
Tyrosine $HOC_6H_4CH_2CH(NH_2)COOH$	<0.5	—	trace–3	tr.–0.5
Phenylalanine $C_6H_5CH_2CH(NH_2)COOH$		—	—	0.1–0.9
Histidine $C_3H_3N_2CH_2CH(NH_2)COOH$		?	trace–3	tr.–2.4
Arginine $NH_2C(:NH)NH(CH_2)_3CH(NH_2)COOH$		?	trace–3	0.1–0.6
Proline C_4H_8NCOOH		?	—	0.3–1.4
Methionine $CH_3SCH_2CH_2CH(NH_2)COOH$		—	trace–3	tr.–0.4
Tryptophan $C_8H_6NCH_2CH(NH_2)COOH$		—	trace–3	—
Glucosamine $C_6H_{13}NO_5$		—	trace–3	—

Name of Compound and Chemical Formula	Concentration						Locality
Free amino acids	(a)† µg/l	(b)† µg/l	(c)† µg/l	(d)† µg/l	(e)†	(f)† µg/l	
Cystine [SCH$_2$CH(NH$_2$)COOH]$_2$					detected	—	Norwegian coastal water
Lysine NH$_2$(CH$_2$)$_4$CH(NH$_2$)COOH					detected	0.2–3.1	Pacific off California
Histidine C$_3$H$_3$N$_2$CH$_2$CH(NH$_2$)COOH					detected	0.5–1.7	
Arginine NH$_2$C(:NH)NH(CH$_2$)$_3$CH(NH$_2$)COOH					detected	0.0	
Serine CH$_2$OHCH(NH$_2$)COOH					detected	2.3–28.4	
Aspartic acid COOHCH$_2$CH(NH$_2$)COOH					detected	trace–9.6	
Glycine NH$_2$CH$_2$COOH					detected	trace–37.6	
Hydroxyproline C$_4$H$_7$N(OH)COOH					detected	trace–2.8	
Glutamic acid COOH(CH$_2$)$_2$CH(NH$_2$)COOH					detected	1.4–6.8	
Threonine CH$_3$CHOHCH(NH$_2$)COOH					detected	2.8–11.8	
α-Alanine CH$_3$CH(NH$_2$)COOH					detected		
Proline C$_4$H$_8$NCOOH					detected	0.0	
Tyrosine HOC$_6$H$_4$CH$_2$CH(NH$_2$)COOH					detected	trace–5.0	
Tryptophan C$_8$H$_6$NCH$_2$CH(NH$_2$)COOH					detected	—	
Methionine CH$_3$SCH$_2$CH$_2$CH(NH$_2$)COOH					detected	—	
Valine (CH$_3$)$_2$CHCH(NH$_2$)COOH					detected	0.3–2.7	
Phenylalanine C$_6$H$_5$CH$_2$CH(NH$_2$)COOH					detected	trace–2.4	
Iso-leucine CH$_3$CH$_2$CH(CH$_3$)CH(NH$_2$)COOH					detected	—	
Leucine (CH$_3$)$_2$CHCH$_2$CH(NH$_2$)COOH					detected	0.5–5.5	

Free compounds

	Pacific coast near La Jolla		Pacific off California
Uracil	detected		
NHCONHCOCH:CH			
Iso-leucine	detected		
CH$_3$CH$_2$CH(CH$_3$)CH(NH$_2$)COOH			
Methionine	detected		
CH$_3$SCH$_2$CH$_2$CH(NH$_2$)COOH			
Histidine	detected		
C$_3$H$_3$N$_2$CH$_2$CH(NH$_2$)COOH			
Adenine	detected		
C$_5$H$_3$N$_4$NH$_2$	detected		
Peptone			
Threonine	detected		
CH$_3$CHOHCH(NH$_2$)COOH			
Tryptophan	detected		
C$_8$H$_6$NCH$_2$CH(NH$_2$)COOH			
Glycine	detected		
NH$_2$CH$_2$COOH			
Purine	detected		
C$_5$H$_4$N$_4$			
Urea			
CH$_4$ON$_2$			

III. ALIPHATIC CARBOXYLIC AND HYDROXY-CARBOXYLIC ACIDS

	mg./l. (0–200 m.)	mg./l. (200–600 m.)	Coastal waters of Gulf of Mexico mg./l. (>600 m.)
Lauric acid CH$_3$(CH$_2$)$_{10}$COOH	0.01–0.32	0.01–0.28	0–0.28
Myristic acid CH$_3$(CH$_2$)$_{12}$COOH	0.01–0.10	0.01–0.05	0–0.07
Myristoleic acid CH$_3$(CH$_2$)$_3$CH:CH(CH$_2$)$_7$COOH	traces–0.02	0.01–0.03	0–0.05
Palmitic acid CH$_3$(CH$_2$)$_{14}$COOH	0.01–0.17	0.03–0.42	0–0.38
Palmitoleic acid CH$_3$(CH$_2$)$_5$CH:CH(CH$_2$)$_7$COOH	0.02–0.16	0.02–0.16	0–0.21
Stearic acid CH$_3$(CH$_2$)$_{16}$COOH	0.04–0.09	0.02–0.13	0–0.10
Oleic acid CH$_3$(CH$_2$)$_7$CH:CH(CH$_2$)$_7$COOH	0.01	0.02	0
Linoleic acid CH$_3$(CH$_2$)$_4$CH:CHCH$_2$CH:CH(CH$_2$)$_7$COOH	0.01	0.01	0

Name of Compound and Chemical Formula	Concentration	Locality
Fatty acids with:	mg/l (1000–2500 m.)	Pacific Ocean coastal water
12 C-atoms	0.0003–0.02	
14 C-atoms	0.0004–0.043	
16 C-atoms	0.0027–0.0209	
16 C-atoms + 1 double bond	0.0003–0.003	
18 C-atoms	0.0037–0.0222	
18 C-atoms + 1 double bond	0.0083	
18 C-atoms + 2 double bonds	0.0000–0.0029	
20 C-atoms	traces–0.0081	
22 C-atoms	traces–0.0014	
Acetic acid CH_3COOH	mg./l. <1.0	Pacific Ocean
Lactic acid $CH_3CH(OH)COOH$		
Glycollic acid $HOCH_2COOH$		
Malic acid $HOOCCH(OH)CH_2COOH$	0.28	Atlantic coastal water
Citric acid $HOOCCH_2C(OH)(COOH)CH_2COOH$	0.14	
Carotenoids and brownish waxy or fatty matter	2.5	North Sea English Channel

IV. BIOLOGICALLY ACTIVE COMPOUNDS

Compound	Value	Location
Organic Fe compound(s)	3.4–1.6 mμg./l.	Deep sea water
Vitamin B_{12} (Cobalamine) $C_{63}H_{90}O_{14}N_{14}PCo$		Long Island Sound
Vitamin B_{12}	0.2 mμg./l. (summer)	Oceanic surface-water
	2.0 mμg./l (winter)	
Vitamin B_{12}	0.2–5.0 mμg./l.	North Pacific Ocean
Vitamin B_{12}	0–2.6 mμg./l.	Sargasso Sea 0—$.05$ m.
Vitamin B_{12}	0–0.03 mμg./l.	Surface water, possibly from land drainage
Thiamine (Vit. B_1) $C_{12}H_{18}ON_4SCl_2$	0–20 mμg./l.	North Sea near Scotland
Plant hormones (auxins)	3.41 mμg./l.	

V. HUMIC ACIDS

Compound	Value	Location
"Gelbstoffe" (Yellow substances) Melanoidin-like		Coastal waters

VI. PHENOLIC COMPOUNDS

Compound	Value	Location
p-Hydroxy-benzoic acid HOC_6H_5COOH	1–3 μg./l.	Pacific off California
Vanillic acid $CH_3(HO)C_6H_3COOH$	1–3 μg./l.	
Syringic acid $(CH_3O)_2(HO)C_6H_2COOH$	1–3 μg./l.	

VII. HYDROCARBONS

Compound	Value	Location
Pristane; (2, 6, 10, 14 tetramethylpentadecane)	trace	Cape Cod Bay

tr. = trace
— = not detected
? = possibly present
† Each column represents determinations by different authors
* From Duursma, 1965

may vary between oceans and also may change seasonally at one geographical site. Populations of animals are present in this layer despite the low concention of oxygen. At times, this zone overlaps the deep scattering layers (DSL), a sound-reflecting layer recorded on echo-sounders which is found at fluctuating depths in open waters. Apparently this layer represents a great concentration of organisms which reflect sound waves to the surface.

In anoxic regions of the world seas, the concentration of hydrogen sulfide (H_2S) tends to be high. When the uppermost boundary of a sulfide-producing region comes into contact with the free atmosphere, a distinctive black layer is formed. The boundary layer between H_2S and O_2 fluctuates, depending on factors such as water circulation, and when this layer rises close to the surface, the result is apt to be mass mortality of benthic organisms.

The well-developed buffering system of sea water apparently prevents variations in CO_2 concentrations from being critical to marine animals. In areas of high oxygen utilization, the CO_2 level is high, while the level is low where plant assimilation rates are high. CO_2 content is obviously important to the marine carbonate system. When CO_2 content is low, calcium carbonate precipitates most readily; high temperature and high salinity also favor precipitation of calcium carbonate, and with increased depths the solubility of calcium increases. Precipitation rates are high in surface waters, especially the tropical shallow-water environments. The net ecological effect of these relationships is that more calcium is in solution in deeper waters, and as a result, skeletal structures of deep-sea animals are less calcified than those of shallow-water organisms. Various detailed reviews of the carbon dioxide-carbonate system are available for readers who require additional information (Revelle and Fairbridge, 1957; Skirrow, 1965; Spencer, 1965; Pytkowicz, 1968).

C. PHYSICAL FACTORS

1. Temperature

Of all the physical characteristics of sea water, probably more data are available on temperature than on any others. This abundance of information reflects both the ease and precision with which thermal measurements can be taken and the influence of temperature on many biological and physical processes.

Temperature may vary with time at one geographical location, with latitude, and with depth. Some marine environments, such as the cold-water abyssal zone or the warm surface waters of the tropics, characteristically have stable thermal regimes. Other habitats, such as the intertidal zone of the temperate zone, experience large daily and seasonal thermal fluctuations. The worldwide temperature range of all oceanic waters is from about $-2°C$ to $+30°C$; in contrast, coastal and estuarine waters show a wider range, from about $-3°C$ to $+44°C$. Intertidal zone organisms may experience a particularly great variation in temperatures throughout the year. North temperate zone animals, for example, survive temperatures as low as −20°C. (Kanwisher, 1955).

In contrast to terrestrial environments, where daily changes in temperature may be in excess of 30°C, sea-water temperatures change slightly, partly because of the specific heat of water. Estuarine waters show greater daily and seasonal fluctuations than either coastal or oceanic waters. The annual range of surface temperature fluctuation is less than 2°C in tropical waters, and about 4°C in north polar waters, and about 8°C in the middle northern latitudes. At a given latitude large water masses with different thermal characteristics may come in dynamic contact with each other, and thus distinctive zoogeographic regions are formed. One example of this phenomenon is the merging of the south-flowing cold water of the Virginia coastal current and the north-flowing warm water of the Gulf Stream off the North Carolina coast of the United States. The lowest surface temperature of the Virginian waters was recorded to be about 4°C; the Gulf Stream water measured 19°C near the surface, and the lowest near-bottom temperatures were 5°C and 16°C, respectively (Stefansson and Atkinson, 1967).

The temperature of a water column may change with depth, as illustrated in the tropics where surface temperature may be 28°C, while at a depth of 4,000 meters the temperature may be 3°C. A sharp thermal difference may also occur between two water layers (a *thermocline*), but its depth tends to vary with time and geography. In shallower coastal waters significant vertical gradients may not be observed or may appear only seasonally (Stefansson and Atkinson, 1967).

Microenvironments are created when marked temperature differences occur in separated regions of a habitat; for example, the upper surface of a rock exposed to direct sunlight may be 20°C higher than its shaded undersurface. Temperatures at the surface and at various depths in the soil of the intertidal zone may also vary, and microthermal stratification may be observed at the surface of bodies of water.

Temperature not only directly influences the physiological capability of an organism, but it also affects various physical and chemical processes, such as the solubility of dissolved gases and the viscosity of sea water, both of which vary inversely with temperature. Chemical reactions are temperature dependent, and typically an increase of 10°C will increase reaction rates by 2–3 times. Because temperature is such a significant environmental parameter that influences various physiological processes of marine animals from different habitats, a discussion of its manifold effects will permeate a considerable part of the following sections of this book.

2. Pressure

For every depth increase of 10 meters, hydrostatic pressure is increased by 1 atmosphere. Although the physiological basis is far from being well understood, organisms are markedly influenced by pressure. *Stenobathic* organisms, those restricted to a narrow range of pressure, respond to slight changes in pressure (less than 1 atmosphere), and frequently die at increased pressures. Others, such as plankton, fishes, and cephalopods, appear to adjust rapidly to the wide changes in hydrostatic pressures which they experience on their extensive diurnal migrations. Some species that do not migrate vertically on a diurnal basis live in shallow waters in polar latitudes, but are found in deeper waters at lower latitudes. This phenomenon has been named *equatorial submergence*. Organisms that tolerate a wide range of pressures are called *eurybathic*, while those that apparently are restricted to the deep sea, such as deepwater bacteria which require high pressures in order to grow (ZoBell, 1952), are called *barophilic*.

3. Tides

The distinctive ecological region known as the intertidal zone has been created by rhythmic fluctuations in sea level known as tides. The temporal pattern of tides varies in different parts of the world: when high water occurs once a day (averaging 24.8 hours) the tidal change is called a *diurnal tide;* when high waters occur twice daily at intervals averaging 12.4 hours, it is a *semi-diurnal tide*. The magnitude of tidal change varies geographically, ranging from a few inches in some areas to more than 40 feet in the Bay of Fundy. Other tidal fluctuations that may occur in one geographical region include *spring tides,* which are unusually high tides resulting at times of the new and full moon, when the sun, moon, and earth are directly in line. When the moon, earth, and sun are at right angles to each other, unusually low tides called *neap tides* result. The gravitational pull of the moon and sun are additive during spring tides, but tend to cancel out each other during neap tides.

In regions where tidal changes are conspicuous, the intertidal zone has been subdivided into various subunits according to different classifications. One system is based on the height of water above a fixed vertical plane called a *datum,* which is arbitrarily selected so that few tides fall below it. On nautical charts the datum is approximately the level of the Mean Low Water Spring tide (M.L.W.S.) (Macmillan, 1966). Other common tidal terms are listed below:

Mean High Water Spring (M.H.W.S.): average height of high water at spring tide.

Mean High Water Neap (M.H.W.N.): average high at neap tide.

Mean Low Water Neap (M.L.W.N.): average low at neap tide.

Mean Tide Level (M.T.L.): mean of mean high and low waters determined over a long period.

Mean Sea Level (M.S.L.): average level of sea above datum determined from a long series of equally spaced observations.

The intertidal zone has also been classified on the basis of the dominant organisms present within each subunit. In some intertidal zones tidal activity has created special microhabitats such as tidepools.

As water level changes, various other environmental factors may also change. Organisms at the higher reaches of the intertidal zone generally are exposed to the greatest fluctuations in intensity and duration of temperature, salinity, sunlight, wave action, currents, and especially desiccation. There are obvious physiological differences in the ability of organisms from different parts of the intertidal zone to respond to these environmental changes. Marked changes in biotic interrelationships also occur during a tidal cycle. For example, some organisms stop feeding when the tide is out, whereas other organisms actively seek food only during the low tide period.

4. Salinity

Salinity of the open ocean, expressed as parts of solid material per 1,000 parts of water (0/00), averages about 35 0/00 and is relatively constant, although minor variations between oceans have been reported. The salinity of other marine environments may range from at least 155 0/00 in tropical shore ponds to less than 1 0/00 in estuarine and coastal regions following excessive rain and run-off from river systems. Rhythmic salinity changes, common to coastal and particularly to estuarine waters, result from diurnal tidal changes and from seasonal changes associated with regularly occurring rainy seasons. Floods from irregular rainstorms bring about arhythmic salinity changes.

Microenvironmental gradients of salinity have been noted in some habitats. For example, Johnson (1967), who found seasonal differences in salinity content in various parts of a sandy beach, observed that the higher regions of the beach showed the greatest variation. Not all habitats, however, exhibit salinity gradients; the salinity of interstitial water of mud-bottoms changes little with tidal flux, although the sea water above varies with the phase of tidal cycle (Sanders et al., 1965).

Salinity was long defined as the amount, in grams, of dissolved solid material in a kilogram of sea water after all bromine has been replaced by an equivalent amount of chlorine, all the carbonate converted to oxide, and all the organic matter destroyed (Knudsen, 1901). However, determinations based on this definition of salinity are technically difficult to perform, and recently salinity has been redefined on the basis of seawater conductivity (Cox et al., 1967; Wooster et al., 1969). The various techniques and problems in estimating salinity have been reviewed by Johnson (1969).

Salinity and osmotic pressure of fluids are obviously interrelated, since reduced salinity results in reduced osmotic pressure. If the salinity of sea water changes, the internal fluids of an organism may either change in order to conform with the new external concentration of salts (*osmoconformer*), or the internal fluids may remain approximately at the original value (*osmoregulator*).

Specific gravity, another important physical characteristic of sea water, is influenced by the level of salinity, since increased salinity raises the specific gravity of sea water. Temperature increase, however, lowers the specific gravity

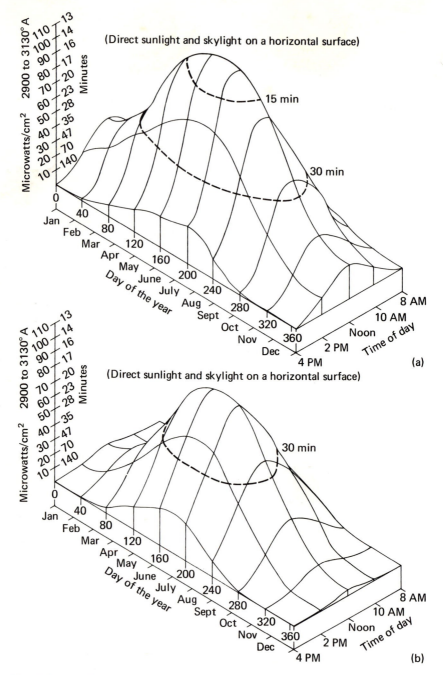

Fig. 2-3. (a) Diurnal and seasonal changes in sunburning ultraviolet radiation at 30°N latitude. (From Daniels, 1962.) (b) Diurnal and seasonal changes in sunburning ultraviolet radiation at 45°N latitude. (From Daniels, 1962.)

of a fluid. Any change in specific gravity may influence the position of an organism in the water column. A high specific gravity is usually characteristic of benthic organisms. Many flotation mechanisms are found in various organisms, including gas and swim bladders, which help maintain an animal at a specific depth.

Because of its general usefulness, and despite its shortcomings, the Venice system was adopted in 1958 for the classification of marine waters according to salinity (Table 2.6). The zonation of biota in a particular region, however, may only approximate the Venice classification. This classification also deals with average salinity conditions, but salinity fluctuations on a tidal and seasonal basis obviously influence the physiology of marine organisms to a considerable extent.

5. Light

Least constant of all the abiotic factors associated with sea water, light changes almost continuously in quality and intensity. These changes may be both ryhthmic (i.e., daily and seasonal) and arhythmic (i.e., the result of differences in cloud cover). The profound influence of light on all levels of bio-

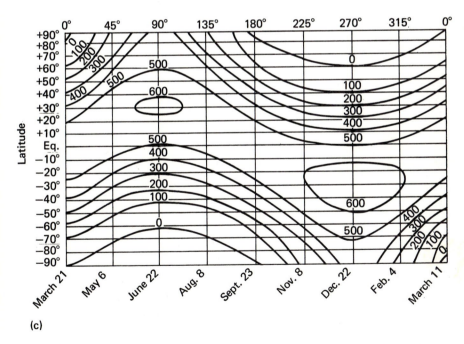

(c)

Fig. 2-3. (c) Daily insolation in calories per square centimeter at the earth's surface when the atmospheric transmission coefficient is 0.7. (From Haurwitz, 1941.)

TABLE 2.6

Venice System for the Classification of Marine Waters According to Salinity

Zone	Salinity (0/00)
Hyperhaline	$> \pm 40$
Euhaline	$\pm 40 - \pm 30$
Mixohaline	$(\pm 40) \ \pm 30 - \pm 0.5$
Mixoeuhaline	$> \pm 30$ but $<$ adjacent euhaline sea
(Mixo-) polyhaline	$\pm 30 - \pm 18$
(Mixo-) mesohaline	$\pm 18 - \pm 5$
α-mesohaline	$\pm 18 - \pm 10$
B-mesohaline	$\pm 10 - \pm 5$
(Mixo-) oligohaline	$\pm 5 - \pm 0.5$
α-oligohaline	$\pm 5 - \pm 3$
B-oligohaline	$\pm 3 - \pm 0.5$
Limnetic (freshwater)	$< \pm 0.5$

logical organization is not fully understood, for it is not always known whether light affects the organism, directly, indirectly, or coincidentally through other environmental changes. Without light, photosynthetic activity would be impossible, and the earth's ecosystem would be destroyed. In addition, light directly influences animal behavior and physiology. The rapid speed of light makes it valuable in the perception of environmental change. Other aspects of light also influence the physiology of organisms, including intensity, wavelength, polarization, angle of reception, and duration of daylight.

Thorson (1964) has estimated that 90 percent of the sea bottom is without daylight. Most benthic species live at depths of less than 200 meters and therefore are subject to the influence of light. Specific examples of the biological implications of light will be presented later; certain physical characteristics of light in the sea need attention here.

Because the sea receives much of the solar radiation reaching the earth, it is important to consider this factor in determining the energy budget of the earth. The quantity and quality of solar radiation vary with time and latitude (Fig. 2.3). The sun emits radiation of different wavelengths ranging from X-rays (short) to radio waves (long). The ionizing radiation, X-ray, and short ultraviolet do not enter the earth's atmosphere, and approximately half of the radiation that does reach the surface of the earth is infrared; the other half is visible and ultraviolet combined.

The amount of radiation penetrating the sea depends in part on two factors:

(1) the amount of radiation immediately above the sea, which is largely determined by the angle of incidence of radiation; and (2) the surface reflectivity.
When penetrating sea water, the beam of light is both scattered and absorbed,
so that radiation decreases with depth (the *extinction coefficient*). The percentage of light penetrating various depths is not the same in different bodies
of water (Fig. 2.4). The spectral distribution of light also changes with depth
(Fig. 2.5), and the degree of change differs in various water columns (Fig.
2.6). Suspended matter, organic substances and their breakdown products and
the presence of kelp or plankton blooms, for example, are only a few factors
that may influence the extent of light penetration.

The angular distribution of light in the sea may be biologically significant,
since it has been suggested that organisms use this clue in orientation. Similarly,
the polarization pattern of light in the sea, which changes with the sun's position, may be another aid to animal orientation.

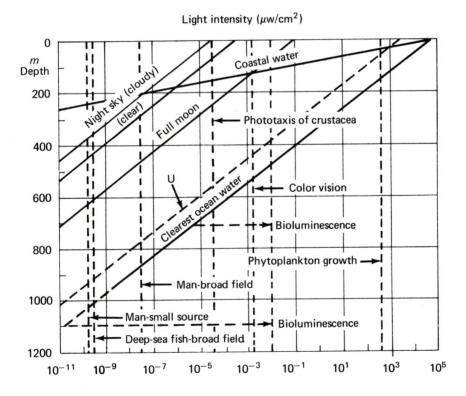

Fig. 2-4. Percentage of light penetrating various depths in different bodies of
water. (From Clarke and Denton, 1962. John Wiley and Sons, Limited.)

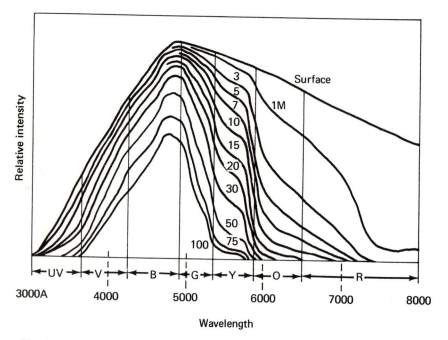

Fig. 2-5. Spectral distribution of light with depth. (From Clarke, 1939. American Association for the Advancement of Science.)

6. Geomagnetism

The earth may be viewed as a magnetized sphere in which the north magnetic pole has probably been stable for at least the past 10,000 years. In the Northern Hemisphere the magnetic field is to the north and downward, and in the Southern Hemisphere it is to the north and upward. However, this relationship is not a stable one, for it is known that the earth's magnetic field has been reversed nine times in the past 3.6 million years (Cox et al., 1967). The reversals of polarity recorded in oceanic rock formations have been an excellent source of evidence for the spreading of the sea floor. As the sea floor cracks and begins spreading, a strip of magnetized rock is produced. Whenever reversal of polarity takes place, the next rock layer formed by sea-floor cracking is magnetized, but in the reverse direction. This sequence of reversals and pattern of spreading is recorded in the magnetized ocean-floor rocks and can be read by a magnetometer from a ship.

During the period when the polarity is reversing the magnetic shielding effect against cosmic rays is lost or altered. Uffen (1963) suggested that when this happened the dramatic increase in radiation dosages could result in increased mutation rates, with the net effect that the rate of evolution would increase.

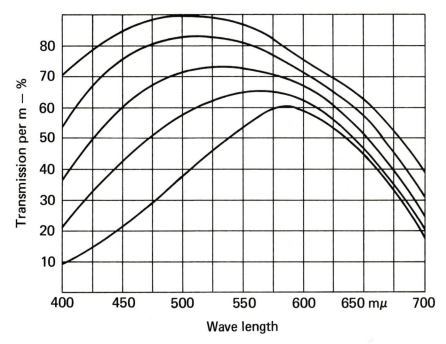

Fig. 2-6. Normal transmission curves for various types of coastal water at solar altitude 45°. (From Jerlov, 1951. Swedish Natural Science Research Council.)

The earth's magnetic field is not uniform and varies in an unpredictable way. Although differences in magnetic intensity (expressed as Gauss Units) are relatively slight between various geographical regions, organisms apparently can detect and are physiologically responsive to them (Brown, 1962).

7. Sound

Recent emphasis on the influence of underwater sound on marine organisms is due in no small measure to military interests, and complex devices to analyze this phenomenon have been developed (Moulton, 1964). Sound travels faster through water than through air; its velocity may be affected by salinity, temperature, and pressure (Fig. 2.7). Various parts of the total sound spectrum have different effects on organisms, and the specific aspects of this interaction will be presented later.

8. Substratum

Although the substratum in marine habitats may be highly variable, its importance must not be overlooked, since it may serve as a source of food to marine organisms or as a protection from predators and marked environmental fluctuations. Two principal types of substrata are recognized arbitrarily:

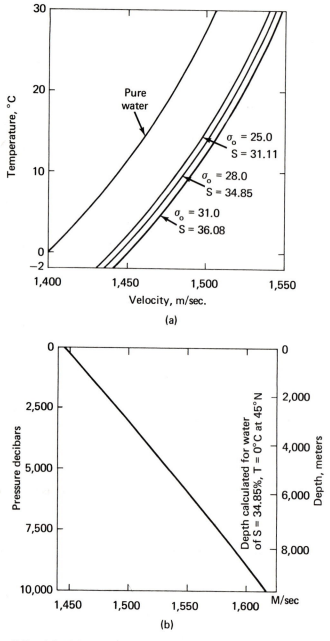

Fig. 2-7. (a) Velocity of sound in pure water and in sea water at atmospheric pressure as a function of temperature and salinity. (b) Effect of pressure upon the velocity of sound in sea water 34.85 0/00 at 0°C. (From Sverdrup, 1942. Prentice-Hall, Inc.)

solid and particulate. Solid substrata include rocks, wood, shells, and beer cans, for example, while particulate substrata are sand, mud, and mixtures of each. According to Zinn (1968), the words substratum and substrate are confused frequently, but *substratum* properly refers to the foundation in which an organism lives, whereas *substrate* is the base on which an organism is found.

Organisms that are attracted to a particular substratum exhibit positive *thigmotaxis,* while those repelled show negative thigmotaxis; the thigmotactic response may change with the stage of the life cycle. Organisms associated with a substratum face a different environmental complex than those organisms living a planktonic existence, with the result that physiological systems may function differently.

The different types of substrata and the correlated physiological diversity of the organisms living there become more apparent in the following examples. Others will be discussed later. Mud flats seldom exist in the presence of significant wave action and generally have a high colloidal content with smaller quantities of poorly circulated interstitial water. As a result, environmental conditions become relatively constant and anerobic conditions characteristically develop. In contrast, rocky shores typically experience heavy wave action along with other factors such as sunlight, desiccation, and temperature which may vary markedly on a diurnal and seasonal basis. The physiological capabilities of the animals living in these two habitats reflect the prevailing environmental conditions.

Many animals show great specificity for a particular substratum, while others are less restricted. For example, the larvae of the wood-boring *Teredo* actively seek wood to complete their life cycle, while size of sand grains determines the distribution of many meiofaunal species.

D. SUMMARY

The vastness and dynamic nature of the sea makes it the most challenging environment for the scientist. Various chemical and physical parameters interact to produce a variety of distinctive environmental complexes, ranging from the cold, dark, unfluctuating depths of the ocean to the rapidly changing condition of an estuary. The physiological capacity of an animal reflects adaptation to its habitat, and to analyze these functional adaptations, the environmental physiologist must know the total environment.

REFERENCES

Brown, F. A., Jr. (1962). Response of the planarian *Dugesia* and the protozoan *Paramecium* to very weak horizontal magnetic fields. *Biol. Bull.,* **123**:264–281.

Chow, T. J., and E. D. Goldberg (1960). On the marine geochemistry of barium. *Geochim. Cosmoch. Acta,* **20**:192–198.

Clarke, G. L. (1939). The utilization of solar energy by aquatic organisms. Problems in Lake Biology. *Publ. Am. Assoc. Advance Sci.,* **10:**27–38.

—— and E. J. Denton (1962). Light and animal life. The Sea: Ideas and Observations, Sec. 4, pp. 456–468. London: Interscience Publishers.

Cox, R. A., F. Culkin, and J. P. Riley (1967). The electrical conductivity/chlorinity relationship in natural sea water. *Deep-Sea Res.,* **14:**203–220.

Cox, A., G. B. Dalrymple, and R. R. Doell (1967). Reversals of the earth's magnetic field. *Sci. Am.,* **216:**44–54.

Culkin, F. (1965). The major constituents of sea water. *In* Chemical Oceanography. J. P. Riley and G. Skirrow, eds. Academic Press, London. **1:**121–162.

Daniels, F., Jr. (1962). Physical factor in sun exposures. *Arch. Dermatol.,* **85:**358–361.

Duursma, E. K. (1961). Dissolved organic carbon, nitrogen and phosphorus in the sea. *Netherl. J. Sea Res.,* **1:**3–135.

—— (1965). The dissolved organic constituents of sea water. *In* Chemical Oceanography. J. P. Riley and G. Skirrow, eds. Academic Press, London. **1:**433–477.

Ekman, S. (1953). Zoogeography of the Seas. Sidgwick and Jackson, Ltd., London.

Goldberg, E. D. (1963). The oceans as a chemical system. *In* The Sea. M. N. Hill, ed. Interscience Publishers, New York. **2.**

—— (1965). Minor elements in sea water. *In* Chemical Oceanography. J. P. Riley and G. Skirrow, eds. Academic Press, London. **1:**163–196.

Guillard, R. R. L., and P. J. Wangersky (1958). The production of extracellular carbohydrates by some marine flagellates. *Limnol. Oceanogr.,* **3:**449–454.

Haurwitz, Bernhard (1941). Dynamic Meteorology. McGraw-Hill, New York. 365 pp.

Hedgpeth, J. W. (1957). Classification of marine environments. *In* Treatise on Marine Ecology and Paleoecology. J. W. Hedgpeth, ed. Geol. Soc. Am. Mem. No. 67, New York. **1:**17–28.

Jerlov, Nils G. (1951). Optical studies of ocean waters. *Swed. Deep Sea Exped. Reports, Physics and Chemistry,* **3:**1–59.

Johnson, R. G. (1967). Salinity of interstitial water in a sandy beach. *Limnol. Oceanogr.,* **12:**1–7.

Johnson, R. (1969). On salinity and its estimation. *Oceanogr. Mar. Biol. Ann. Rev.,* **7:**31–48.

Kanwisher, J. (1955). Freezing in intertidal animals. *Biol. Bull.,* **109:**56–63.

Knudsen, M. (1901). Hydrographical Tables. G. E. C. Gad., Copenhagen. 63 pp.

Kuenen, P. H. (1950). Marine Geology. John Wiley and Sons, Inc., New York.

Macmillan, D. H. (1966). Tides. American Elsevier Publishing Co., Inc., New York.

Mare, M. F. (1942). A marine benthic community, with special reference to the microorganisms. *J. mar. biol. Ass. U.K.,* **25:**517–554.

Menard, H. W. (1964). Marine Geology of the Pacific. McGraw-Hill Book Co., New York.

Moulton, J. M. (1964). Underwater sound: biological aspects. *Oceanogr. Mar. Biol. Ann. Rev.,* **2:**425–454.

Nicholls, G. D., Herbert Curl, Jr., and V. T. Bowen (1959). Spectographic analyses of marine plankton. *Limnol. Oceanogr.,* **4:**472–478.

Pickard, G. L. (1968). Descriptive Physical Oceanography. Pergamon Press, London.

Pytkowicz, R. M. (1968). The carbon dioxide-carbonate system at high pressures in the oceans. *Oceanogr. Mar. Biol. Ann. Rev.,* **6:**83–136.

Remane, A. (1933). Verteilung und Organisation der benthonischen Mikrofauna der Kieler Bucht. *Wiss. Meeresunters. Kiel.*, **21**:161–221.

Revelle, R., and R. Fairbridge (1957). Carbonates and carbon dioxide. *In* Treatise on Marine Ecology and Paleoecology. J. W. Hedgpeth, ed. Geol. Soc. Am. Mem. No. 67, New York. **1**:239–295.

Sanders, H. L., P. C. Mangelsdorf, Jr., and G. R. Hampson (1965). Salinity and faunal distribution in the Pocasset River, Massachusetts. *Limnol. Oceanogr.,* **10**(Suppl. Nov.): R216–R229.

Skirrow, G. (1965). The dissolved gases—carbon dioxide. *In* Chemical Oceanography. J. P. Riley and G. Skirrow, eds. Academic Press, London. **1**:227–322.

Spencer, C. P. (1965). The carbon dioxide system in sea water: A critical appraisal. *Oceanogr. Mar. Biol. Ann. Rev.,* **3**:31–57.

Stefansson, U., and L. P. Atkinson (1967). Physical and Chemical Properties of the Shelf and Slope Waters off North Carolina. Technical Report, Dec. 1967. Duke University Marine Laboratory, Beaufort, North Carolina. Pp. 1–71.

Sverdrup, H. U., M. W. Martin, and R. H. Fleming (1942). The Oceans. Their physics, chemistry and general biology. Prentice-Hall, Inc., New York.

Tatsumoto, M., and C. C. Patterson (1963). The concentration of common lead in sea water. *In* Earth Science and Meteoritics. J. Geiss and E. D. Goldberg, eds. North Holland Publishing Co., Amsterdam. Pp. 74–89.

Thorson, G. (1957). Bottom communities (sublittoral or shallow shelf). *In* Treatise on Marine Ecology and Paleoecology. J. W. Hedgpeth, ed. Geol. Soc. Am. Mem. No. 67, New York. **1**:461–534.

—————— (1964). Light as an ecological factor in the dispersal and settlement of larvae of marine bottom invertebrates. *Ophelia,* **1**:167–208.

Uffen, R. J. (1963). Influence of the earth's core on the origin and evolution of life. *Nature,* **198**:143–144.

Von Arx, W. S. (1962). An Introduction to Physical Oceanography. Addison-Wesley Publishing Co., Inc., Reading, Pa.

Weiss, R. F. (1970). The solubility of nitrogen, oxygen and argon in water and sea water. *Deep-Sea Res.,* **17**:721–735.

Wooster, W. S., A. J. Lee, and G. Dietrich (1969). Redefinition of salinity. *Limnol. Oceanogr.,* **14**:437–438.

Zenkevitch, L. (1963). Biology of the Seas of the U.S.S.R. George Allen and Unwin Ltd., London.

Zinn, D. J. (1968). A brief consideration of the current terminology and sampling procedures used by investigators of marine interstitial fauna. *Trans. Am. Mic. Soc.,* **87**:219–225.

ZoBell, C. E. (1952). Bacterial life at the bottom of the Philippine Trench. *Science,* **115**:507–508.

An intertidal zone mud flat at the Belle W. Baruch Coastal Research Institute field station near Georgetown, South Carolina (Courtesy of Dr. P. J. DeCoursey).

CHAPTER III

THE INTERTIDAL ZONE

The intertidal zone is a narrow fringe of land between the ocean and land, which is alternately exposed to air and covered by water. It is not homogeneous in structure or in physical, chemical, or biotic characteristics. Many schemes of classification have been proposed for the intertidal zone, but probably the most universally accepted and most logical system is based on the nature of the substratum. Because intertidal zone organisms are intimately associated with the substratum, they are often identified as being attached to, or burrowed into, a particular bottom type. In general three major types of habitats are recognized: rocky shores, sandy beaches, and mud-flats. These habitats are found on the edge of the open ocean and bordering protected regions such as harbors and estuaries. All major types may be found close to one another and show varying degrees of intermixing. For example, a rocky shore may be adjacent to a sandy beach, and the transition area between may consist of rocks ranging from large boulders to small stones to various-sized large sand grains. Because microhabitats occur within each microenvironment, a complex terminology and system of classification of intertidal subunits has evolved. This subject, however, is not central to the physiological theme of this book, and the reader is referred to the following references for a review of the subject: rocky intertidal zone, Stephenson and Stephenson (1949); Doty (1957); Southward (1958); and Lewis (1964); sandy beaches, Hedgpeth (1957); Swedmark (1964); and Jansson (1967 a, b); and mud-flats, Carriker (1967); and Postma (1967).

Although there are many biotypes within the general classification of intertidal zones and each has a different microenvironment, certain examples can serve to demonstrate the general environmental characteristics of the intertidal zone.

Because of alternate exposure to water and air, the various abiotic components in the intertidal zone fluctuate widely with differences in duration of exposure controlled by tidal frequency and amplitude which vary geographically (see Chapter II). In regions of noticeable tidal change, sections of the intertidal zone are exposed for varying lengths of time to the aerial environment; the upper region of the intertidal zone particularly is exposed to air for longer periods of time. This time differential in exposure has a profound effect on the vertical distribution of organisms, and organisms near the high tide mark tend to be more resistant to environmental stress than those found near the low tide mark.

There are often marked annual and diurnal environmental changes in the intertidal zone as illustrated in studies by Jansson and his coworkers on sandy beaches in Sweden. Over a 24-hour period during the summer, the temperature was measured in air and at various depths. The greatest change over 24 hours was observed at the surface of the sand where temperatures varied as much as 28°C. At a depth of 70 cm, however, no diurnal change was observed (Fig. 3.1). The surface temperature of different parts of a beach measured at the

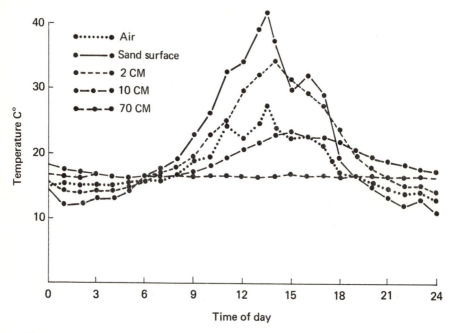

Fig. 3-1. Diurnal variations in temperature at different depths during the summer at a beach in Sweden. (After Jansson, 1967b.)

same time also varied least near the water (Fig. 3.2). During winter the daily thermal amplitude was small, and temperatures were uniform at depths below a few centimeters. In spite of low air temperatures, the temperatures below a few centimeters in depth never fell below −2 to −3°C. Although the beaches studied bordered the low salinity Baltic Sea, there were salinity gradients with distance from the sea (Fig. 3.3). Vertical differences in salinity were also recorded, and the magnitude of change varied with distance from the sea. When

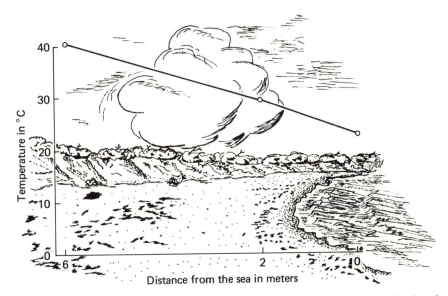

Fig. 3-2. The horizontal distribution of surface temperature on a Swedish beach during the summer at 1 p.m. Insolation: 62,000 lux, wind: 0 m/sec. (After Jansson, 1967b.)

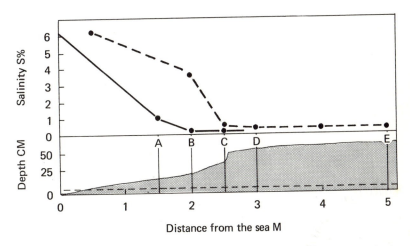

Distance from the sea M

Fig. 3-3. The horizontal distribution of salinity at a Swedish beach during the summer on two different days. Day 1 (-----) was cloudy, with winds of 2–4 m/sec, and small waves. The ground water table is indicated by the dashed line near the bottom. Day 2 (——) was sunny, with winds of 0–2 m/sec, and small waves. The ground water table is identical with the abscissa. A through E represent different sampling sites. (After Jansson, 1967b.)

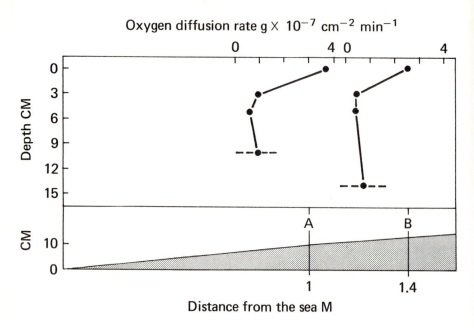

Fig. 3-4. Horizontal and vertical distribution of oxygen on a Swedish beach. (After Jansson, 1967a. North-Holland Publishing Company.)

the sea-water salinity was 6.4 0/00, the salinity at a distance of 10 cm from the sea and a depth of 0.5 cm was 11.4 0/00; at 3 cm depth it was 7.9 0/00. In contrast, at a distance of 40 cm from the sea the salinity was only 0.9 0/00 and did not appreciably change with depth. Marked changes in salinity were observed with heavy rains, fluctuations in ground water pressure, and evaporation from the sand surface, which is influenced by temperature and wind. Similar general findings have been reported for a sand beach in California (Johnson, 1965, 1967).

An analysis of the concentration and distribution of oxygen determined in sandy beaches in Sweden by Jansson (1967a) demonstrated the following trends. Along a transect parallel to the beach, the oxygen content at corresponding depths was similar. Although the highest oxygen concentrations occurred near the water, horizontal and vertical distribution of oxygen showed no clear trend, with the maximum appearing at various depths (Fig. 3.4). The availability of oxygen is governed in great measure by the physical and chemical properties of the sand, such as porosity, permeability, and percentage air volume. The flow of interstitial water in the beach and the dynamic impact of wave action greatly increase the availability of oxygen in adjacent areas. One example from the work of Jansson (1967a) demonstrates the relative influence of oxygen, temperature, and salinity on the vertical distribution of the marine fauna inhabiting sandy beaches: at a depth of 1 cm the oxygen content was 5.64 mg/

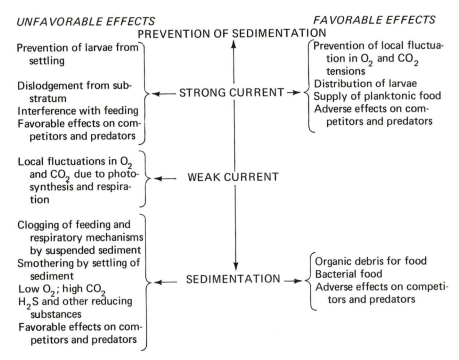

UNFAVORABLE EFFECTS   *FAVORABLE EFFECTS*

PREVENTION OF SEDIMENTATION

Prevention of larvae from settling

Dislodgement from substratum
Interference with feeding
Favorable effects on competitors and predators

← STRONG CURRENT →

Prevention of local fluctuation in O_2 and CO_2 tensions
Distribution of larvae
Supply of planktonic food
Adverse effects on competitors and predators

Local fluctuations in O_2 and CO_2 due to photosynthesis and respiration

← WEAK CURRENT

Clogging of feeding and respiratory mechanisms by suspended sediment
Smothering by settling of sediment
Low O_2; high CO_2
H_2S and other reducing substances
Favorable effects on competitors and predators

← SEDIMENTATION →

Organic debris for food
Bacterial food
Adverse effects on competitors and predators

Fig. 3-5. The influence of water currents on other environmental factors. (From Lilly *et al.,* 1953. Blackwell Scientific Publications Ltd.)

liter, the temperature 21.5°C, and the salinity 7.3 0/00; but at a depth of 10 cm the oxygen content was 0 while the temperature and salinity were relatively unchanged. This drop in oxygen content influenced the vertical distribution of nematodes and turbellaria. At 10 cm only 1 nematode per 10 ml was found; at depths of 1–2 cm, 11–19 nematodes and 1–4 turbellarians per 10 ml were found. The type of substrate also has a profound effect on the distribution of animals inhabiting sandy beaches. Faunal assemblages occupying siliceous sand are quite different from those of sublittoral shell-sand bottoms. Furthermore, chemical composition of sand grains and size of sediment particles influence the distribution of organisms. Swedmark (1964) has reviewed in detail the interstitial fauna of marine sand.

Water movement influences the biota not only of sandy shores but also of the rocky shore zone. Two aspects of water movement in particular affect the organisms in these areas: water flow and wave crash. The violent and irregular crashing of waves exerts direct physical force on organisms inhabiting rocky surfaces, while wave action and water flow influence sediments, physiochemical properties, and the transport of food and reproductive stages. The influence of water currents is summarized in Fig. 3.5. Frequently, in high wave action regions littoral organisms are found higher above the high water level than organisms living on wave-sheltered shores (Fig. 3.6).

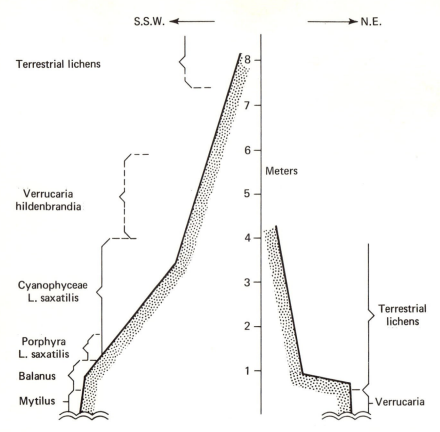

Fig. 3-6. The influence of high wave action on the zonation of littoral organisms in the littoral zone at adjacent sites in Norway. The southern area is fully exposed, the northern one very sheltered. (From Lewis, 1965. Pergamon Publishing Company.)

Although all intertidal habitats have general characteristics in common, each major type has its own distinctive features. On rocky shores the nature of the substrate inhibits burrowing, but some species living here seek crevices or the undersurface of boulders or rocks to escape environmental extremes. The hard substrate on rocky shores is favorable for the attachment of sessile organisms and also provides a hard surface on which other organisms can move. The sessile animals in these regions are especially subject to environmental extremes. The effects of sunlight may act directly on rocky shore species or influence the expression of other physical parameters, such as salinity, humidity, and temperature. The intensity of different factors varies on temporal and geographical bases; for example, temperate zone animals are exposed to wide seasonal temperature ranges. The generally pronounced wave action common to this rocky shore environment greatly influences many of these environmental factors.

Mud shores are absent in regions of marked wave action or major water movements. Many of the animals living here are burrowing forms since the nonburrowing species typically do not need a hard surface for attachment or mobility. Organic content is high, providing a food-rich habitat for detritus feeders. Anaerobic conditions with high concentrations of H_2S are common.

Although sand beaches vary in chemical composition and size of sand particles, they are somewhat intermediate between mud beaches and rocky shores in terms of environmental characteristics. The substrate is sufficiently firm to support crawling organisms but sand is unstable, particularly on unprotected wave-pounded beaches. Thus, few organisms exist at the surface of these sand beaches, and most are burrowing forms. Protected sand beaches show gradation from pure sand to various amounts of mixture with mud. As in mud beaches, environmental factors fluctuate less at a depth of a few centimeters than at the surface.

Sulfides and H_2S, which are present in the substratum when oxygen is absent, have been shown to influence intertidal animals. Organisms from a mud habitat are more resistant to H_2S than those from a mussel-bank community, and organisms exhibiting a high degree of locomotor activity are less resistant to H_2S than less active or sessile animals (Jacubowa and Malm, 1931; Theede et al., 1969).

Although excessive CO_2 may never be a limiting factor for animals living on sandy beaches (Bruce, 1928), some data indicate that CO_2 dissolved in sea water influences the survival of marine animals. The shore crab *Carcinus maenas*, which may inhabit isolated shallow pools, tolerates higher CO_2 concentrations than many other decapod crustaceans (Arudpragasan and Naylor, 1964), and bottom-dwelling fish survive high CO_2 concentrations better than more pelagic species (Shelford and Powers, 1915).

Although the above studies have emphasized the lethal effects of a single environmental factor, it should be emphasized that the environmental complex consists of multiple factors acting in concert on the animals in their environment. See Chapter V for further discussion of the interaction of multiple factors.

A. RESISTANCE ADAPTATIONS

Since intertidal zone organisms are subject to widely fluctuating environmental factors, it is not surprising to find that these animals can tolerate more extreme conditions than other marine organisms. Intertidal zone organisms are generally more eurythermal, euryhalinic, and more resistant to desiccation than animals occupying other marine habitats. Furthermore, toleration limits can be correlated with the position the animal occupies in the intertidal zone: generally, animals living in the higher regions of the intertidal zone are more resistant to environmental fluctuations than those living closer to the subtidal zone.

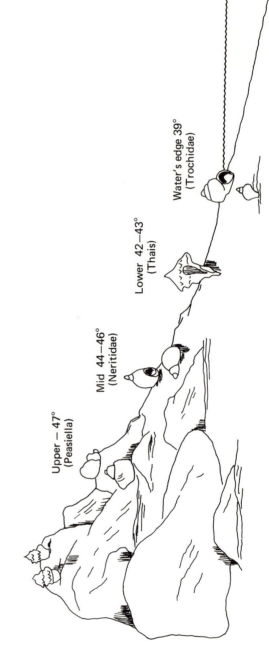

Fig. 3-7. Temperature tolerances of molluscs living in different parts of the intertidal zone. (Based on data from Fraenkel, 1966.)

Intertidal Zone

Spray zone 47–48.5°
(Nodilittorina)

Upper – 47°
(Peasiella)

Mid 44–46°
(Neritidae)

Lower 42–43°
(Thais)

Water's edge 39°
(Trochidae)

1. Temperature

Tolerance to high temperatures is one of the chief characteristics of intertidal zone organisms, especially those living in the upper areas of this region.
For example, Fraenkel (1966) found that molluscs living in the upper part of
the intertidal zone could tolerate temperatures of 49–50°C, but molluscs living
in the lower part of the intertidal zone could not resist temperatures above
37°C (Fig. 3.7). Thermal tolerance at different tidal levels may also be illustrated by the difference between the intertidal zone amphipods *Haustorius
canadensis* and *Neohaustorius biarticulatus,* which can survive temperatures up
to 41°C, and the subtidal species *Acanthohaustorius millsi, Protohaustorius
deichmannae,* and *P. longimerus,* which die at 39°C (Sameoto, 1969).

Some intertidal zone animals distributed over a wide latitudinal range are
resistant to very low temperatures as well as high temperatures. One of the best
examples of wide thermal tolerance can be found in the intertidal mollusc
Modiolus demissus. This animal, which is found from Canada to Brazil, can
survive temperatures ranging from −22° to between 34° and 40°C (Lent,
1968).

There has been considerable speculation about whether thermal tolerances
shift in latitudinally separated populations of the same species. Ushakov (1964)
has argued that resistance to high thermal stress in each species changes little
except for slight adaptation to local and seasonal conditions. Thus, the thermal
lethal limits of a widely distributed species are approximately the same throughout its geographical range. Fraenkel (1968) has pointed out that the Littorinidae are the most representative world-wide family in the upper intertidal
zone, and all are resistant to high temperature. Similar tolerance to high temperatures has been observed in the widely distributed intertidal zone fiddler
crabs (genus *Uca*), which do not vary significantly in the upper thermal lethal
limits throughout their geographical range from Massachusetts to Brazil (Vernberg and Vernberg, 1967). Although both the Littorinidae and species of *Uca*
occur intertidally throughout their range, some widely distributed animals shift
from an intertidal to a subtidal position as habitat temperatures change. For
example, the sea urchin *Strongylocentrotus purpuratus,* ranging from Alaska
to Mexico, has an upper thermal limit of 23.5°C. Distribution of this species in
the marine environment reflects its lack of ability to adapt to higher temperatures, because these urchins are found intertidally at higher cold-water latitudes,
but subtidally at lower, warm-water ones (Farmanfarmaian and Giese, 1963).

Lower thermal tolerances may be markedly different in latitudinally separated species. In *U. rapax,* which is distributed transequatorially along the
eastern coast of the Americas in subtropical and tropical areas, the lower
thermal lethal limits can be correlated with the thermal habitat of a particular
population. Thus, subtropical species tolerate significantly lower temperatures
than do tropical zone populations of crabs (Fig. 3.8). Furthermore, these differences in response to low temperature appear to be genotypic, for the lower

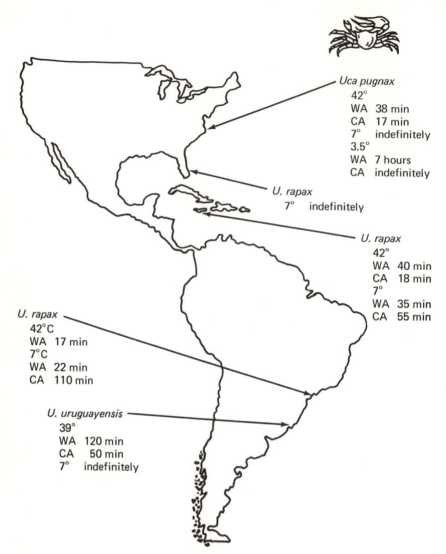

Fig. 3-8. Temperature tolerances of temperate and tropical zone species of fiddler crabs (genus *Uca*). Each LD_{50} given in minutes. WA-warm-acclimated, CA-cold-acclimated. (Based on data from Vernberg and Tashian, 1959; Vernberg, F. J. and W. B. Vernberg, 1966a.)

thermal limits of the tropical zone species cannot be shifted significantly (Vernberg and Vernberg, 1967).

Upper thermal limits may be modified by decreases in salinity, and for many marine animals the upper lethal temperatures decrease as salinity decreases (Kinne, 1964). For example, survival time in the tube-building polychaete worm *Clymenella torquata* at 37°C is markedly less at low salinities; at low temperatures, however, the worms can tolerate low salinities for prolonged periods of time (Kenny, 1969). The increased tolerance to low salinities at decreased temperature can be correlated with the normal environmental conditions *C. torquata* encounters in nature. The estuary where this study was made, the Newport River estuary in North Carolina, is a winter low salinity system. Thus, periods of low salinity usually coincide with periods of low temperature.

What are the mechanisms that enable these intertidal zone organisms to tolerate thermal stress? Although intertidal zone invertebrates cannot regulate body temperature at a high constant level as most birds and mammals can, they do have a limited ability to thermoregulate at high temperatures. Evaporative cooling, which is one method of thermoregulation, has been studied in three species of tropical intertidal marine animals, a barnacle, *Tetracleta squamosa,* a limpet, *Fissurella barbadensis,* and a gastropod, *Nerita tesselata* (Lewis, 1963). It was found that at high temperatures all three species had body temperatures lower than their environment, and none absorbed radiation as did inanimate or black bodies. Tissue temperatures of the barnacles were 4.9°C lower than that of the black bodies, limpet body temperatures were lower by 5.6°C, and temperatures of gastropod tissue were lower by 8.0°C. These differences led Lewis to suggest that the lower body temperatures were due to evaporative cooling. There also seems to be a correlation between position in the intertidal zone and the ability to regulate body temperature. The most effective regulator is *Nerita,* which lives at the high tide level, while the barnacle and limpet, living at mid-tide levels, did not regulate as effectively. Evaporative cooling has also been reported in fiddler crabs, genus *Uca* (Edney, 1961, 1962). At high temperatures, body temperatures of fiddler crabs always are lower than that of their habitat temperature, a fact that Edney (1962) attributes to transpiration and the convection of heat.

Rather than using physiological strategies to avoid environmental stress, some animals have evolved behavioral adaptations to avoid elevated temperatures. By clustering on a falling tide, the Australian gastropod *Cerethrium clypeomorus moniliferum* can decrease environmental temperature. Temperatures on the sand surface average 30.6°C, but in the center of the cluster the temperature is only 27.8°C (Moulton, 1962). The bivalve mollusc *Modiolus demissus,* which is thigmokinetic, tends to assume clumped distribution patterns buried in marsh mud up to its siphons. This pattern of distribution has two advantages: relative humidity is 100% and temperature is 5–10° cooler in the middle of the clump of bivalves than it is in the air (Lent, 1968). The diurnal phototactic rhythm reported for *Uca pugnax* is another example of a behavioral thermoregulatory mechanism (Palmer, 1962). In their daily rhythm these crabs are attracted to light in the morning, but avoid light during the hottest part of the day. When temperatures reach a certain high point, the fiddler crabs will return to their burrows, which are several degrees cooler than at the surface. In this manner the crabs can lower their body temperature appreciably.

At low temperature, however, the body temperature of poikilotherms does not differ significantly from the ambient temperature. Although tropical zone animals cannot long survive temperatures below 10°C, adaptation to low temperature has been clearly demonstrated in the case of temperate and polar zone intertidal animals. Kanwisher (1955) reported that Arctic animals may be frozen for six months. Intertidal animals from Massachusetts survive prolonged exposure to −20°C, and the percent of body water frozen at −15°C in some of these animals ranged from 54 to 67. Histological studies revealed large pockets of intercellular ice with shrinkage and distortion of surrounding cells (Kanwisher, 1959). The metabolic rate dropped markedly (Q_{10} values may be as high as 50), quite possibly as a result of increased salinity in the tissues.

The question of the biochemical adaptations to both high and low thermal stress has not yet been resolved. It is known that some enzyme systems in high intertidal molluscs remain stable at very high temperatures. Aspartic glutamic transaminase from *Modiolus demissus,* for example, showed no reduction in activity after 12 minutes at 56°C (Read, 1963). Although similar thermostability for this enzyme system was reported in intertidal molluscs, Read suggested that perhaps this enzyme could be more important in adaptation to anaerobic conditions than in temperature adaptation per se.

Usually cold-acclimation enhances enzymatic activity. One key enzyme system which has been implicated in thermal acclimation is cytochrome *c* oxidase, the terminal oxidase of the electron transfer system in molluscan and anthropod tissues (Bliss and Skinner, 1963). When cytochrome *c* oxidase activity was determined in muscle tissue and in the supraoesophageal ganglion (SEG) of widely separated populations of *Uca,* marked differences were observed. The enzymatic activity in muscle tissue from *U. pugnax* from North Carolina and *U. rapax* from Florida was significantly increased after exposure to cold. In contrast, enzymatic activity in muscle tissue from a tropical population (Puerto Rico) of *U. rapax* did not show significant alteration following cold exposure. Enzymatic activity in the SEG was enhanced after cold only in the North Carolina population. In the SEG of crabs from Florida enzymatic activity did not change after cold, and in the Puerto Rican population enzymatic activity decreased (Vernberg and Vernberg, 1968a).

2. Salinity

Intertidal zone animals tend to tolerate wider ranges in salinity than do subtidal and open ocean organisms, and generally animals living in the upper intertidal zone area are more tolerant of salinity fluctuations than are those in the lower intertidal zone. Some animals can even survive extremely hypersaline conditions in tidal pools cut off from ocean waters. For example, fiddler crabs, *Uca rapax,* are commonly found living on the salt flats of Puerto Rico in salinities as high as 90 0/00. The strong ability to hyporegulate is evident in two species of crabs, *Pachygrapsus crassipes* and *Hemigrapsus oregonensis,* which are known to thrive in a hypersaline lagoon (66 0/00) cut off from the sea. Some species of intertidal zone fish that inhabit protective rocky shores show remarkable tolerance to high salinities. Along the Texas coast where salinities in tide pools may reach very high levels, fish have been found living in salinities as high as 142.4 0/00 (Gunter, 1967). At the other extreme, prolonged rains can lower salinities markedly in the intertidal zone region, and, although some intertidal organisms migrate during low salinity stress, many can survive for prolonged periods of time in low salinity waters. Tolerance to low salinities in intertidal zone animals has been linked to several mechanisms. It has been suggested, for example, that the ability of certain polychaetes to adapt to low salinity situations may be due to a number of factors: (1) active transport of salts by the body surface from the medium to the body fluids; (2) a reduction of the permeability of the body surface to salts, to water, or to both; and (3) possible production of hypoosmotic urine (Oglesby, 1969). Certain intertidal bivalves may lose amino acids in response to low salinities; these changes in amino acid concentrations are believed to have mainly an osmotic function (Hammen, 1969). It is quite probable that similar mechanisms would be operative in other intertidal organisms (see p. 126 for a more detailed discussion of osmoregulatory mechanisms).

3. Desiccation

Most marine organisms are not confronted with the problem of desiccation, and it is not a serious problem for mobile intertidal organisms. However, it may be a major problem for sessile animals living near the high tide mark in the intertidal zone. Many of these animals are able to tolerate desiccation for long periods of time, and their resistance to desiccation appears to be related to vertical zonation (Fig. 3.9). Brown (1960) analyzed three factors that influence desiccation in relation to vertical distribution of six species of intertidal gastropods: (1) rate of water loss, (2) amount of water loss the molluscs could survive, and (3) length of time the animals could tolerate various saturation deficiencies. The first two factors correlated to some extent with vertical zonation, but zonation correlated almost completely with the third factor (Fig. 3.10). Another study suggests that the upper level of distribution of intertidal organisms is determined by interplay between rate of loss of water and the time required to regain the water lost when the organism is again covered by the tide (Davies, 1969). The mollusc *Patella vulgata* inhabits both high and low level areas, whereas a closely related species, *Patella aspera,* is confined only to those lower intertidal regions that do not dry out during tidal exposure. Davies attributed the ability of the high level limpet *P. vulgata* to limit the rate of water loss better than low level limpets to its relatively small shell circumference, thus presenting less surface area of tissue from which water may be lost. He also suggested that permeability to water of the mantle tissues could be a factor in water loss.

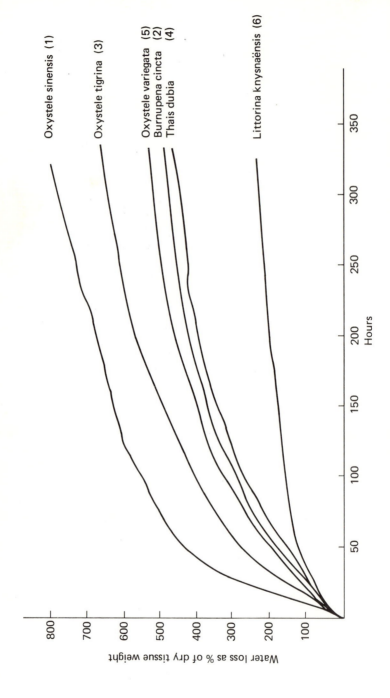

Fig. 3-9. Rate of loss of water of six species of gastropods when dried over calcium chloride at 19°C. The relative positions of the species on the shore, from low to high levels, are indicated by the figures in brackets. (From Brown, 1960.)

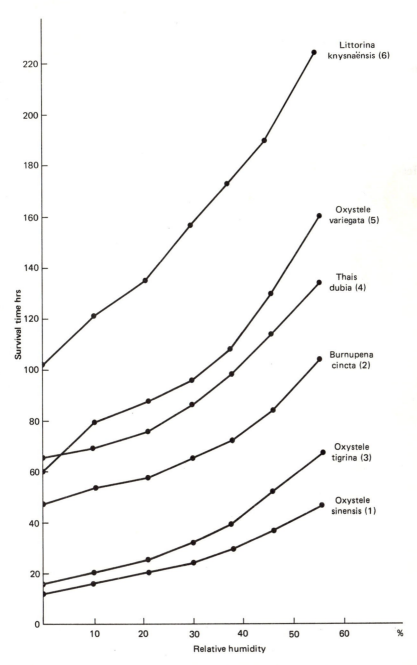

Fig. 3-10. Survival times of six species of gastropods at low humidities at 19°C. The relative positions of the species on the shore, from low to high levels, are indicated by the figures in brackets. (From Brown, 1960.)

To assess the relative success of sessile species in occupying different parts of the intertidal zone, Foster (1971) studied the desiccation resistance of various life stages in the following species of barnacles: *Chthamalus stellatus,* from high tidal levels; *Elminius modestus* and *Balanus balanoides,* residents of the middle tide levels; and *Balanus crenatus,* usually found near the low tide level. Both the adults and the spat of *Chthamalus* lose water at a slower rate and survive longer under comparable conditions than the other three species; *B. crenatus* was the most susceptible to desiccation. Small organisms are less resistant to water loss because of greater surface area to volume ratios. Foster suggested that when the tide is out, small animals may be susceptible to desiccation at normal temperatures and low humidities, while large barnacles may be prone to death from high temperature.

Some slower moving animals also must be able to resist desiccation when the tide recedes and they are exposed to air. Certain intertidal zone fish, for example, can survive out of water for periods up to several days (Gibson, 1969). Their survival depends on keeping the body wet, particularly respiratory surfaces, and some have morphological modifications that aid in keeping body surfaces moist. Two species of *Adamia,* for example, have fine grooves in their skin which hold moisture when they are out of water. Other species rely on behavioral patterns, burrowing in moist substrate. Survival time in some of these fish is greatly reduced in completely dry air, although survival time is not affected when relative humidities are 35% or higher.

4. Gases

Many intertidal animals experience periods of oxygen stress. For example, inhabitants of a soft substratum may be exposed to low oxygen tensions or anaerobic conditions in their burrows, and some sessile species, such as barnacles and clams, may close themselves off from the external environment during the low tide period. During these times such animals are effectively anaerobic organisms.

The fiddler crab, *Uca pugnax,* a semiterrestrial salt marsh species, is active at low tide, but lives in burrows during high tide. These crabs can survive anoxia for 24 hours at 21°C (Teal and Carey, 1967), whereas rapid swimming portunid crabs die quickly under these conditions. Diversity in survival ability can also be shown by the permanent and temporary inhabitants of large brown seaweeds, where anaerobic conditions commonly develop during low tide. The permanent residents of this vegetation tend to be slow-moving animals capable of surviving anaerobiosis for as long as 16 hours at 25°C. The temporary animals, which are generally mobile species capable of leaving their seaweed habitat, are much more susceptible to anaerobiosis, some dying within minutes (Wieser and Kanwisher, 1959). Within one habitat the normal locomotor pattern of a species can be correlated with the ability to withstand anaerobiosis. Cercariae of two trematode species emerging from the same species of snail can illustrate this point. The very active, tailed cercariae of *Himasthla quissetensis* survived anaerobic conditions for 6 to 8 hours, but, in contrast, the tailless, non-swimming cercariae of *Zoogonus lasius* did not die until after 12 hours of oxygen deprivation (W. Vernberg, 1963). Some mud-dwelling species are harmed by relatively high concentrations of oxygen. *Scoloplos armiger*, a mud-dwelling polychaete worm, survives better in poorly aerated water (4%) than in fully aerated water (21% oxygen) (Fox and Taylor, 1955).

Although the above examples emphasize the survival of the individual organism under anoxic conditions, oxygen lack also may influence the reproductive capability of a species without harming the individuals, yet effectively killing the species in terms of population continuity. The work of Reish and Barnard (1960) on the estuarine polychaete *Capitella capitata* demonstrates this point. When little or no dissolved oxygen was present, the test animals stopped eating and died after several days; when the median oxygen content rose to 2.9 p.p.m. (parts per million), animals began eating; however, it was not until the oxygen level rose to 3.5 p.p.m. that animals would reproduce and complete their life cycle.

The lethal oxygen level for a species is not easily determined since a number of variables may be operative. Tolerance to anoxia may vary with different stages in the life cycle; differences may be extremely pronounced in those species where the various stages occupy widely divergent habitats. For example, the free-living cercariae of trematodes are much more sensitive to anoxia than the endoparasitic adults (W. Vernberg, 1969). A number of other variables may be important, but, as yet, it is difficult to generalize on their influence since all species do not respond similarly. Some of the known variables include sex, body size, parasitism, temperature, pollutants, molting stage, and previous exposure to level of oxygen concentration (Vernberg, 1971).

The respiratory pigments of some animals may function as an oxygen reservoir during anoxia, releasing oxygen as needed for general oxidative processes (Manwell, 1960). This oxygen may be released slowly. Hemoglobin of the pogonophorian *Siboglinum atlanticum*, for example, was not completely deoxygenated even after an anaerobic period of 24 hours (Manwell et al., 1966). However, the adaptive role of respiratory pigments in anaerobiosis may not be a general phenomenon. On the basis of analyses of water and blood samples of the worm *Arenicola marina*, Eliassen (1955) concluded that during the summer, oxygen content was high enough in the Wadden Sea to eliminate any need for hemoglobin to function as an oxygen reservoir. In addition, Fox (1955) could find no increase in hemoglobin in *Arenicola* when it was subjected to low oxygen tensions.

The amount of hemoglobin an animal has apparently is related to its habitat preferences. In a study on a number of species of clams from different habitats, Manwell (1960) found that the species living in open sandy beaches, the pismo clam *Tivela stultorum*, had the highest concentration of tissue hemoglobin. It was also the most sensitive to oxygen lack. Generally, respiratory pigments of animals inhabiting oxygen-poor habitats have a greater affinity for oxygen than those of animals from oxygen-rich environments. A more detailed discussion on the interaction of the environment and respiratory pigments follows (p. 118).

In the absence of oxygen, an organism may utilize metabolic processes which do not require oxygen. Two main methods may be used: (1) the incomplete breakdown of glucose to CO_2 and water with the formation of intermediate compounds resulting from reactions not utilizing oxygen; and (2) the utilization of alternate metabolic pathways, especially the hexose monophosphate shunt. These schemes of metabolic pathways were discussed in Chapter I. When the animal returns to an oxygen-rich environment, the accumulated intermediates may be oxidized to CO_2 and water. Oxidizing these accumulated intermediates may require additional oxygen resulting in an increase in the metabolic rate of the organism. This increase in oxygen uptake, then, repays the oxygen debt incurred during anaerobiosis. Some anaerobes can excrete the accumulated intermediates without oxygen, but from an energetics viewpoint the incomplete oxidation of glucose during anaerobiosis is very costly. The efficiency of conversion of glucose is approximately ten times less for anaerobic processes than for aerobic ones.

Typically during anoxia, glycogen levels decrease, lactic acid increases, and an oxygen debt is incurred. All three of these physiological properties were observed in the fiddler crab, *Uca pugnax,* which lives in burrows in salt marshes and can survive anaerobic conditions for prolonged periods of time (Teal and Carey, 1967). When comparing a subtidal snail, *Gibbula divaricata,* and a midlittoral species, *Monodonta turbinata,* Bannister et al. (1966) concluded that *G. divaricata,* which typically is exposed to longer periods of anaerobiosis, demonstrated a better capacity to withstand an oxygen debt. This conclusion was based on the higher consumption rate of pyruvate by tissue homogenates of *Gibbula,* since pyruvate is incompletely broken down during anaerobiosis.

Not all marine animals follow this basic anaerobic theme. *Arenicola marina,* an annelid, consumes glycogen during anoxia but lactate or pyruvate do not accumulate, nor is there an apparent oxygen debt (Dales, 1958). A further modification is demonstrated by another worm, *Owenia fusiformis,* which is very resistant and can tolerate at least 21 days of anaerobiosis. This species does not deplete its glycogen reserve, but instead reduces metabolic demands to a minimum by becoming quiescent (Dales, 1958).

Metabolic studies on the oyster *Crassostrea virginica* illustrate how one species can cope with anaerobic conditions. The rate of oxygen uptake by oysters when their valves are open is relatively high for a sessile bivalve mollusc, but when the valves are closed the oxygen uptake rate becomes zero and remains at this level until the valves open (Galtsoff, 1964). Thus, the oyster shuts itself off from the external environment by closing its shell and under these conditions essentially becomes anaerobic. Under adverse situations, such as prolonged exposure to fresh water, this species can remain tightly closed for periods up to two weeks (Pearse and Gunter, 1957).

Lund (1957) found that temperature can influence survival under anaerobic conditions. During the winter, oysters survived anoxia for 18 days when the temperature varied from 14° to 23°C, but summer temperatures of 26.8° to 31.4°C reduced survival time to only 7 days. The oyster's rich supply of glycogen (1–8% by weight) must serve as a metabolic resource during the anaerobic phase of glucose breakdown. Glycolysis, as measured in the oyster's mantle tissue, produces more succinic than lactic acid, a result earlier correlated with a low activity of lactate dehydrogenase (Simpson and Awapara, 1966). Recently, Hammen (1969) proposed the following metabolic scheme: pyruvate, which results from the degradation of glucose by the classical glycolytic pathway, is converted through a reversal of the citric acid cycle to fumarate and then to succinate. The production of more succinate than lactate during anaerobiosis could be advantageous in that succinic acid, being a weaker acid than lactic, might produce less dissolution of shell if it diffused out of the tissue. Upon return to aerobic conditions, excess succinate could be oxidized readily through the Krebs Cycle. Since most of the intermediates and enzymes required for complete oxidation of glucose have been found in species of *Crassostrea,* Hammen (1969) concluded that the Krebs Cycle is operative in the oyster.

B. CAPACITY ADAPTATIONS

1. Perception of the Environment

The sensory modalities that enable animals to survive and compete in the intertidal zone can be correlated with habitat and mode of life. A polychaete that lives burrowed in sand, for example, does not depend on a well-developed visual sense in the search for food or a mate, but a highly mobile, semiterrestrial fiddler crab does. Furthermore, since many intertidal zone animals have pelagic larval stages, the sensory modalities important during larval life may not be significant during the adult stage. For a comprehensive discussion of sensory receptors in marine invertebrates, readers are referred to Laverack's (1968) review of the various sense organs. This section will be more concerned with the manner in which an intertidal zone animal acquires sensory information from the external environment and relays it to the central nervous system to aid in locating a particular habitat, searching out a specific type of food, detecting enemies, and finding a mate.

a. Sensory modalities of larvae

The pelagic larval stage common to most intertidal zone animals is advantageous in perpetuating the species, for it transports the larvae into new areas and thereby increases chances of interbreeding. On the other hand, this stage necessitates the search for a suitable substrate for settling in a favorable environment where survival is possible. This problem is particularly critical for sessile intertidal forms. The solution is not left entirely to chance, since many interacting factors insure that larvae are in the right place at the right time for settling. Among the most significant are phototactic, hydrostatic, and geotactic responses.

The behavior of the larval stage of many intertidal zone animals is controlled by their response to light. In some larvae, light response can be correlated with habitat preferences of the adults, as illustrated by the two species of intertidal sponges *Mycale macilenta* and *Haliclona* sp. As in the majority of intertidal sponges, the larvae of these two species are liberated at an advanced stage and have a free-swimming period lasting only a few hours. Adult *M. macilenta* are cryptic species occurring under stones in the lower tidal regions, and their larvae are photonegative throughout the swimming period. On the other hand, adults of *Haliclona* sp. live in a higher tide region in more exposed conditions, and the larvae swim actively for 9–10 hours after they have been released. Their strong photopositive response keeps them swimming near the surface of the water until they are ready to settle (Berquist and Sinclair, 1968). Response to light may also vary from stage to stage in the life cycle. Consider, for example, larvae of the bivalve mollusc, *Mytilus edulis*. The trochophore stages show no response to light, young veligers are photonegative, the straight-hinge veliger stages will concentrate toward light, the veloncha larvae do not respond to light, and the eyed-veliger stages are positively phototactic. Then at the time of settlement, the larvae become photonegative. The adaptive value of light response can be illustrated by these mussels; for example, young larvae do not become photopositive until the swimming and defensive mechanisms, including the ability to retract the velum between the shell valves, have become well developed (Bayne, 1964).

In general, larvae of most species that have been studied are photopositive when first released, starting life by swimming toward light, and becoming photonegative with age. Thus, the larvae are brought to the phytoplankton-rich waters where they grow and develop. When they are ready to settle, they become photonegative and retreat from light to the bottom (Thorson, 1964). Larvae appear to generally prefer a diffuse, not too strong light, and they are more responsive to blue and green wavelengths (460–510 mμ) that penetrate most deeply in the sea (Clarke and Oster, 1934).

Response to light can be modified by such environmental factors as temperature and reduced salinity. High temperatures tend to reduce photopositive response, as illustrated by the eyed-veliger stage of *Mytilus edulis*. Between 7° and 15°C, these larvae concentrate toward light, but when temperatures are raised to 20°C, they become generally distributed and show no photopositive response (Bayne, 1964). A similar response in adult marine animals has been explained by the fact that the respiration rate of these organisms is conditioned in part by temperature and in part by light (Friedrich, 1961). The exchange of gases between animal and environment, which must be in balance, will shift in the same direction with increasing temperature and light. Thus, if respiration is to continue optimally, an animal must avoid light when water temperature rises. Larval responses to light are thought to have the same basis (Thorson, 1964). If surface layers of the water were to become too warm, a reversal of the photopositive response would tend to remove the larvae to deeper, cooler water where they can better adapt metabolically.

Reduced salinity caused by heavy run-off of fresh water from rivers or by prolonged periods of rain also tends to lessen the photopositive response in larvae. During these periods the water will become stratified with lower salinity water staying near the surface. To avoid reduced salinity, the larvae become photonegative, seeking out higher salinity waters at greater depths. Larvae of intertidal zone animals, however, are much less affected by low salinity waters than are larvae of pelagic animals. The photopositive response of the oldest larval stages of intertidal zone animals is so strong that it takes a very bright light and very high temperature, or a much reduced salinity, to drive the larvae from the surface waters. This reaction is considered to be a natural ecological response, since the larvae must stay at the surface if they are to come in contact with an intertidal area (Thorson, 1964).

A photopositive response to light is not the only factor acting to keep young larvae in the surface waters. For example, during much of the larval life of *Mytilus edulis* when they are indifferent to light, larvae are kept at the surface of the water by a combination of hydrostatic and gravity responses; they respond to increased water pressure by swimming upward and show a strong negative geotaxis (Bayne, 1963, 1964).

As larvae approach metamorphosis and settlement, their response to light, gravity, and hydrostatic pressure often changes. They tend to become photonegative and geopositive and they no longer respond to increased pressure by swimming. This combination of responses effectively takes them out of the surface water and places them in contact with the substrate.

Once intertidal zone larvae reach the settling stage in their development, they must choose the proper substrate. Although the larvae are photonegative at the time of settlement, this photonegativity must not encourage them to settle during the night. They might settle in a position directly exposed to sunlight; direct sunshine is often injurious to intertidal zone animals, and ultraviolet light will either severely injure or kill them. Because total darkness will postpone or even prevent settling in many animals, it appears that there is a direct correlation between response to light at the time of settling and the position the organism maintains in the intertidal zone. In three species of barnacles it has been shown that larvae of the barnacle *Chthamalus stellatus*, living in the upper part of the tidal zone, settled most abundantly in direct sunlight; larvae of *Balanus amphitrite*, which lives lower in the intertidal zone, settled abundantly in bright but not direct sunlight; and larvae of *Balanus tintinnabulum*, inhabiting a still lower part of the intertidal zone, settled most abundantly during dusk and at daybreak (Daniel, 1957). Numerous factors associated with light affect settling behavior. The color of the substratum influences where the larvae will set, and most intertidal larvae find shaded or dark areas and dark surfaces the most attractive. Working with *Balanus amphitrite* and *Balanus improvisus,* Smith (1948) found that these barnacle larvae could distinguish between large black and white squares, for they settled much more frequently on black squares than on white. They seem to be influenced not by directional light, however, but by the intensity of diffused light from a distance; when small black and white squares were used, barnacles settled in equal numbers on both.

The alternation of crawling and swimming seems to indicate that the larvae are searching for the proper substrate. There is evidence that some searching larvae discover suitable substrata through chemoreception. Scheltema (1961) was able to demonstrate that for *Nassarius obsoleta* metamorphosis-inducing properties of the substratum probably are water soluble and are carried into adjacent waters. Therefore, this suggests that the infauna probably emit pheromones that cause larva settlement in the close vicinity of adults of the same species.

The settling responses of sessile intertidal animals seem to be affected by the "gregarious" response, which was first demonstrated in the study of Cole and Knight-Jones (1949) on *Ostrea edulis.* Pairs of similarly sized and shaped shells were placed in tanks containing settling oyster larvae. One of the pair was permitted to accumulate oyster spat continuously, while the other was cleaned daily and the oyster spat removed. Since many more larvae settled on the uncleaned shells than on the cleaned ones, it was suggested the small oysters were secreting some substance that attracted larvae to set. In a further study, Knight-Jones (1952) showed that more larvae settled in well-stocked oyster grounds than in grounds where there were few adult oysters. Crisp (1967a) experimentally exposed settling larvae of *Crassostrea virginica* to shells which had been cleaned in hot water (all visible traces of muscle tissue were removed but the periostracum was left intact) and to shells whose inner surface had been treated with hypochloride. When the organic layers on the surface of the oyster shells were destroyed, larvae settled almost exclusively on shells which had retained at least a part of the protein covering. Crisp inferred that larvae could recognize layers of conchiolin on the surface of the shells. Hidu (1969), also working with *Crassostrea virginica,* suggested that the initial settling of oysters was spontaneous, but as spontaneous settling proceeded, the new spat on the shell stimulated more and more larvae to set near this same site. Because the larvae were stimulated to set on a clutch mixed with two-month-old spat placed inside larval proof plankton mesh bags, a water-borne pheromone was thought to be involved (Hidu, 1969).

Although larvae of some animals apparently are able to detect a diffusion gradient in the water, in other species the sensory basis for settlement is more complex. The cyprid larvae of barnacles which spend long periods of time sinking through the water column or lying on the bottom offer a case in point (Crisp and Meadows, 1962; Crisp, 1965a). Upon contact, the larvae crawl over various surfaces, attaching temporarily from time to time. The times between these exploratory migrations are occupied by periods of swimming, but once a desirable substrate is found, the larvae will settle permanently within one hour. The specific recognition of a suitable substrate does not depend upon diffusion of a specific chemical through the water, for behavior of the cyprid larvae is the same in untreated sea water as in water containing dilute solutions of tissue extract of barnacles. However, when suitable settling areas were treated with integumental proteins of the same species, cyprids would settle almost immediately. Comparable untreated surfaces did not induce settlement. Thus, the larvae apparently did not detect the metamorphosis-inducing substance until contact was made with the surface. These studies led Crisp (1965a) to hypothesize that specific recognition of a suitable surface by the cyprids occurs only when the larvae come in contact with a protein that is held at a surface in a particular configuration. Bayne (1969) has reported similar findings in the settlement behavior of larvae of the oyster *Ostrea edulis*.

b. Sensory modalities of adults

Intertidal zone animals and their pelagic larvae, particularly sessile organisms, do not necessarily utilize the same sensory modalities to carry out life functions. For example, at the time of settlement, barnacles orient vertically upward toward the light, and they have only a very weak orientation to water currents. After settlement, however, adult barnacles are oriented by water currents rather than light, and they change from a predominately vertical to a nearly horizontal position; generally the carinae are directed away from the current source, and the cirral net faces the current (Crisp and Stubbings, 1957).

Many sessile organisms do not have well-developed photoreceptors, but they do have dermal light receptors. Although specific dermal light receptors have not been identified, it has been suggested that the light response probably results from accumulation of small amounts of photosensitive substances in the cytoplasm (Steven, 1963). Some sessile organisms having a dermal light sense will orient toward light by bending the body. For example, the sea anemone *Metridium* will bend when illuminated from one side; maximum sensitivity to light occurs at wavelengths of 490–520 mμ (North and Pantin, 1958). Other animals, such as many sessile polychaetes and bryozoans, orient toward the light by the action of their tube-secreting glands.

A dermal light sense is also responsible for the shadow reflexes observed in many sessile animals, such as barnacles, tubiculous worms, and lamellibranchs. Although some can detect either a decrease or increase in light intensity, usually they respond more vigorously to decreased illumination. This response, functioning as a protective mechanism, has great adaptive value to an animal. Barnacles, for example, will withdraw their cirri and close their valves when shaded.

Response to light can also serve to maintain zonation of intertidal zone organisms. For example, the polychaete *Protodrilus symbioticus*, which lives beneath the surface on sandy beaches, avoids high light intensities. This avoidance response serves to keep *P. symbioticus* within a narrow depth range, from 2.8 to 5.1 mm, beneath the surface of the sand (Gray, 1966).

Because vision is vital in many of the life-supporting activities of more mobile intertidal organisms, many, such as intertidal zone fish, have very well-developed photoreceptors. Most of these fish have large eyes with a foveated retina that causes images reaching the retina to be magnified. Both eyes can converge independently in temporary coordination onto an object in the narrow anterior field for binocular vision (Brett, 1954). Many of these fish rely primarily on vision in perception of their environment (reviewed by Gibson, 1969). Using visual cues, several species of *Blennius* are able to perceive, learn, and remember specific features of the topography and find their way back when displaced from their environment. The majority of intertidal zone fish search for and locate their food by sight. Visual signals are also important in reproductive behavioral patterns, although acoustical cues are also used (Tavolga, 1958).

Visual cues are also important in the life of the fiddler crabs (i.e., *Uca pugilator*). These crabs are active and completely exposed in the intertidal region during periods of low tide, moving about over the beach to different microenvironments to feed, reproduce, wet their gills, escape from predators, or release larvae. Thus, many of their activities have directional components that are adaptive. For example, to escape the approach of a predator when the crabs are some distance from their burrow, they may run landward and enter burrows or vegetation, or they may run offshore. In either event, the crabs are able to orient and return to the beach, guided primarily by visual mechanisms (Herrnkind, 1968). The primary cues for guidance are sun position and plane of polarized light, but the crabs can also use landmarks to supplement celestial cues. Thus, they are able to return to the beach if they are displaced either inland or offshore.

Fiddler crabs use not only visual but also acoustical signals during courtship. The displays of the males, which involve the waving of the large major cheliped, are especially conspicuous, and have attracted the attention of many workers. In a series of studies of crabs from throughout the world, Crane (1941, 1943, 1957) found each species of *Uca* possessed a characteristic waving display. Recent investigations have revealed that sound production is also an important component of courtship behavior that is distinctive for each species (Salmon and Atsaides, 1969). In *U. pugilator,* when females are absent, the crabs wave during the day and produce sound at night. If a female is nearby during the day, the male waves more rapidly, and if she approaches closer to the male's burrow, the waving is followed by sound production. At night a male produces sounds at low rates, but will increase sound rate when touched by a female (Fig. 3.11). Females have been observed in the field to orient and move toward males at night in response to sound from distances as great as 25 cm (Salmon and Atsaides, 1969).

The sound reception of fiddler crabs, and probably in many other intertidal

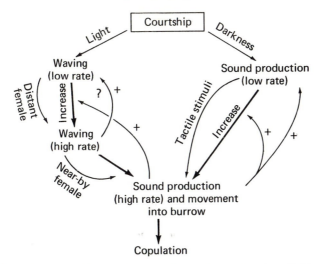

Fig. 3-11. Courtship in *Uca pugilator* during the day (left) and at night (right). Normal sequences of response by male to female are indicated by the thick arrows. Arrows to left of normal sequence indicate effect of stimuli provided by female upon the male. Arrows to right, indicate effects of signals produced by male upon behavior of adjacent males. (From Salmon and Atsaides, 1968. American Zoologist.)

zone species, is primarily a response to sonic energy carried through the substrate rather than to an airborne component (Salmon and Atsaides, 1969). In studies on the rapping sounds of the male, Salmon (1965) found that most of the energy of the substrate-borne component was below 300 Hz, and that females showed the greatest sensitivity to substrate displacements between the frequencies of 60 and 2,400 Hz. Thus the female's sensitivity to frequencies containing most of the acoustic energy enables her to detect the sounds of the male.

Although vision and, to a lesser extent, sound are probably the sensory modalities most widely used by many of the more mobile adult intertidal zone animals, chemoreception also helps some animals perceive their environment. *Nassarius obsoletus* locates food, for example, primarily by chemoreception (Carr, 1967a and b), for the snails apparently depend on waterborne odors to detect food. Snails submerged below the tide mark, where slight currents flow, were observed to move directly upcurrent from distances of two or three feet directly to freshly killed crabs, shrimp, or oysters. As the snail moved toward the food, the siphon was waved horizontally, and once the snail was within a few centimeters of the fresh tissue, the previously unexposed proboscis was extended. This same proboscis-search reaction occurred even when food was absent, if a sufficient concentration of certain tissue extracts was present. Chemical analyses indicated that the stimulatory components possessed physical properties similar to those of amino acids and certain other nonvolatile, nitrogenous compounds of low molecular weight.

Many molluscs are able to detect potential predators through chemoreception. Responses to predatory starfish are mediated through contact chemoreception, but other predators, such as some predatory gastropods, are detected through distance chemoreception (Ansell, 1969).

Sessile animals also may use chemoreception in locating live prey. The pedunculate barnacle, *Lepas anatifera,* lives attached to pieces of wood and feeds on living plankton obtained from the surrounding sea water. Crisp (1967b) demonstrated that these barnacles are able to detect amino acids, and he suggested that they use this sensitivity to determine whether their prey is living. The barnacles crush the plankton with the sharp spines on their cirri and, if the plankton is living, release stimulating substances, setting in motion the responses that cause the living prey to be eaten.

Many intertidal animals are very sensitive to chemical changes in the water, such as changes in oxygen tension, carbon dioxide content, or the presence of pollutants. The burrowing shrimp *Callianassa,* for example, can detect oxygen concentrations. During low tide, the shrimp are buried deep where oxygen tensions are very low, but as oxygen tensions rise with the incoming tide, the shrimp, detecting the increase, move upward toward the burrow openings and hyperventilate.

One of the most interesting examples of specific recognition of a chemical factor can be found in the work of Ross (1960) on the association between the sea anemone, *Calliactis parasitica,* and the hermit crab, *Eupagurus bernhardus. C. parasitica* is commonly found on mollusc shells that are inhabited by *E. bernhardus;* however, there is no attraction between these two animals. Instead, the two species occur together because they respond independently to some organic constituent in the periostracum of the molluscan shells.

Since it is a constant mechanical force in the environment and its direction of pull or intensity does not vary in any given location, gravity serves as the basic plane of reference for many animals, enabling them to maintain a definite attitude in space. Animals may have an orientation response toward or away from the gravitational force. Many of the burrowing intertidal zone animals, for example, show an orientation toward the earth's center of gravity (positive geotaxis), and the force of gravity orients the direction of digging (Schone, 1961).

Some intertidal zone animals which have a very simple nervous system display complex behavior patterns that are difficult to explain on the basis of any one sensory modality. In studying homing behavior of the chiton *Acanthozostera gemmata,* Thorne (1968) ruled out sight, odor of homesite, and celestial navigation as orienting mechanisms. These molluscs do not appear to be completely dependent on a landmark recognition, since a breaking up of their surroundings or the introduction of new features into the environment does not seriously disrupt their homing abilities. Yet these chitons, which have no cerebral ganglia and little differentiation of the cerebral commissures, are able to return over distances up to 20 cm to a definite homesite after each nocturnal feeding excursion.

c. Rhythms in intertidal zone animals: the right place at the right time

In addition to sensory modalities, many animals rely on a biological or internal "clock" mechanism to aid in finding the right place at the right time. In most terrestrial animals such temporal patterns recur rhythmically, coordinated with the time of day, with periods of activity separated by periods of inactivity. If these rhythmic functions persist in the absence of daily temperature and light cues with approximately 24-hour periodicity, they are called circadian rhythms. The name was derived from the Latin *circa dies* (about a day). Under constant environmental conditions, the physiological rhythm free-runs and gradually loses phase with the natural photoperiod (DeCoursey, 1961). Exposure to a new light-dark cycle will reentrain the rhythm by controlling both phase and frequency, thereby synchronizing the function to a specific time of day. Persistence of a rhythm under these constant conditions has led many workers in the field to believe that such rhythms are endogenously controlled. However, this viewpoint has been challenged by Brown (1968), who has pointed out that environmental factors other than light and temperature—including barometric pressure, humidity, and cosmic radiation—vary during the day. Therefore, Brown suggested that until these environmental factors are controlled in experiments, endogeneity cannot be proven beyond all doubt.

One of the most characteristic phenomena of the intertidal zone area is the ebb and flow of the tide. Thus, it is not surprising that physiological and behavioral patterns of intertidal zone animals reflect the tidal rhythmicity; in fact, the tidal rhythm is often more pronounced and pertinent to the life cycle of these animals than the circadian rhythms which are so important in terrestrial animals.

Often the peak of activity coincides with periods of high tide. The swimming rhythm of the isopod, *Eurydice pulchra,* for example, shows a definite correlation with tidal conditions (Jones and Naylor, 1970). At low tide this isopod burrows in the sand, but with the incoming tide period, it leaves the sand, feeds actively during the high tide period, and then burrows in the sand again as the tide begins to ebb. This synchronizes the end of the swimming period with the falling tide, allowing the animals to maintain a favorable position in the intertidal zone. As the tide begins to ebb, the isopod burrows in the sand at approximately the same level from which it emerged, and thus avoids being carried out to the sea by the ebb tide.

These tidal cycles of locomotor activity also persist under constant conditions in the laboratory, as illustrated by a study on three species of fiddler crabs, genus *Uca* (Barnwell, 1966). The crabs are highly mobile, moving throughout the exposed part of the intertidal zone during low tide. When the crabs were brought into the laboratory and placed in the natural light cycle, the tidal rhythm persisted in all three species for prolonged periods of time. In one species, *U. minax,* the rhythm was apparent for at least 46 days. When *U. minax* was kept under constant illumination, activity peaks occurred at approximate tidal frequency, but the free-running period of the timing mechanism was slightly longer than actual tidal cycles (Fig. 3.12).

Locomotor activity is not the only function that is linked to a tidal rhythm. Persistent tidal rhythms have been described for oxygen consumption in *Littorina littorea* and *Urosalpinx cinereus* (Sandeen et al., 1954), in *Uca pugnax*, *U. minax*, and *U. pugilator* (Brown et al., 1954; Barnwell, 1963, 1966), and in *Carcinus maenas* (Arudpragasan and Naylor, 1964); for color changes in *U.*

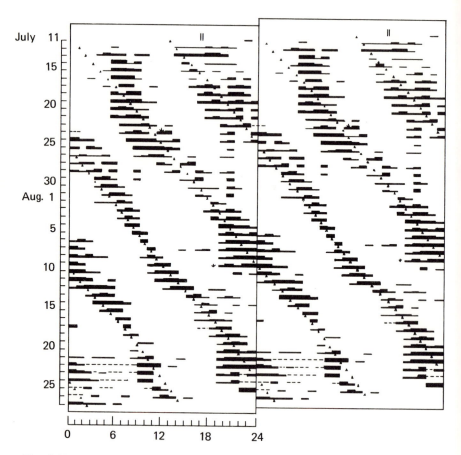

Fig. 3-12. Activity pattern of a male *U. minax* in LD. The graph has been reproduced twice, with the right-hand graph displaced upward by one day, in order to facilitate the visualization of the drift of activity peaks across the solar day. The height of the blocks indicates the percent of the hour during which activity was recorded, either 25, 50, 75, or 100 percent. Parallel vertical lines at the beginning of the record mark the time of collection. Small triangles indicate the predicted times of high tide at the beach of collection. Small arrows point to times when water in the tipping pans was changed. Broken lines indicate failure of the recording system. (From Barnwell, 1966.)

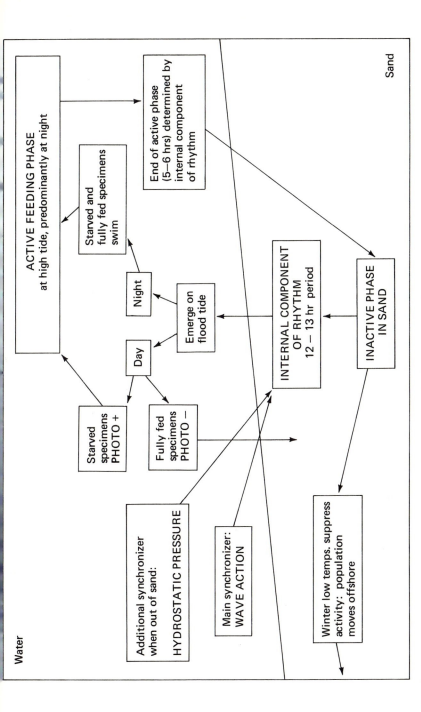

Fig. 3-13. Summary of interaction between environmental variables and the endogenous component of tidal periodicity in *Eurydice pulchra*. (From Jones and Naylor, 1970. North-Holland Publishing Company).

pugnax (Brown et al., 1953) and *Callinectes sapidus* (Fingerman, 1955); and for filtration rate in *Mytilus* (Rao, 1954).

Enright (1968) has suggested that the tidal rhythm is physiologically and evolutionarily similar to the circadian rhythm, but modified to permit entrainment predominantly by stimuli other than light. Evidence at hand would tend to support this viewpoint. The tidal rhythm shows the same basic characteristics as a circadian rhythm, i.e., it persists under constant conditions, with the free-running periods slightly longer than actual tidal cycles. The mechanisms of entraining tidal rhythms can be linked to the habitat preferences of the organisms. For example, wave action, with the accompanying changes in hydrostatic pressure, is an important entraining stimulus for activity rhythms in animals that burrow in the sand, such as the isopods *Excirolana chiltoni* and *Eurydice pulchra* (Enright, 1965; Jones and Naylor, 1970); but in the nonburrowing crab *Carcinus maenas* the most effective synchronizer proved to be a combination of immersion in water and associated temperature changes (Williams and Naylor, 1969).

There has been some question whether intertidal zone animals simultaneously show both circadian and tidal rhythms. Some species, such as *Hemigrapsus edwardsi* and *Orchestia mediterranea,* undergo peak activity during the nocturnal high-tide period (Williams, 1969; Wildish, 1970); other species, such as *Blennius pholis,* are more active when light period and high tide coincide (Gibson, 1969). However, these species showed only two peaks of locomotor activity, both associated with the high-tide periods. If there was a circadian component, then an additional peak of locomotor activity would be expected when high tides do not coincide with a particular time of day. Thus, although environmental variables without doubt influence the behavior of the animals, the actual tidal rhythm is not affected by environmental factors. Jones and Naylor (1970) have graphically illustrated this interplay between environmental variables and the apparently endogenous component of tidal periodicity in one species of intertidal zone animal, the beach isopod *Eurydice pulchra* (Fig. 3.13).

Some intertidal zone species, however, apparently do have both a circadian and tidal rhythm. In a study on activity rhythms in the fiddler crabs *Uca pugilator, U. pugnax,* and *U. minax* Barnwell (1966) found both daily and tidal rhythms, and Palmer (1967) reported a nocturnal peak of locomotor activity in addition to the two peaks associated with the high tides in the crab, *Sesarma reticulatum* (Fig. 3.14). It is of interest that many of the intertidal zone species in which only a tidal rhythm was observed are active primarily at high tide. The species showing both tidal and circadian components tend to be active and completely exposed at low tide; thus they are essentially terrestrial animals during this period of time. It would, therefore, be particularly advantageous to them to have both circadian and tidal rhythms.

2. Feeding

Energy is derived from the food an animal eats. Since this energy is necessary to sustain the metabolic machinery of the organism, it is not surprising

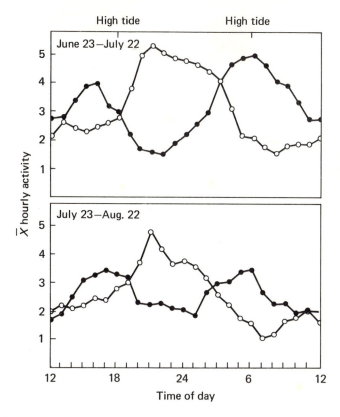

Fig. 3-14. Dual components in the persistent rhythmic activity of *Sesarma reticulatum*. Mean solar day rhythm is indicated by open circles. Solid circles represent the bimodal lunar day rhythmic component. Maxima in this rhythm correspond to the times of high tide (midpoints of high tide are indicated at the top of the figure). (From Palmer, 1967.)

that morphological-physiological mechanisms have evolved to ensure the extraction of necessary energy from the environment. Both food choice and feeding patterns reflect the position in the intertidal zone and way of life of intertidal zone animals. According to the classification of Yonge (1928), the feeding mechanisms of intertidal zone animals usually fit one of two categories: small particle or large particle feeders.

a. Small particle feeders

In general, sessile animals in the intertidal zone and those animals with limited locomotor ability are filter feeders. The modes of feeding, even within this group, are by no means identical. Barnacles, for example, use their long-haired thoracic appendages as a net to sweep the water for food. In bivalves the feeding organs are the mantle, gills, and mouth palps. Filter feeders generally do not discriminate beyond particle size, and consequently some of the filtered material has no energetic value for the organism. Since the aquatic environment is generally a highly dilute medium of organic material, filter-feeding organisms must either be able to filter a great deal of water, or, alternatively, to concentrate molecules against a gradient from a dilute medium.

Not all filter feeders, however, are sessile animals. A good illustration is *Donax incarnatus,* a small tropical bivalve that lives in India (Ansell and Trevallion, 1969). This bivalve has evolved mechanisms enabling it to use the continuous disturbance of the surf zone to its advantage. Washed out by the surf, *D. incarnatus* uses the energy of the waves to move up and down the beach following the tide. Between waves, it burrows rapidly and opens an elaborate system of tentacles which prevents the entry of sand grains into the inhalant siphons, but allows water containing food to be filtered.

Small particle feeders that eat detritus, material scraped from sand grains or fresh and decaying flesh of other marine organisms, are often highly mobile. Again, the method of feeding can be correlated with both food availability and habitat selection. For example, fiddler crabs capture food material in spoon-tipped "hairs" on the mouth parts. One of the species of fiddler crabs, *Uca pugilator,* lives and feeds on sandy beaches where food is not abundant and the substratum is heterogeneous. Material from the beach is picked up, passed to the buccal cavity, and sorted by a flushing of respiratory water from the gill cavities back and forth over the ingested material. Much of the small organic matter is retained by the spoon-tipped hairs, and the larger inorganic particles are discarded. In contrast, *Uca minax,* which lives in tidal marshes where abundant food material is present on the surface of the silty substratum, has fewer specialized, flat-tipped hairs on the mouth parts (Miller, 1962).

b. Large particle or mass feeders

(1) *Inactive food*

Sluggish and/or burrowing forms generally utilize inactive food. They tend to be omnivorous, feeding on mud, sand, or bottom deposits, where the organic material usually consists of feces and microorganisms. The animals swallow everything indiscriminately by means of extroversible gullets, pushing tentacles, or other "shoveling" devices. Some evidence, however, suggests that among these animals there may be diversification in either the feeding process or the digestive process in closely related sympatric species occupying the same area (Mangum, 1962). Maldanid polychaetes obtain food by ingesting sand grains through the lower end of their vertical tubes approximately 20 cm below the water-substratum interface. Two species, *Clymenella torquata* and *Axiothella mucosa,* show a striking dichromatism, with populations being either green or orange. The color of a population inhabiting a particular area is always consistent. The green pigment proved to be mesobiliverdin, which probably reflects the density of algal food in the sediment; the orange pigment was not identified. Species diversity, however, is noted in a third closely related species, *Euclymene collaris,* which feeds in the same manner but is always found in the orange form, even when it occurs sympatrically with the green forms of the other two species.

Many intertidal gastropods have jaws or radulae that enable them to eat encrusting animals or plant species. In these animals the radula is a movable, elongated structure containing numerous teeth. When active, the radula is worked back and forth over the surface of the food, and the particles ripped off are swallowed.

(2) Active food

The story of predatory feeding is an intriguing one. Wood (1968) has documented the relationship betwen the predatory oyster borer, *Urosalpinx cinerea,* and its prey in different areas ranging from Massachusetts to South Carolina. *Urosalpinx* utilizes three main groups of prey: barnacles, oysters, and mussels. In attacking all of these, the borer first softens its prey's shell by secretions from its accessory boring organ located in the foot. Then it uses its radula to rasp through the weakened shell (Carriker et al., 1967). The prey choice of *U. cinerea* in different geographical areas is linked to intertidal zonation and relative abundance of prey (Fig. 3.15). In Massachusetts, *U. cinerea* is found primarily with *Balanus balanoides,* and the barnacle is the chief prey species. In Maryland, however, the barnacles are found living above the bulk of the predator population, and the distributional limits of *Urosalpinx* now overlap that of the mussel, *Mytilus edulis,* which served as the prey choice. Thus, where prey species are separated from one another by intertidal zonation, then prey selection more or less depends on the coexistence of the predator with whatever particular prey species is dominant in its zone. However, further south, where the prey populations overlap extensively, the oyster *Crassostrea virginica* is the prey choice. In these mixed prey habitats, the prey is selected on the basis of the relative density of the prey species (Wood, 1968).

Another predatory gastropod species, *Murex fulvescens,* attacks its prey in a manner very different from *Urosalpinx. Murex,* which attains large size, feeds on oysters, clams, and mussels. Oysters are opened by pulling the valves apart using the lip as a brace; clams and mussels are opened by grinding the bivalve shell across the outer lip. In the laboratory *M. fulvescens* shows a very strong preference for oysters over four other species of bivalves. The interesting point about this relationship is that *Crassostrea virginica* is not the dominant species in the area where *M. fulvescens* normally lives. Yet in a choice situation, eight specimens of *M. fulvescens* ate 134 *Crassostrea virginica* in contrast to only six *Ostrea equestris,* the dominant species in the natural habitat area of *M. fulvescens* (Wells, 1958).

c. Feeding stimuli

Stimuli which cause the release of feeding patterns can be related to both the internal state of the animal and habitat preference. Pelagic larvae of a number of prosobranch gastropods will feed almost continuously if concentra-

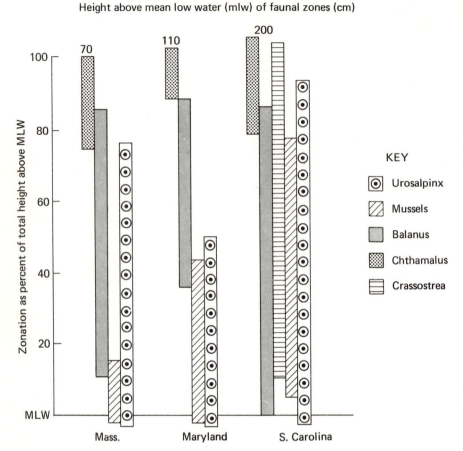

Fig. 3-15. Prey choice of *Urosalpinx cinerea* in different geographical areas. (From Wood, 1968.)

tions of food are low. On the other hand, if concentrations of digestible food are high, larvae will fill their stomachs in a few minutes and then stop feeding until digestion is under way (Fretter and Montgomery, 1968). Larvae apparently are not only able to discriminate between digestible food and nondigestible food but can also select digestible foods. When larvae of the mud-flat snail, *Nassarius obsoleta,* were offered three different species of algae, *Cyclotella, Phaeodactylum,* and *Dunaliella,* young veligers showed a definite preference for *Cyclotella* over the other two species. As the veligers aged, food preference changed and *Phaeodactylum* was selected more often. All sizes, however, preferred *Cyclotella* over *Dunaliella,* and the selection against *Dunaliella* occurred regardless of what food the larvae had been presented during their early stages (Table 3.1). Selection of food by the larvae of *N. obsoleta* would seem to de-

TABLE 3.1   *Preference of* Nassarius obsoletus *Veliger Larvae Among Three Different Species of Algal Food**

| Species of Algae | Percentage of experiments | | |
	First Choice	Second Choice	Third Choice
Cyclotella	*81.3*	18.8	0.0
Phaeodactylum	12.5	*62.5*	25.0
Dunaliella	6.3	18.8	*75.0*

* From Paulson and Scheltema, 1968.

pend on size, concentration, and perhaps a chemotactile sense (Paulson and Scheltema, 1968). However, if concentrations of food are too great, even when it is a preferred food, feeding is inhibited. When larvae of the intertidal zone clam, *Venus mercenaria,* were held in heavy concentrations of algae cells, their stomach contained less food than those held in lighter food concentrations (Loosanoff et al., 1953).

The rate of feeding can also be correlated with the vertical position of a species in the intertidal zone. In laboratory studies on feeding in *Balanus balanoides,* Ritz and Crisp (1970) found that animals living near the high tide mark consumed more food than low tide level animals. They suggested that the higher feeding rate in the high tide level animals is correlated with the short amount of time they have to feed in nature. This difference appears to be an acclimation phenomenon rather than a genetic difference. When barnacles from the high tide region were brought into the laboratory and given an unlimited supply of food for long periods of time, they took in large amounts of food at first, but, after a few days, they adapted to a lower rate of feeding comparable to that of lower shore animals (Table 3.2).

Feeding activity of many sessile forms is stimulated by water movement. A good example can be found in the study by Crisp (1965b) on species of barnacles inhabiting different areas in the intertidal zone. Surf zone species, such as *Chthamalus tetraclita* and *Balanus cariosus,* and those inhabiting areas exposed to strong bottom currents, required a much faster flow of water to induce feeding behavior than did the forms that live in sheltered silted areas, such as *Elminius modestus* and *Balanus crenatus.* Feeding activity in barnacles is also stimulated by the presence of food particles in the water.

Feeding behavior is the result of the interaction of a number of factors in the hermit crab *Clibanarius vittatus.* Once feeding has been elicited by chemical stimuli, visual cues serve to orient the crab and to initiate the grasping response (Hazlett, 1968).

The stimulus to start feeding appears to be mechanical in other organisms. Two species of interstitial amphipods, *Bathyporeia pilosa* and *B. sarsi,* live on food material or matter adhering to sand grains. In observing the feeding be-

TABLE 3.2 *Comparison of Feeding Rates of* Balanus balanoides *at Daily Intervals after Transferring to Experimental Conditions in the Laboratory**

Days after Transferring	Rates in thousands of nauplii/g barnacle/day. Mean feeding rate of barnacles from			
	High Water	Mean Tide	Low Water	Mean
1	3.37	2.16	1.25	2.26
2	2.93	2.24	0.99	2.06
3	2.12	1.93	0.85	1.64
4	1.54	1.43	1.15	1.37
5	1.74	1.73	1.08	1.52
Mean	2.34	1.90	1.06	1.77†

Days after Transferring	Mean feeding rate at			
	4°C	10°C	15°C	Mean
1	1.35	2.27	2.33	1.98
2	1.16	2.21	2.02	1.80
3	1.06	1.59	1.64	1.43
4	0.96	1.62	1.36	1.31
5	1.01	1.55	1.59	1.38
Mean	1.11	1.85	1.79	1.58†

* From Ritz and Crisp, 1970.
† Grand mean.
Cambridge University Press.

havior of these animals it was noted that some of the sand grains were not treated by the mouth parts, but others were thoroughly worked over. To determine if feeding was initiated chemically or mechanically, a number of animals were allowed to burrow in sand mixed with particles of an inert fluorescent pigment which adheres to the sand grains. After feeding for a short while, the animals were removed, then examined. Since the foregut of all the animals was completely filled with particles of the compound, it may be that the attractiveness of a sand grain is determined by the amount of material adhering to the grain rather than through any chemical cue (Nicolaisen and Kanneworff, 1969).

d. Environmental modification of feeding

Numerous environmental factors can modify both feeding behavior and food preferences. Temperature probably is the most important factor in poikilotherms. If temperature is lowered, most rate functions in poikilotherms including that of feeding are reduced. Reduced energy requirements usually accompany reductions in rate function. Many animals will feed only within a certain temperature range, and most poikilotherms have an optimum temperature for feeding. However, this optimum temperature need not necessarily be the same for geographically separated populations of a single species. For example, the pumping rate of the clam *Mytilus californianus* is the same at 6.5°C from latitude 48° 27′ north, as that of a population situated more southerly at 12°C latitude 38° 31′ north, or of a population from a latitude of 34° north at about 14°C (Rao, 1953). Although natural thermal differences are experienced by these widely separated populations, the rate of pumping water is about equal at their respective environmental temperatures. A similar type of response has been recorded for populations of this same species of clam occupying different regions of the tidal zone at one latitude. The pumping rate was measured at a common temperature, and it was found that the animals from the cooler subtidal zone pumped water faster than clams from the warmer regions of the intertidal zone (Segal et al., 1953).

The temperature at which blockage of feeding response is observed can also be correlated with the distribution of different species. In a study on feeding response of two polychaetes, Mangum (1969) compared the temperate zone *Nereis succinea* with the boreal *N. vexillosa* from Alaska. *Nereis succinea* was collected at temperatures of 12–14°C, *N. vexillosa* at 10–13°C. The temperate zone species gave a positive feeding response at 15° and 20°C. Although they could tolerate temperatures of 5° and 10°C, feeding at this temperature was completely inhibited. In contrast, *N. vexillosa* presented quite a different picture. Ninety percent of the worms had a positive feeding response between 10° and 12°C, 80% responded at 5°C, and 20% at −2.5°C. *Nereis succinea* is a cosmopolitan species ranging northward to the Gulf of St. Lawrence, but *N. vexillosa* is found only in the cold waters of the north Pacific. Therefore, Mangum concluded that feeding responses are correlated not only with the environmental temperature, but also probably with geographic range.

In a study of the effect of temperature on the barnacle *Balanus balanoides,* Ritz and Crisp (1970) measured throughout the year the fecal output of the barnacles fed on larvae of *Artemia*. There were two main yearly peaks of feeding, one from March to May, the other in October. The March-to-May increase in feeding was channeled into gonad production, and the October peak contributed to the increased nutrient needed for completion of maturation of eggs and sperm. At the onset of breeding in November, there was a very sharp decline in feeding; in late November and December there was virtually no food uptake. Feeding was not equally influenced by temperature at all seasons of the year. Before March and after May, feeding rates were much lower regardless of temperature, and in January when feeding activity was at a minimum, low temperature was without effect on food intake. However, when animals were feeding actively, higher temperatures up to 18°C resulted in a greater intake in the mid- and high-water animals, but low-water animals did not seem to be much influenced.

It has also been suggested that the optimum temperature for feeding behavior is a significant factor in the competition between species for food. Two species of barnacles, *Balanus balanoides* and *Chthamalus stellatus,* are in direct competition over a large part of the intertidal zone in southwest Britain. Competition for food between these two species is great, but competition is reduced since temperature differentially influences their feeding behavior. The optimum temperature range for *Balanus balanoides* lies between 0° and 18°C, for *C. stellatus* between 5° and 30°C (Southward, 1955). When either of the barnacles are outside these temperature ranges, the species cannot feed efficiently; *B. balanoides* is more efficient below approximately 15°C, *Chthamalus* above 15°C. Competition for food and for space is keen between adults, but it is especially acute for young spat surrounded by older individuals, for the older barnacles will feed on newly hatched nauplii of their own or of other species. Thus, the low temperature at which the spat and adults of *B. balanoides* are growing and feeding lessens the interspecific competition to which they will be exposed. Above 17°C, *Balanus balanoides* shows reduced activity but *Chthamalus* has increased activity. Hence, the balance between the two species is maintained by the effect of temperature on feeding behavior throughout the year (Southward and Crisp, 1965).

Salinity and pH also influence the feeding behavior of intertidal zone animals. The stomachs of subtidal oysters, *Crassostrea virginica,* contained food irrespective of tide or time of day, indicating that these oysters feed almost continuously (Loosanoff and Nomejko, 1946). The decisive factor, however, may be salinity gradient. In Delaware Bay, oysters living in parts of the bay where the salinity is relatively constant, fed during both low and high tide, but in regions with a pronounced salinity gradient, they fed only during high tide when the salinity was relatively high (Chestnut, 1946, cited from Korringa, 1952). In the laboratory *C. virginica* maintained normal pumping rates over a salinity range of 25–30 0/00, but stopped pumping water effectively at salinities below 13 0/00 (Hopkins, 1936). In *C. virginica* a drop in pH from 7.75 to 6.75 resulted in a temporary increase in pumping rate, followed by a rate that was lower than the original value, while pH as low as 4.25 greatly reduced the pumping rate (Loosanoff and Tommers, 1947). In suspension feeders the rate of feeding is generally independent of concentration and quality of suspended particles (Jørgensen, 1966). However, high concentrations result in clogging of the filter devices, and in these instances filtering rate is reduced. Whether or not silt depresses the feeding rate of oysters tends to vary with the population and its geographic location. Adult oysters living in relatively silt-free waters are adversely influenced by a small amount of silt, whereas populations normally found in habitats where there is more silt can withstand higher silt concentrations (Loosanoff and Tommers, 1948).

e. Energy budgets

In recent years it has become increasingly important to know not only the energy requirements of a species and how energy is partitioned between various physiological processes, but also the rate of energy flow between various components of the ecosystem. One way of expressing the results of energy balance studies is to use the concept of the energy budget (computing energy input into a living organism and balancing it against total utilization and distribution of the energy).

A detailed energy budget has been determined by Hughes (1970) for a tidal-flat population of the bivalve *Scrobicularia plana*. This bivalve, which is a deposit feeder, lives in permanent burrows, and thus population estimates are not complicated by immigration or emigration. The approximate life history table for this bivalve is as follows:

October–March	No growth, mortality only
April–July	No mortality, only growth
August–September	No mortality (except slight in September)
	No growth
	Gamete release

The equation used to determine the energy budget of a population may be summarized as:

$$C = P + R + F + U,$$

where $P = P_r + P_g$ and $P_g = \triangle B + E$.

Each component may be measured in kilocalories per annum. C is the energy content of the food consumed by the population; P, the total energy produced as flesh or gametes; P_g, the energy content of the tissue due to growth and recruitment; B, the net increase in energy content of standing stock; E, elimination, or energy content lost to the population through mortality; P_r, the energy content of the gametes liberated during spawning; R, the energy lost due to metabolism (respiration); F, the energy lost as feces; and U, the energy lost as urine or other exudates.

Energy flow (Smalley, 1960) or gross production (Englemann, 1966) is the proportion of ingested energy assimilated by the population $(C - F - U)$, where $C - F$ is that part of the food absorbed into the body wall through the wall of the alimentary canal. Another expression of energetics is energy flow $= P + R$. *Production* $= C - (F + R)$; this has been termed net production by Englemann (1966). In this study U is assumed to be negligible, and assimilation and absorption become synonymous; hence, assimilation $= C - F$.

Each component in the energy equation was estimated for populations of *S. plana*. Total growth, $P_g = \triangle B + E$, was estimated from data on rate of shell growth, shell length, dry flesh weight ratios, calorific value of the tissue, and size-frequency distribution. The amount of energy liberated as spawned gametes (P_r) was estimated from changes in the ratio of shell length to dry flesh weight during the breeding season. Mortality estimates (E) were made by comparing density and size-frequency structure of the *Scrobicularia* population during the month of November in two consecutive years. Metabolic heat loss (R) was calculated by measuring the monthly oxygen consumption of different size classes of bivalves in the laboratory under thermal conditions approximating field temperatures. Defecation (F) rates were determined in terms of dry weight of feces produced per animal per day, and the weights of feces produced/M² were computed for each month. *Scrobicularia* is a deposit feeder, and large amounts of bottom sediment are taken in by its long mobile inhalant siphon, then sorted on the palps. Only a small amount of organic material is actually ingested, and the rest is expelled as pseudofeces. To estimate ingestion (C), it was necessary to determine caloric content of the sediment and the rate it was taken up by the inhalant siphon, the caloric content and rate of pseudofeces production, and the caloric content of the ingested sediment and the ingestion rate. From these data, energy budgets were calculated for populations inhabiting both high and low areas in the intertidal zone.

The energy flow ($C - F - U$ or $P + R$) was much higher through the lower sampling area where the population size was greater than in the higher sampling area, 600 kcal/m²/year in the former, 72.1 kcal/m²/year in the latter. The average flow of both populations was 336 kcal/m²/year. These values, when compared with energy flow results for other intertidal zone animals, indicate that *Scrobicularia* is a very important component of the tidal-flat community. Net growth efficiency (production × 100/assimilation) decreases as the animals grow. This tendency for a decrease in net growth efficiency has been observed in other continuously growing invertebrates. Ecological efficiency, which is the ratio of any of the various parameters of energy flow between trophic levels, was only 2.4% in the *Scrobicularia* population. Therefore, only a small proportion of the production was passed on to the next trophic level, the oyster catcher, which is the only known predator of *S. plana*. A number of energy budget components and efficiency ratios which have been determined for other marine invertebrates are listed in Table 3.3.

3. Respiration

A wide range of physiological strategies enable intertidal zone animals to adapt to the markedly fluctuating conditions characteristic of this zone. Although no single physiological parameter reflects the total fitness of an organism for a given habitat, basic metabolic-environmental interactions are reflected in oxygen utilization. The energy expended by most animals to meet the demands of this changing external environment can be measured in terms of oxygen uptake since the rate of oxygen uptake is influenced by many intrinsic and extrinsic factors.

a. Adaptive mechanisms

The various adaptive mechanisms utilized by intertidal animals in response to environmental change may be arbitrarily divided into three categories: physiological, morphological, and behavioral. However, it must be kept in mind that this classification is artificial and proposed only for purposes of discussion. The survival of an organism depends upon a complex series of integrated responses; function and structure are interrelated, and overt behavior has both a physiological and structural basis.

Animals originated in the sea and have successfully invaded land following

TABLE 3.3

Assimilation, Production, Respiration as Per Cent Assimilation, Assimilation Efficiency, Net Growth Efficiency and Ecological Efficiency of Different Invertebrates from Data of Several Authors

Species	A Assimilation = Energy Flow or Gross Production (kcal/m²/yr)	P Production = Net Production (kcal/m²/yr)	Rx100/A Respiration Assimilation (0/0)	Ax100/C Assimilation Efficiency (0/0)	Px100/A Net Growth Efficiency (0/0)	Ex100/C Ecological Efficiency (0/0)	Habitat	Authority
Scrobicularia plana	336.1	70.8	79	61	21.0	2.4	Tidal mud-flat	Hughes (1970)
Littorina irrorata	290.0	40.6	86	45	14.0	—	Salt marsh	Odum & Smalley (1959)
Modiolus demissus	56.0	17.0	70	—	13.5	—	Salt marsh	Kuenzler (1961)
Mytilus edulis	—	—	—	—	73–11	—	—	Jørgensen (1952)
Crassostrea virginica	—	—	40–92	—	60–8	—	Inter-tidal	Dame (1971)
Orchelimum fidicinium	30.8	10.4	66	36	33.8	—	Salt marsh	Odum & Smalley (1959)
Prokelisia sp.	275.0	—	—	—	—	6.8	Salt marsh	Teal (1962)
Nematodes	85.0	21.0	75	—	24.7	6.8	Salt marsh	Teal (1962)

different routes. One avenue includes movement from the sea to estuaries to fresh water and finally to land; subsequent radiation into various terrestrial and aerial habitats has then occurred. A second route to land is from the sea across the intertidal zone. Following either route, the organism faces a new enrivonmental complex, and new mechanisms of metabolic adaptation are to be expected. An insight into this problem can be gained by reviewing studies based on crustaceans ranging from subtidal to intertidal to terrestrial habitats. Whereas aquatic and terrestrial animals face different environmental conditions, many of the intertidal zone crabs lead an amphibious life and must contend with the vicissitudes of both aquatic and aerial modes of life.

Aquatic and land environments differ in certain features which affect respiration. The oxygen content of an aquatic environment may vary markedly, from supersaturation to anaerobiosis, whereas oxygen levels in air are relatively constant. In contrast, temperature fluctuations are more rapid and more extreme on land than in water. Furthermore, as animals leave the water and migrate to land, they may experience greater desiccation which, in turn, tends to dry out their respiratory membranes. Other functional processes which are reflected in metabolic changes are also influenced by this transition from water to land; for example, in some animal groups (but not crabs) a marked shift from ammonotelism to uricotelism is observed.

Numerous morphological differences exist between animals occupying these different habitats. Some of the best examples of morpho-physiological adaptations to habitat are exhibited by crabs. The number and the total volume of gills of crabs living on land are less than in aquatic species (Pearse, 1929). In a more comprehensive, interspecific study of crabs ranging from the sublittoral zone to land, Gray (1957) reported that the gill area per unit body weight tended to be reduced in species occupying more terrestrial habitats. This trend could be correlated with the increased availability of oxygen in an aerial habitat. However, this correlation of the reduction of gill surface with terrestrialism is not simple, as illustrated by certain land crabs that have a reduction in gill surface and highly vascularized gill chamber walls which could have a respiratory function. Not only has the total amount of gill surface been altered, but also the structure of the gills has undergone modification to compensate for a terrestrial existence. In the ocean, the gill leaves, subunits of the gills, could be held apart by the influence of water currents, whereas on land they would tend to adhere together with a resultant decrease in surface area for respiratory exchange. To overcome this problem, the gills of some intertidal animals have become highly sclerotized with the subunits being rigid and supported. The gills also may be at right angles to the gill bar, and thus the gill leaves are not easily closed (van Raben, 1934).

The behavior of crabs affords a degree of protection to their respiratory system against environmental stress, such as desiccation and thermal extremes. Adaptive mechanisms have been discussed in an earlier section (Zone of Resistance).

When comparing the oxygen consumption rate of crabs from the subtidal, intertidal, and terrestrial zones, a tendency for an increased rate with an approach to terrestrialism was observed (Table 3.4). As is to be expected, interspecific variation exists in each habitat type, and it can be attributed in part to differences in the level of locomotor activity—the more active the species, the higher the metabolic rate. These differences are not only attributable to

TABLE 3.4

Oxygen Consumption of Gill Tissue and Mid-gut Gland of Nine Species of Decapod Crustacea from Three Ecological Zones, Determined at 27°C*

Species	Gill Tissue		Mid-Gut Gland	
	Mean	S.E.	Mean	S.E.
Terrestrial zone:				
Ocypode	15.231	1.051	12.862	0.993
Sesarma	15.186	1.035	5.943	0.572
Intertidal zone:				
U. pugilator	9.165	0.496	6.007	0.511
U. minax	6.208	0.303	6.380	0.561
Menippe	5.482	0.292	10.386	0.676
Panopeus	5.321	0.471	8.935	0.445
Below low tide:				
Callinectes	5.853	0.564	12.225	0.610
Clibanarius	5.414	0.458	10.368	0.855
Libinia	3.572	0.331	5.761	0.313

* From Vernberg, 1956.
University of Chicago Press.

increased locomotor behavior, which would require a greater expenditure of energy, but also reflect certain basic cellular and enzymatic differences between species. The Q_{o2} values of gill tissue of these crabs show significantly the same metabolic trend with terrestrialism as noted for the intact organism (Table 3.5). Comparative enzymatic studies on subtidal, intertidal, and semiterrestrial crabs demonstrate the same trend in that the cytochrome *c* oxidase activity in gill tissue is progressively higher in species approaching land (Fig. 3.16).

In contrast to relatively stable levels in aerial land environments, the oxygen content in an aquatic marine environment may vary sharply. This difference in oxygen levels appears to have differentially influenced the metabolic response to oxygen lack of animals from these two habitats. Marine amphipods and isopods, for example, are sensitive to a decrease in dissolved oxygen and respond

TABLE 3.5

Comparisons of Oxygen Consumption of Whole Animal and Q_{O_2} of Gill Tissue*

Species	Average Oxygen Consumption of Whole Animal ($\mu l/Gm/Min$)	Q_{O_2} (Gill)
Ocypode	2.35	15.23
Sesarma	2.21	15.19
U. pugilator	2.03	9.17
U. minax	1.28	6.21
Clibanarius	1.28	5.41
Callinectes	1.14	5.85
Panopeus	0.93	5.32
Menippe	0.51	5.48
Libinia	0.42	3.57

* From Vernberg, 1956.
Ibid.

by increasing their respiratory movements (rate of pleopod beating). However, the respiratory movements of semiterrestrial species do not change appreciably with fluctuation in oxygen tension (Walshe-Metz, 1956). Terrestrial animals tend to be more sensitive to increased CO_2 levels than to decreased oxygen concentration. This pattern of response to oxygen levels was studied by Gamble (1970), who compared the effect of low oxygen tension on a free-swimming amphipod with that of three tubicolous species of crustaceans from different microhabitats. At low tide very low oxygen tensions are found in the tubes of these intertidal burrowing animals. After exposure to low dissolved oxygen, the ventilation rate increased for the free-swimming species, but remained unchanged for the tubicolous amphipods. Moreover, a marked change in the ventilation rhythm was found; at air-saturated oxygen tensions, only the tubicolous amphipods exhibited an intermittent rhythm of ventilation, but when the oxygen levels dropped, a continuous rhythm resulted. These findings coupled with Gamble's work on anaerobic resistance suggest that tubicolous amphipods are more resistant to anoxia than other crustaceans of similar size, and that they do not attempt to hyperventilate at reduced oxygen tension. Because of the viscous drag resistance of the tube, water movement costs the organism energy, and hyperventilation would be energetically expensive. However, a minimal ventilation rate apparently is maintained since the organisms show a continuous ventilation rhythm.

Other studies on respiratory adaptation of amphipods show that the critical tension at which oxygen consumption shifts from oxygen independence to oxygen dependence is higher in semiterrestrial species and lower in brackish and fresh-water species of amphipods (Walshe-Metz, 1956).

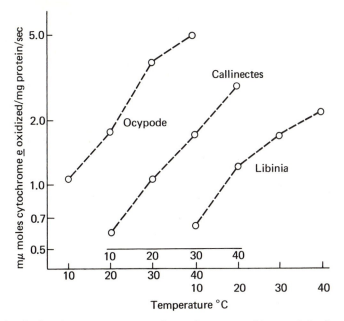

Fig. 3-16. Comparative rates of cytochrome *c* oxidase activity in gill tissue from *O. quadrata, C. sapidus,* and *L. emarginata.* (From Vernberg and Vernberg, 1968. American Zoologist.)

It has also been shown that crustaceans from different habitats show different metabolic responses when the oxygen tension decreases. A study of seven species of crabs from different habitats (Van Weel et al., 1954) demonstrated that the rate of oxygen uptake of a mud-dwelling crab was independent of ambient oxygen tensions until a low level was reached; at this point the respiration pattern became oxygen dependent. Species from well-aerated waters, however, exhibited an oxygen-dependent response over a wide range of oxygen tensions. All species showed a steep increase in percentage oxygen utilization at oxygen concentrations less than 2.5 ml of oxygen/liter. This response was interpreted as an attempt by the organisms to insure a minimum amount of oxidation. Similarly, fiddler crabs, *Uca pugilator* and *U. pugnax,* living in burrows in sandy-muddy substrates, may experience low oxygen tensions when the tide is in, since they apparently do not pump water through their burrows. These species are not only relatively resistant to anoxia, but also the critical oxygen tension is low: 1–3% of an atmosphere for inactive and 3–6% for active crabs. These crabs continue to consume oxygen down to a level of 0.4% of an atmosphere; in contrast, the nonburrowing wharf crab *Sesarma cinereum* stops respiring at a somewhat higher value (Teal and Carey, 1967).

Metabolic diversity is further illustrated in other groups of animals represented in the intertidal zone only during certain stages of their life cycle. The developmental stages of these animals may occur in habitats with very different oxygen regimes, and an excellent example is to be found in certain parasitic platyhelminths, the trematodes. The free-living cercarial stage emerges from intertidal snails and inhabits waters with widely fluctuating oxygen levels, while the parasitic stages live internally in their host where oxygen tension levels are generally very low. The oxygen consumption of the cercaria of *Himasthla quissetensis* decreases proportionately when the ambient oxygen tension is lowered from 5% to 0.5%, but the metabolism of the redial stage is oxygen-independent over this range of tensions (W. B. Vernberg, 1963).

b. Metabolism and environmental stress

Faced with living in a rapidly changing environment, it is not surprising that intertidal animals exhibit some degree of metabolic adaptation to meet environmental stress. Although numerous physical factors affect intertidal organisms, we will restrict our discussion to the influence of temperature, salinity, and photoperiod on the metabolism of one of the dominant intertidal macrofauna groups, the crustaceans. The adaptive principles demonstrated by this heterogeneous group are applicable to other intertidal animals.

(1) Temperature

By definition poikilotherms have a variable body temperature and apparently do not expend energy on thermal regulation (Vernberg and Vernberg, 1970). This inability to attain homeostasis by having a constant body temperature has serious ecological implications, since the metabolic machinery of such an organism is more vulnerable to environmental change, particularly temperature fluctuation. However, some intertidal zone organisms have achieved a modicum of freedom from thermal fluctuations by means of metabolic adaptation. Naturalists have long observed that animals from polar regions function normally at low temperatures which would be lethal to tropical animals. Metabolic adaptations to temperature have evolved to permit biological success in just such diverse thermal environments. These metabolic adaptations have been studied at such various levels of biological organization as the intact organism, tissues, and subcellular units, including enzymes and hormones.

Since crustaceans are heterothermic, their metabolic rate will in general vary directly with temperature—low temperature will lower the rate of oxygen uptake and increased temperature will increase the metabolic rate. However, the metabolic rate of some fiddler crabs is thermally insensitive over a relatively wide temperature range (Fig. 3.17) which appears to coincide with the range that occurs during this animal's period of annual activity. During the colder months of the year, metabolic rate is depressed by low temperature, and the animals remain inactive in their burrows. The Q_{10} values reflect both of these metabolic-temperature response patterns. Over the thermal range of 20° to 30°C the Q_{10} value is 1.2, and the animals exhibit a high degree of locomotor activity. When temperatures drop, a marked change in Q_{10} values is observed; for example, between 10° and 15°C the Q_{10} value is 6.3, indicating a

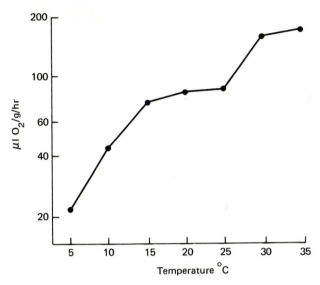

Fig. 3-17. Metabolic-temperature response of the fiddler crab, *Uca pugilator,* from a Massachusetts population. Note that the metabolic rate is thermally insensitive over a relatively wide temperature range. (Modified from Vernberg, F. J., 1969. American Zoologist.)

marked change in metabolic rate. The thermal range of metabolic-temperature insensitivity is not the same for animals occupying different biogeographical zones—for tropical animals this thermal zone is shifted toward higher temperatures than for temperate zone species. In addition, the critical low temperature range at which high Q_{10} values are observed varies with latitude—for tropical fiddler crabs this range is 12–15°C and for temperate zone species 7–12°C (Vernberg, 1959a). The influence of temperature on the standard and active metabolism of intertidal zone animals has also been shown to influence metabolism only slightly (Newell and Pye, 1971). Q_{10} values close to 1 over a range of about 17°C were reported for the standard metabolism of an anemone (7.5–25°C), a polychaete (3–20.5°C), a cockle (6.5–23°C), and a snail (5–35°C). Both the study by Newell and Pye (1971) and one by Halcrow and Boyd (1967) on *Gammarus oceanicus* demonstrated that the rate for active metabolism was more influenced by temperature than the rate of standard metabolism, while Mangum and Sassaman (1969) reported both rates to be equally influenced by temperature in the polychaete *Diopatra cuprea.*

Despite obvious exceptions, certain generalizations can be made on the basis of the results from various metabolic-temperature acclimation studies on intertidal animals, but it is expected that additional studies will modify present knowledge about this subject. Much of the earlier literature on the general problem of acclimation has been summarized (Bullock, 1955; Prosser, 1955;

1958, 1967; Vernberg, 1962; Dill *et al.*, 1964); some papers have emphasized intertidal organisms (F. J. Vernberg, 1969; W. B. Vernberg, 1969; and Newell, 1970).

It is generally accepted that animals experiencing fluctuating thermal environments tend to metabolically acclimate to temperature. Moreover, the temperature range over which acclimation is expressed may be correlated with the ecology and biogeography of the population being studied. For example, temperate zone fiddler crabs are more metabolically labile at lower temperatures than a tropical species which demonstrates a greater degree of lability at higher temperatures (Fig. 3.18).

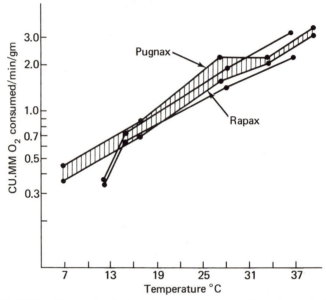

Fig. 3-18. Genotypic limits of oxygen consumption of tropical and temperate zone fiddler crabs. (*Uca rapax,* tropical, *U. pugnax,* temperate.) (From Vernberg, F. J., 1963. Van Nostrand Reinhold Company.)

Because ambient temperature changes seasonally in many marine environments, it is logical to ask whether organisms metabolically acclimate or passively submit to these changes. Various studies have demonstrated a variety of responses, each of which has been correlated with the ecology of the species under study. It has been shown, for example, that seasonal acclimation is characteristic of animals which are active throughout much of the year, while temperate zone fiddler crabs from the Carolinas had a higher metabolic rate at temperatures between 7° and 28°C during the winter months than during the summer period; a closely related tropical species, *U. rapax,* did not exhibit seasonal variation. Since the tropical species does not normally experience seasonal changes in temperature, this lack of metabolic machinery to compensate is not unexpected. As temperature decreases with the approach of winter, the ability of temperate zone animals to be more metabolically active with this temperature drop imparts a certain amount of organismic independence of environmental changes. However, this metabolic-temperature independence is not complete, since temperate zone fiddler crabs remain inactive in their burrows when the air temperature is low.

Acclimation to a high temperature may involve mechanisms different from acclimation to low temperature. Evidence to support this generalization is illustrated by the following examples. Fiddler crabs acclimated to a low temperature and exposed to a high temperature may exhibit a different acclimation pattern than when acclimated to a high temperature and then subjected to low temperatures (Table 3.6). Differences may be noted not only for the response of the whole organism, but also at the tissue and enzyme levels (Table 3.7).

TABLE 3.6 *Types of Thermal Acclimation by Geographically Isolated Populations of* Uca *(after the classification of Precht)**

Species and location of population	Original acclimation temperature	
	High	Low
Uca rapax		
Santos	4	3
Salvador	4	3
Jamaica	4	4
Florida	3	5
Uca pugnax		
North Carolina	3	3
Uca uruguayensis		
Torres	3	4
Santos	5	3

* From Vernberg & Vernberg, 1966a
Pergamon Publishing Company.

Different mechanisms of metabolic-temperature control apparently have evolved in geographically separated populations of one species. That different control mechanisms exist at both ends of the thermal scale and at different levels of biological organization should emphasize the complexity of acclimation phenomena.

When an intertidal organism is subjected to changes in temperature, the initial metabolic response may change with prolonged exposure to this new temperature. For example, in a north temperate zone species of fiddler crab, *Uca pugnax,* the metabolic rate initially decreased and then gradually increased with time (Fig. 3.19). An apparent plateau was reached after 6–8 days. If ani-

TABLE 3.7 *Type of Thermal Acclimation Pattern of Cytochrome-c Oxidase in Tissues of Isolated Populations of* Uca *Based on the Classification of Precht (1958)**

Type of tissue	Species and location of population	Original acclimation temperature	
		High	Low
Muscle	*U. pugnax* North Carolina	4	3
	U. rapax Florida	2	2
	U. rapax Puerto Rico	4	3
Supraoesopha- geal ganglion	*U. pugnax* North Carolina	2	2
	U. rapax Florida	4	4
	U. rapax Puerto Rico	4	5

Type 2—100 per cent acclimation.
Type 3—Partial acclimation.
Type 4—No acclimation.
Type 5—Inverse acclimation.
* From Vernberg and Vernberg, 1968a
Pergamon Publishing Company.

mals were fed, they continued to show a greatly increased metabolic rate (30–40% increase), whereas the metabolic rate of unfed animals decreased to about the level of the control animals (Vernberg, 1959b). Therefore, food is obviously important in metabolic-temperature adaptation, although its role is unclear.

Four distinct populations of a tropical fiddler crab, *Uca rapax,* exhibited diversity in their type of M-T responses (Fig. 3.20). After one day at 15°C two populations showed a positive acclimation response (the metabolic rate had increased by 12–35%), and two populations showed a negative acclimation response (the rate decreased by 10–24%). By day 7 the three populations normally living in thermally stable environments showed no statistically significant change in metabolic rate, but the metabolic rate increased by 45% in the population from Santos, Brazil, where temperatures fluctuate seasonally. Such a response is typical of temperate zone species. Unlike studies on the north temperate zone *U. pugnax,* which were conducted over a 21-day period, the experiments on the tropical animals were discontinued after 7 days owing to an increase in mortality levels.

Once metabolic-temperature patterns were established for the intact organism, then the question could be posed: Do all of the tissues respond in a similar

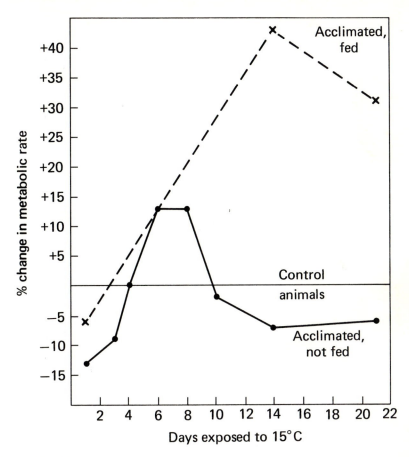

Fig. 3-19. Metabolic response of the fiddler crab, *Uca pugnax,* subjected to changes in temperature. (From Vernberg, 1959b.)

manner over these same temperatures? If so, the metabolic response of the intact organism would appear to be the summation of the oxygen demands of the individual tissues. If not, then some overall metabolic regulating mechanism (possibly endocrine or nervous in origin) would be suspected of integrating the metabolic response of the separate tissues to give a total organismic response which is adaptive to a changing environment.

Gill tissue from *U. pugnax* showed positive acclimation, as did the intact organism, but the mid-gut gland, heart, and supraoesophageal ganglion (SEG) showed negative acclimation, and the muscle tissue was unchanged (Fig. 3.21). *Uca rapax* from Jamaica, the West Indies, demonstrated no metabolic acclimation, as did gill, muscle, and heart tissues, but mid-gut gland showed positive acclimation, and the SEG showed inverse acclimation. From these studies one can conclude that the metabolic-temperature responses of the various tissues

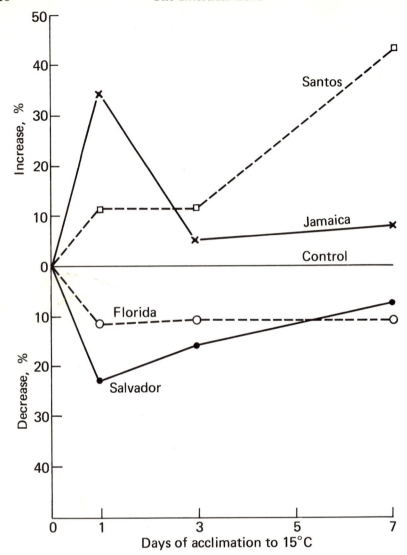

Fig. 3-20. The influence of exposure to 15°C for various periods of time on the respiratory metabolism of four populations of *Uca rapax* determined at 28°C. Percentage change in rate of oxygen consumption relative to control animals as 0 percent change. (From Vernberg, F. J. and W. B. Vernberg, 1966a. Pergamon Publishing Company.)

are not the same, nor do they all correspond with the response of the intact organism. Because detailed studies on the possible integrative role of the endocrine and nervous system on temperature-metabolic adaptation of fiddler crabs are lacking, we can only suggest that integrative mechanisms are of importance.

Also temperature adaptation at the enzyme level has been demonstrated for fiddler crabs by measuring cytochrome *c* oxidase activity in muscle tissue and the supraoesophageal ganglion (SEG) from cold- and warm-acclimated crabs from Florida and Puerto Rico (*Uca rapax*) and North Carolina (*Uca pugnax*) (Vernberg and Vernberg, 1968a). After a period of exposure to low temperature (15°C), cytochrome *c* activity in SEG from *U. pugnax* was enhanced over that in preparations from animals acclimated to 25°C. Cytochrome *c* activity for SEG from Floridian *U. rapax* was the same for cold- and warm-acclimated crabs, whereas cold-acclimation reduced this enzyme activity in animals from Puerto Rico. For muscle tissue, enzyme activity was higher for cold-acclimated animals from North Carolina and Florida, but it was relatively unchanged in Puerto Rican fiddler crabs. These results are consistent with the general theory of metabolic-temperature adaptation which states that the activity rate of a physiological function is higher in a cold-acclimated animal than a warm-acclimated animal. Cytochrome *c* activity of the temperate zone *U. pugnax* demonstrated this response while the activity of tropical crabs from Puerto Rico either was not influenced or was inhibited.

The interspecific diversity between Floridian and Puerto Rican populations is an example of biochemical evolution. Cytochrome *c* oxidase from the muscles of Florida crabs demonstrated a marked acclimation effect when crabs were exposed to cold, whereas enzymatic activity of the more tropical Puerto Rican population remained unchanged. However, the enzyme activity of SEG (the "brain") of Florida crabs was not influenced by temperature acclimation, while the activity of the enzyme from Puerto Rican animals was relatively reduced by exposure to cold. It is postulated that the Florida population has partly adapted to the slightly colder environment of Florida, but the Puerto Rican population lacks this adaptational response. Florida crabs, then, are intermediate in response to cold between tropical and temperate zone species. The failure of the metabolic machinery of the "brain" tissue to respond to cold, as shown both by the oxygen consumption rate and this enzymatic activity, is strong evidence to support the view that the inability of tropical species to live in cold environments may be correlated with the poor development of homeostatic integrative mechanisms such as the "brain."

Since more detailed studies have been made on enzymatic activity and metabolic pathways in sublittoral species, a more thorough discussion of this phase of the problem will be presented in Chapter VI.

Data presented in the preceding paragraphs dealt with comparative responses of latitudinally separated populations of animals. If we now consider results of studies on animals from one geographical region, it becomes apparent that the correlation of metabolic acclimation phenomena and temperature is not a simple one, but reflects the complex interaction of numerous intrinsic and extrinsic factors.

One prominent intertidal zone species along most of the eastern United States is the mud-flat snail *Nassarius obsoleta*. In addition to playing a role in

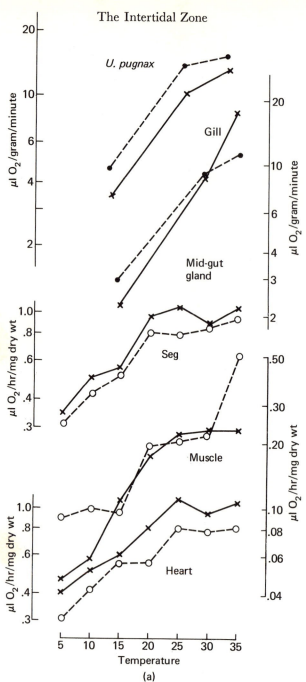

Fig. 3-21. (a) The metabolic-temperature curves of various tissues from cold- and warm-acclimated *Uca pugnax*. (b) The metabolic-temperature curves of various tissues from cold- and warm-acclimated *Uca rapax*. (—— = warm-acclimated, - - - - - = cold-acclimated). (From Vernberg, F., and Vernberg, W., 1966b and Vernberg, W., and Vernberg, F., 1966.)

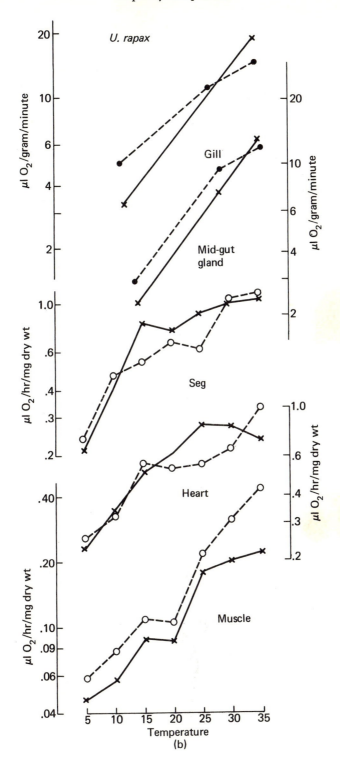

(b)

the ecological energetics of the mud flats, this species serves as an intermediate host for various trematodes. Parasitized snails are not only less resistant to temperature extremes than nonparasitized animals, but also metabolically acclimate differently to temperature (Fig. 3.22). Further, parasitism has been shown to influence the metabolic-temperature response at the enzyme level. For example, the cytochrome c oxidase activity of digestive gland preparations from infected and noninfected animals acclimated to different temperatures

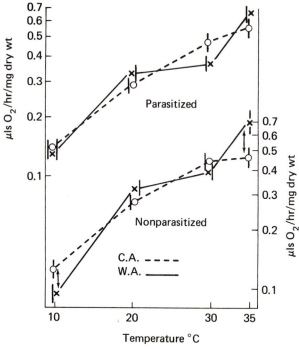

Fig. 3-22. The effect of parasitism on the thermal metabolic acclimation patterns of *Nassarius obsoleta* (——— = warm-acclimated, 30° C; ----- = cold-acclimated, 10°C). (From Vernberg and Vernberg, 1967.)

was not only quite different, but also the level of enzymatic activity was dependent upon which species of parasite was infecting the digestive gland (Fig. 3.23).

Intertidal animals do not show the same response seasonally to fluctuating temperature, as demonstrated by studies on two species of crabs which occupy the same beach. The mole crab *Emerita talpoida*, which consumes oxygen at a much higher rate during the winter than in summer at temperatures below 20°C, is active during the winter and continues growing throughout the year. In contrast, the beach flea *Talorchestia megalophthalma*, shows no metabolic acclimation with season and it becomes dormant during the winter (Edwards and Irving, 1943a, b). Another example of metabolic diversity in response to

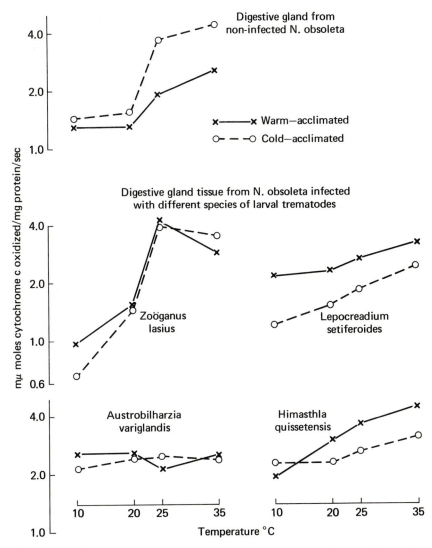

Fig. 3-23. Thermal acclimation patterns of cytochrome *c* oxidase activity in tissue from digestive glands of *Nassarius obsoleta,* infected with different species of larval trematodes (—— = warm-acclimated, 25° C; ----- = cold-acclimated, 10°C). (From Vernberg, W., 1969. American Zoologist.)

temperature is illustrated by the enzymatic activity of the semiterrestrial ghost crab *Ocypode quadrata* and the offshore spider crab *Libinia emarginata.* The activity of cytochrome *c* oxidase in the ghost crab, which is dormant in the winter, was relatively unaffected by acclimation temperature, while enzyme activity varied with thermal acclimation (Vernberg and Vernberg, 1968b).

Seasonal differences in respiratory rates reported in barnacles (Barnes et al., 1963a, b) may be correlated with the animal's biochemical composition during the various seasons. The low metabolic rate of *Balanus balanoides* in October may be the result of the increase in relatively inert semen, which may account for 50% of the total body weight. The scarcity of food and the increase in tissue water content during the winter months may account for the continued low rate of oxygen uptake after the barnacle has discharged its semen during copulation in about November. The metabolic substrates for winter animals are proteins and lipids, but their greater dependence on carbohydrates during the summer when food is more abundant may account for the increase in metabolism during the summer.

(2) Salinity

Exposure to low salinity waters generally causes increased oxygen consumption in intertidal animals originally maintained at higher salinities. This type of response has been reported for the crabs *Carcinus maenas* (Schlieper, 1929), *Ocypode quadrata* (Flemister and Flemister, 1951), *Uca* (Gross, 1957a), and *Hemigrapsus nudus* and *H. oregonensis* (Dehnel, 1960), and the polychaete *Nereis diversicolor* (Schlieper, 1929). In contrast, the metabolic rate of the wool-handed crab *Eriocheir sinensis* is unaffected by salinity change (Schwabe, 1933), while the bivalve *Mytilus edulis* consumes less oxygen when moved to a new salinity (cited by Kinne, 1964). Salinity effects on metabolism may depend on the temperature at which the organism is acclimated, but this interaction between temperature, salinity, and respiration is complex and may differ even in two closely related species of intertidal crabs found on the same beach (Dehnel, 1960). For example, *Hemigrapsus oregonensis* has a higher metabolic rate when acclimated to 25% sea water (salinity of 7.97 0/00) than when acclimated to 75% sea water. This response was reported at all acclimation temperatures (5°, 10°, 15°, and 20°C). In contrast, *H. nudus* consumed oxygen faster when acclimated to 75% sea water than when acclimated to 25% sea water, but only when acclimated to higher temperatures.

Some caution must be exercised in interpreting these results involving the intact organism, since a change in salinity (or for that matter any other environmental factor) may alter the locomotor activity of an organism. Thus, observed changes in metabolic rate may reflect behavioral changes rather than the effect of salinity on basic metabolic processes. For example, the increased metabolic rate of the brackish water snail *Potamopyrgus jenkinsi,* when subjected to a new salinity regime, was reported to be due to increased locomotor activity (Duncan, 1966). In contrast, Dehnel (1960) suggested that the greater metabolic rate at low salinity of the crabs he studied did not result from muscular activity. Various workers have suggested that energetic requirements for osmotic work is only approximately 1% of the total metabolism of an organism. Moreover, it was suggested that the change in oxygen consumption with salinity may reflect the change in water content of the tissues (Schlieper, 1935). This change in degree of hydration may influence the hormonal and/or enzymatic relationships within the cells.

Salinity effects on metabolism may be contingent on the "normal" salinity regime encountered by the organism. For example, the bivalve *Mytilus edulis* living in low salinity water consumes less oxygen at higher salinities while a population of this species from high salinity water has a lower metabolic rate at low salinity (cited by Kinne, 1964). The size of an animal may influence metabolic response to salinity stress as in the crab *Sesarma plicatum*. Small crabs consume 50% less oxygen in 50% sea water than when they are in tap water. However, adult crabs have similar metabolic rates in both of these media (Madanmohanrao and Rao, 1962).

(3) Photoperiod

Photoperiod, which may vary seasonally, can influence the metabolism of intertidal animals. For example, *Hemigrapsus oregonensis*, an intertidal crab, had a higher oxygen uptake rate when maintained under a photoperiod of 16 hours darkness–8 hours light than when exposed to either constant darkness or 8 hours darkness–16 hours light (Dehnel, 1958). Although this study utilized animals collected during the summer and determinations were made only at 15°C, the response was interpreted as reflecting a metabolic adjustment to the winter light regime. Presumably this response would be adaptive in that the increased rate of metabolism would compensate for the expected metabolic decrease in winter due to low temperature.

c. Metabolic cycles

In addition to seasonal changes in metabolic rate, shorter-term rhythmic respiratory cycles have been reported. Of particular interest to our discussion of intertidal zone organisms is the occurrence of metabolic changes correlated with tides. A persistent oxygen consumption rhythm which was correlated with tidal frequency has been reported for various intertidal animals: snails, *Littorina littorea* and *Urosalpinx cinereus* (Sandeen et al., 1954); fiddler crabs (Brown et al., 1954; Barnwell, 1968); and the crab *Carcinus maenas* (Arudpragasan and Naylor, 1964). Typically the oxygen consumption rate of these animals is highest during that phase of the tidal cycle when greatest locomotor activity occurs. For example, fiddler crabs are relatively inactive in their burrows when covered by sea water at high tide. Under this circumstance, the oxygen content within the burrows may drop to almost zero. Conceivably a rhythmic drop in metabolism during high tide would have adaptive significance for these animals. Other rhythms, such as circadian (about 24 hours), semilunar, and lunar, have been demonstrated to influence the metabolism of various intertidal species. Since oxygen consumption is but one physiological indicator of cyclic phenomena, the reader is referred to p. 85 for a more detailed discussion of rhythmicity and its biological importance.

4. Circulation

Circulating body fluids of intertidal animals transport various chemical substances necessary for metabolic continuity. These compounds have many functions; they may serve as food, act as a regulating substance, or they may be by-products to be delivered to a disposal site. The organism's environment may directly or indirectly influence the composition of its blood and thus affect various physiological processes. In our present discussion, obvious compositional changes associated with osmoregulatory phenomena, such as ionic composition, will be excluded, but since many physiological processes are interrelated, some of the examples cited below may be partly correlated with osmoregulation.

a. Respiratory pigments

In many animals oxygen transport to respiring tissues is assisted by oxygen transport molecules found in some types of circulatory fluid. Typically these molecules will reversibly combine with oxygen in body regions where the ambient oxygen content is higher than that of the circulating fluid, such as gills or lungs. Oxygen is released to the surrounding tissue media where the oxygen content is reduced. The interactive effects of environmental factors on this vital organismic function has been studied in varying degrees of thoroughness for intertidal animals. Examples from various animal groups will introduce some of the current problems.

(1) Polychaetes

The common Atlantic bloodworm *Glycera dibranchiata,* which lacks a vascular system, has a large pool of hemoglobin in the small nucleated cells of the coelomic fluid. This fluid contains as much hemoglobin, 3.5 g/100 ml, as typical invertebrate bloods. As a result of waves of contraction by body wall muscles and by the beating of cilia lining the coelomic cavity, the coelomic fluid is transported throughout the length of the body, moving through segmentally arranged gills and to tissues. This hemoglobin is thought to help transport the oxygen, since low CO concentrations blocked the functioning of hemoglobin, resulting in oxygen consumption being depressed by about 15% in worms maintained at 20°C (Hoffmann and Mangum, 1970; Mangum, 1970). With standard procedures, it was determined that two different sized hemoglobin molecules are present in the coelomic cells, although a closely related species, *G. americana,* has only one kind of molecule. Oxygen equilibrium curves for the light-weight, heavy-weight, and a mixture of the two were significantly different. The light-weight hemoglobin had the highest affinity for oxygen, while the mixed solution had the lowest affinity, an unexpected result (Fig. 3.24). Although the Bohr effect was observed, its adaptive significance in this species was questioned because a pH gradient has not yet been demonstrated. These workers suggest that hemoglobin may function in this worm as a storage site for oxygen. During periods of intermittent cessation of ventilation, a brief period of oxygen storage could be of adaptive significance. Also, these coelomic cells may be of great value as a source of

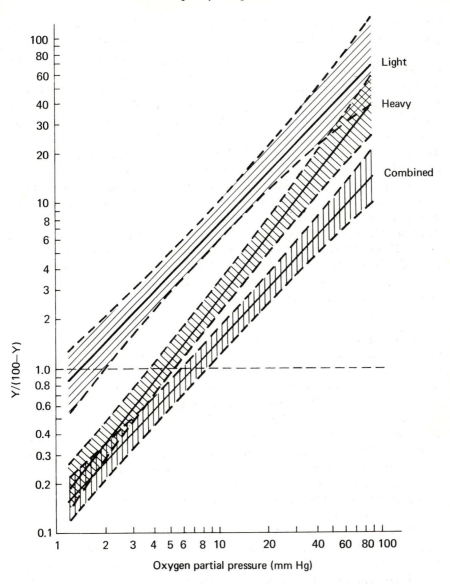

Fig. 3-24. Oxygen equilibrium curves for dissolved hemoglobin from coelomic cells of the bloodworm *Glycera dibranchiata*. Y is the percent of the total that is combined with oxygen. The partial pressure at which Y = 50 is known as the oxygen affinity (P_{50}); a high numerical value of P_{50} indicates a low oxygen affinity and vice versa. (From Hoffmann and Mangum, 1970. Pergamon Publishing Company.)

oxygen to the organism when the tide is out and ventilation stops for a longer period. Hoffmann and Mangum (1970) estimated that this species could draw on its oxygen supply for almost two hours after ventilation is initially inhibited, a time period which could be of significance to this species which reaches its maximum population density near the low water mark. Earlier Jones (1955) determined the oxygen equilibrium curves for vascular and coelomic hemoglobin from the polychaete, *Nephthys hombergii*, and suggested these hemoglobins could function at higher oxygen concentrations, but not when the tide was out and worm burrows collapsed. The P_{50} at a pH of 7.4 for the vascular hemoglobin is 5.5 mm Hg, and this value is 7.5 mm Hg for coelomic hemoglobin. Unlike the vascular hemoglobin, the coelomic hemoglobin demonstrated a reverse Bohr effect, and Jones suggested that oxygen could be transferred from the coelom to the blood. The respiratory pigment chlorocruorin found in sabellid worms had a low affinity for oxygen ($P_{50} = 27$ mm Hg). This pigment could function only in worms living in well-aerated habitats. Although the amount of blood pigment will increase in some crustaceans during a period of anoxic stress it does not in two worms, *Arenicola marina* and *Scoloplos armiger* (Fox, 1955).

(2) Mollusca

Several species of chitons from tropical and cold waters had similar P_{50} values ranging from 20 to 26 mm Hg. Redmond (1962a, b) correlated the low oxygen affinity of the respiratory pigments of these species with their well-aerated sea water habitat, their low oxygen demand, and a relatively large gill area. In contrast, the P_{50} value for hemocyanin from the snail *Busycon canaliculatum* which inhabits muddy salt flats, is 15 mm Hg in the absence of CO_2 and 6 mm Hg in the presence of CO_2. An even more dramatic increase in oxygen affinity is noted in the tissue respiratory pigments of burrowing clams. The tissue hemoglobin from the gills of the mud-dwelling clam *Phacoides pectinatus* has an extremely low P_{50} of 0.19 mm Hg (Read, 1962). The muscle hemoglobin of the clam *Mercenaria mercenaria* also has a very high oxygen affinity (Manwell, 1963).

The inverse Bohr effect has been reported in various species, including *Busycon* and the horseshoe crab *Limulus polyphemus*, and appears to be an adaptive characteristic of organisms living in regions of high CO_2 content and low O_2 concentration. Redmond (1955) proposed that the combination of the inverse Bohr effect would cause about 28% more oxygen turnover coupled with a decreased partial pressure of oxygen in the veins.

(3) Crustacea

Generally, the respiratory pigments of animals living in oxygen-poor habitats have a higher affinity for oxygen than do pigments from animals inhabiting oxygen-rich habitats. For example, the P_{50} values for respiratory pigments of a number of birds and terrestrial mammals are from 25 to 56 mm Hg. The P_{50} values for various aquatic crustaceans range from 7 to 15 mm Hg. There are,

however, always exceptions to generalities, and Redmond (1962a, b) reported a surprisingly low P_{50} value of 3.5 mm Hg for the land crab *Cardisoma
guanhumi* and suggested that this low value may facilitate stable loading pressures during periods of thermal or osmotic stress. However, the response of most
terrestrial crabs is consistent with the general trend that oxygen affinity of the
hemocyanin from terrestrial species is greater than that of aquatic species (Redmond, 1968). The hemocyanin of *Gecarcinus lateralis,* a terrestrial crab, became half-saturated at 16–17 mm Hg O_2 pressure (Fig. 3.25). Also represented

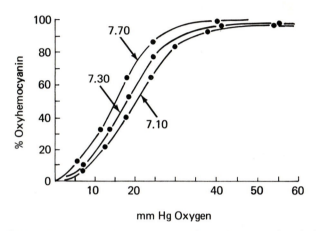

Fig. 3-25. Oxygen-equilibrium curves of the hemocyanin of *Gecarcinus.* Each curve represents determinations at three different pH's at 27–
28°C. (From Redmond, 1968. American Zoologist.)

in Fig. 3.25 are data demonstrating that as acidity is increased, the pigment's
affinity for oxygen is decreased. Redmond (1968), on the basis of direct
measurements of the oxygen tension of blood in the pericardium and the venous
return from the cheliped, reported that the hemocyanin was approximately 87%
oxygen saturated in the gills, and returned from the tissues about 18% saturated, while the pO_2 changed from 32 to 9 mm Hg. A summary of oxygen exchange by the blood of *G. lateralis* is represented in Fig. 3.26. Redmond (1968)
reported that 94% of the oxygen in the blood of *G. lateralis* is in the form of
oxyhemocyanin. In this species the pO_2 is relatively high in both the pericardium (23–32 mm Hg) and the venous blood (9–14 mm Hg) when compared to
the low values in the pericardium (5–8 mm Hg) and venous blood (2–4 mm
Hg) for aquatic species.

b. Blood characteristics

(1) Coagulation of blood

Because various types of adverse clashes may cause the body of an organism
to be punctured, a mechanism is required to plug any resulting hole and

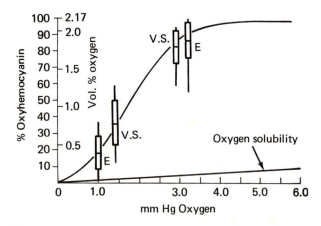

Fig. 3-26. Summary of O_2-exchange by the blood of *Gecarcinus later-alis.* The curve indicates the O_2-equilibrium of *G. lateralis'* hemocyanin at pH 7.40 and 27°C. The vertical bars crossing the curve show the per-cent saturation of the hemocyanin and the volumes percent oxygen found in the blood by Van Slyke analysis (V.S.) and by O_2-electrode (E). The two bars to the right represent pericardial blood; the two bars to the left, the venous outflow of the cheliped. The bars show the range, standard deviation, and average for each set of measurements. Since conditions of pH and temperature were not constant during these measurements, the O_2-pressure corresponding to the percent saturations may not be read from the pH 7.40 O_2-equilibrium curve. The bars are drawn over the average partial pressure of oxygen found in the pericardial and cheliped blood. The O_2-solubility line at the bottom of the graph shows the approx-imate volumes percent oxygen that will be dissolved in the blood as the O_2-pressure varies. (From Redmond, 1968. American Zoologist.)

thereby to prevent excessive loss of body fluid. In more primitive marine ani-mals, the loss of fluid to the sea need not be serious since the composition of their body fluids may not be very dissimilar from sea water. In more complex organisms a variety of mechanisms respond to body damage, from the rather simplistic constriction of the body wall until regeneration occurs, to the more complex coagulation process of the body fluid. Compared to the amount of knowledge about clotting in mammals, little is known about how fluids of in-vertebrates and lower vertebrates coagulate.

In intertidal and terrestrial crabs, clotting is of particular significance since these crabs live out of sea water for significant periods of time. During the molt cycle, the hemolymph of *Gecarcinus lateralis* (a land crab) changes in its capacity to clot, with the most rapid clotting occurring in the premolt period. Unlike that of vertebrates, the blood of *G. lateralis* clots rapidly in the absence of large numbers of cells or a large mass of clotting proteins. The crab's hemo-lymph clots from two to twenty times faster than human lymph (Stutman and

Dolliver, 1968). Temperature also influences clotting. For example, the clotting time for fiddler crab blood acclimated to 10°C is greatly reduced over that of animals acclimated to 30°C; the clotting times are 221.2 seconds for cold-acclimated animals and 56.8 seconds for warm-acclimated crabs. Although a higher level of total blood protein was observed in warm-acclimated crabs, no qualitative difference in the electrophoretic patterns in these two thermally treated groups of crabs was found (Dean and Vernberg, 1966). However, Stutman and Dolliver (1968) reported that a new protein appeared in the blood of *Gecarcinus* during the premolt period, but that its role in the whole clotting mechanism is unknown. No quantitative or qualitative difference in fibrinogen was detected between cold- and warm-acclimated crabs using an electrophoretic technique (Dean and Vernberg, 1966). This type of response might be correlated with the observation that a fibrin-like substance apparently plays a minor role in clotting (Morrison and Morrison, 1952).

(2) *Blood volume*

Blood volume values of many marine species vary markedly, depending in part on the type of body fluid measured and type of circulatory system. At present insufficient data are available to correlate with certainty the relationship of blood volume to environment on an interspecific basis. Crustaceans tend to have values from 17–37% and molluscs from 40–60%; in contrast, the blood-worm does not have blood, but about 34% of its body weight is coelomic fluid. In crustacea, the hemolymph tends to be restricted to the blood vessels and lacunae. The hemolymph volume is less in those species of crustaceans that have a well-developed exoskeleton than in those species in which the body fluid acts as a hydrostatic skeleton (Croghan, 1958).

The influence of environmental change on body fluid volume of an individual animal is not well-known except in the case of an alteration of external salinity, where it has been demonstrated that marked change occurs (see section on osmoregulation). Although the total water content of many animals decreases with decreasing temperature, it has been shown that the blood volumes of fiddler crabs acclimated to 10°C and 30°C were statistically the same. The values for these two groups were 15.5 ± 0.70 vol % and 15.9 ± 0.56 vol % respectively (Dean and Vernberg, 1966). The blood volume of crustaceans will change with the molting stage, and volumes are highest immediately after molt. However, the method by which change in water content occurs in various species may be related to habitat. For example, aquatic species can obtain water from the surrounding milieu during ecdysis, while terrestrial species may have to acquire water in advance and store it in structures, such as the pericardial sacs. At the time of molting, the internal hydrostatic pressure increases and greatly helps to facilitate growth and the shedding of the old exoskeleton. The uptake and redistribution of water in the crab's body during molt appears to have a hormonal basis (Bliss, 1968).

(3) Blood glucose

The concentration of blood glucose varies with a number of factors, but comparison of values published by various investigators is difficult because their experimental procedures may differ. For example, a specific enzymatic method for measuring blood glucose (the glucose oxidase test) developed by Huggett and Nixon (1957) puts in question the comparative value of previous studies, although general trends suggested by earlier workers may still be valid. A number of variables, however, can be recognized. The amount of blood glucose may vary individually in animals directly brought in from the field, but a lower uniform value is obtained after a period of starvation at 30°C. Animals acclimated to 10°C showed little variation in concentration with time. The blood sugar concentration of fiddler crabs varied with time of day the sample was taken, the highest values being obtained between 5:30 p.m. and 9:30 p.m. and a second smaller peak occurring about 9:30 a.m. Although photoperiod acting independently did not appear to markedly affect the blood glucose level, a significant interaction between photoperiod and time of day has been statistically demonstrated. This result indicates that the joint effect of time of day and photoperiod acting simultaneously is different from the sum of their separate effects. An obvious correlation of diurnal changes in blood sugar level with cyclic variation in other physiological parameters does not presently exist; for example, the peaks in oxygen consumption (Brown et al., 1954) are different from the peaks in blood glucose levels. However, these differences may reflect variation in experimental design and population differences, since the separate studies were conducted using either northern or southern populations of crabs.

Nonovigerous female crabs have a lower blood glucose level than those carrying freshly laid developing eggs (called a sponge). With time, the blood glucose level of females carrying a mature sponge drops to a level comparable with that of the nonovigerous female. This response probably reflects the active mobilization of energy reserves to provide nutrients for the eggs, but once the egg mass has been laid and receives no further energy from the female, this phase of the mobilization process is over and the blood glucose level drops (Dean and Vernberg, 1965a).

Blood glucose levels in fiddler crabs also may vary with the temperature at which animals are acclimated (Dean and Vernberg, 1965b). Although the lowest values are at lowest temperatures, the response pattern is not simply correlated with temperature. The concentration of blood glucose is independent of temperature over certain thermal ranges, but a further slight thermal change will alter the concentration markedly. For example, over the ranges 2–10°C and 15–25°C little change is noted with resultant low Q_{10} values, but from 10° to 15°C and from 25° to 30°C the Q_{10} values are over three.

In addition to external environmental factors, intrinsic physiological mechanisms have also been reported to control blood glucose levels. When the eyestalks of fiddler crabs acclimated to 30°C are removed, the blood glucose level drops significantly after one day (Dean and Vernberg, 1965a). In a separate experiment, the injection of eyestalk extracts increased the blood sugar rapidly within one hour, but a gradual decrease to the pretesting level followed (Kleinholz et al., 1950).

c. Heart rate

Various internal and external factors influence the heart rate of intertidal animals. Although reduced temperature typically decreases the heart rate, another factor to consider when comparing latitudinally separated populations is that each might be differentially adapted to different thermal regimes. For example, a species of crab from Plymouth, England, had a heart rate 12% faster than a population sample of the same species from the Mediterranean when measured at common temperatures of 18°C (Fox, 1939). The heart rate of a tropical bivalve, *Isognomum alatus,* living in mangroves reacts rapidly to temperature change, probably as a result of thermoreceptor stimulation rather than direct temperature effect on the heart muscle (Trueman and Lowe, 1971). Oxygen concentrations also influence heart rate. In the clam *Mytilus edulis* an oxygen tension below 1% or an increased CO_2 content causes a gradual decrease in heart rate, 3% oxygen produces an initial increase in rate, but by 130 minutes the heart stops beating. When animals are exposed at low tide, their heart rate is greatly reduced (Schlieper, 1955). Similar results were reported for the cockle *Cardium edule* (Trueman, 1967). However, during exposure to air the heart rate of a tropical bivalve, *Isognomum alatus,* was unaltered and dependent, instead, on the ambient temperature (Trueman and Lowe, 1971). Generally, small individuals have a higher heart rate than larger forms. The rate of beat increases during exercise for most animals which have nervous regulation of the heart. When various related species are compared, the heart rate is lower in sluggish animals than in active ones.

d. Hydrostatic mechanism

The hydrostatic function of body fluids in some intertidal animals gives them a semblance of skeletal rigidity. The entire body, or only specific parts, may become more or less turgid, and this response may aid an animal in activities such as burrowing, horizontal locomotion, feeding, and "swelling up" in a burrow or crevice to inhibit removal by predators. Knowledge of the hydraulics of body fluids during the active process of burrowing has been greatly expanded by recent advances in electron recording techniques (see review of Trueman and Ansell, 1969). It is known, for example, that soft-bodied animals burrow utilizing two types of anchors, penetration anchors and terminal anchors (Fig. 3.27). In contrast, hard-bodied animals generally have numerous appendages which displace the substratum to create a cavity into which the animals passively fall or actively are pushed by other appendages (Fig. 3.28).

The circular body is ideal for burrowing since the entire body wall is in contact with the substratum and all muscles assist in digging. Further, a true

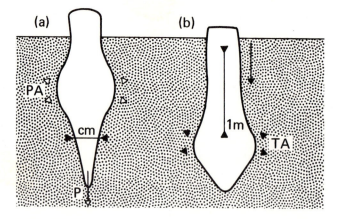

Fig. 3-27. Diagrams showing the two principal stages in the burrowing process of a generalized soft-bodied animal. (a) formation of penetration anchor (PA, hollow arrowheads) that holds the animal when the
distal region is elongated (P) by contraction of the circular (cm) or
transverse muscles. (b) dilation of distal region producing a terminal anchor (TA, solid arrowheads) that allow contraction of longitudinal (lm)
or retractor muscles to pull the animal into the substratum (→). Maximum pressures are developed in the fluid system at this stage. (From
Trueman and Ansell, 1969. George Allen and Unwin Ltd.)

fluid skeleton in the circular animal facilitates powerful movements since muscular force can be transferred to the lower extremity by means of the hydraulic
system of the body cavity. Fig. 3.29 represents recordings of pressures produced
during burrowing activity of the intertidal worm *Arenicola marina* (Trueman
and Ansell, 1969).

5. Chemical Regulation and Excretion

a. Osmoregulatory mechanisms

The extent to which intertidal animals may regulate their blood concentration generally may be correlated with the degree of stress imposed by their
particular environment. Therefore, physiological diversity can be expected between species occupying different parts of the intertidal zone. Lockwood (1962)
proposed that five principal types of osmoregulatory mechanisms are utilized
by animals to physiologically meet the environmental challenges of the intertidal
zone: (1) reduced permeability of the body surface, (2) active uptake or extrusion of ions, (3) regulation of body water volume, (4) conservation of water
or salts by the excretory organs, and (5) regulation of cellular osmotic concentration.

The effects of body surface permeability can be readily compared in that
the exoskeleton of terrestrial crabs is less permeable to water and salts than
that of semiterrestrial species, and in turn the exoskeleton of these species is less

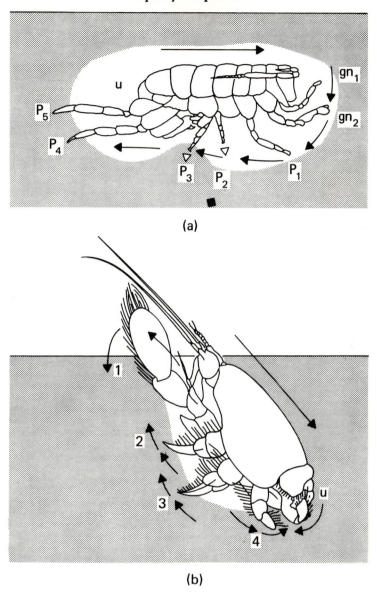

(a)

(b)

Fig. 3-28. Burrowing in animals with a hard exoskeleton. (a) The position of the amphipod, *Talorchestia deshayesii* in its burrow; gnathopods 1 and 2 (gn 1, 2) and periopod 1 (p_1) pull sand down from in front of the animal and sweep it back below the flexed urosome (u); peripods 2 and 3 (p_2, p_3) brace the body while the sand is pushed out behind by the urosome. Periopods 4 and 5 (p_4, p_5) push the animal forward as the sand is removed in front. (After Reid, 1938.) (b) the position of the mole crab *Emerita talpoida* in the sand, with arrows to indicate the direction of the power stroke during burrowing of the first to the fourth thoracic legs(1–4) around the uropod (u). (From Pearse, Humm, and Wharton, 1942.) (Cited in the review of Trueman and Ansell, 1969. Ibid.)

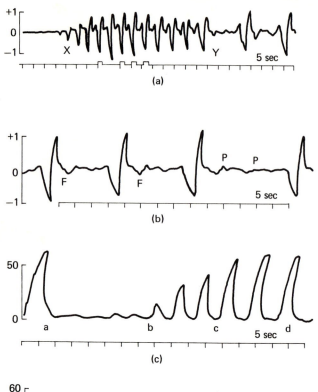

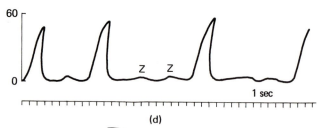

Fig. 3-29. Recordings of the pressures (cm of water) produced by *Arenicola marina* externally in sand a and b and in the coelom c and d during burrowing. (a) sequence from the commencement of burrowing (extreme left) showing increase in amplitude of the negative pulses as penetration proceeds (at X) and the reduction of their frequency (at Y) after six branchial segments have passed into the sand. Visual observation of dilation of the anterior trunk segments is marked above the time trace. (b) three digging cycles when burrowing against glass with the trunk almost completely beneath the surface of the sand, flanging (F) and proboscis extrusion (P) being marked by direct observation. (c) pressure changes in the coelom during the commencement of burrowing; a, head down on sand, followed by repeated proboscis eversion, b, c, and d, 2, 3, and 4, chaetigerous annuli respectively beneath the surface. The flat top of the last two pulses is due to limitation of travel of the pen. (d) coelomic pressure recording after 2 min of burrowing showing increased duration of the digging cycle, minor fluctuations in pressure occurring (Z) between high pressure pulses, corresponding to a peristaltic wave arriving in the anterior segments and proboscis eversion. (From Trueman and Ansell, 1969. George Allen and Unwin Ltd.)

permeable than that of subtidal crabs. In one study, water loss, chiefly through the exoskeleton, was about half as much per unit time for four species of semi-terrestrial crabs as in three intertidal species, and only one-third to one-fifth as much for two aquatic species (Herreid, cited by Bliss, 1968). A similar trend in decreased exoskeletal permeability of animals occupying more terrestrial habitats has been reported for both salt loss (Gross, 1957b) and rates of sodium loss (Shaw, 1961): subtidal species lost more salt and/or sodium than more terrestrial or freshwater species. Within one species, *Carcinus maenas,* the rate of water loss varied with external salinity, being highest at 75% and lowest at 30% seawater (Smith, 1970). Whether this change in permeability reflects reduced circulation, integumental changes, or another factor is not understood.

That organisms in more terrestrial habitats tend to regulate their blood concentration better than subtidal animals has been illustrated by numerous workers (Gross, 1964; Lockwood, 1962; Potts and Parry, 1964; Vernberg and Vernberg, 1970). Results of many studies have shown that most terrestrial species hypo- and hyperosmoregulate, while many intertidal and estuarine species can maintain their blood hyperosmotic to reduced salinity with various degrees of success (Fig. 3.30 and Table 3.8). This response has obvious adaptive signifi-

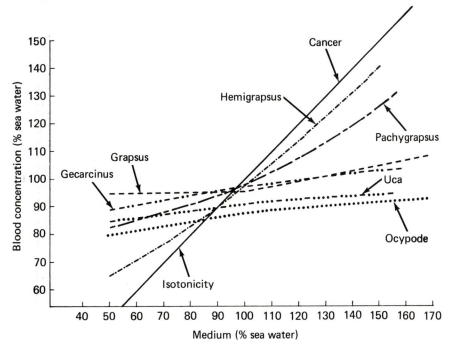

Fig. 3-30. Comparative osmoregularity of crabs showing different degrees of terrestrialness. *Cancer* is an aquatic species; *Hemigrapsus* and *Pachygrapsus,* low intertidal zone; the other species are terrestrial or high intertidal zone inhabitants. (From Gross, 1964.)

TABLE 3.8   *A Comparison of Differences in Osmoregulation for Species of Terrestrial, Intertidal, and Estuarine Crabs**

PROBABILITY VALUES

A. Blood Osmotic Concentrations	Concentration of Medium (% sea water)			
	50	100	150	168
Hemigrapsus v. *Pachygrapsus*	<0.001	not sig.	<0.001	
Pachygrapsus v. *Grapsus*	<0.01	<0.01	<0.01†	
Grapsus v. *Uca*	<0.01	not sig.	not sig.†	
Uca v. *Ocypode*	<0.05	not sig.	—	
Grapsus v. *Ocypode*	<0.001	<0.01	—	<0.02
Gecarcinus v. *Grapsus*	<0.05	not sig.	—	
Gecarcinus v. *Uca*	not sig.	not sig.	<0.01	
Gecarcinus v. *Pachygrapsus*	not sig.	not sig.	<0.001	
Gecarcinus v. *Ocypode*	<0.001	<0.01	<0.02†	

† Crabs in 150% sea water compared with *Grapsus* and *Ocypode* in 168% sea water.

B. Na U/B values	50% SW v. 100% SW	100% SW v. 150% SW
Hemigrapsus	<0.001	<0.001
Pachygrapsus	<0.001	<0.01
Uca	**	**
Gecarcinus	not sig.	not sig.

	100% SW v. 115% SW
Cancer	<0.05

** U/B values for Na were not determined on individual specimens of *Uca*.
* From Gross, 1964.

cance in view of the limited water supply from which semiterrestrial and terrestrial species can replace losses due to transpiration or during molting. Also, the only water available to certain species may be a seawater pool, which often is concentrated owing to water evaporation.

The time taken for an organism's blood to osmotically adjust to a new salinity is to some extent a function of both the amount of change encountered and other environmental/organismic factors, such as temperature and time. Dehnel (1962) reported that during the summer about one-half of the changes in blood concentration of crabs *Hemigrapsus oregonensis* maintained at 95% sea water and then exposed to new external salinities, occurred within three hours. At the end of 24 hours, the blood concentration appeared to be in equilibrium with the new salinity over the range of 12% to 100% sea water, but by 48 hours it was not clear if the blood of animals subjected to high salinity (125 and 150% sea water) or low salinity (6% sea water) had adjusted. The response of winter animals was somewhat different: blood concentration appeared to stabilize more rapidly in that little change occurred between 24 hours and 48 hours. Also, summer animals died after a three-hour exposure to 175% sea water, while winter animals lived at least 48 hours; by this time their blood concentration was about 190% sea water. However, this species is known to inhabit salt lagoons which have a salinity of 175% sea water. Gross (1963), using longer term, controlled salinity acclimation procedures, reported that *H. oregonensis* showed some hypo-osmotic regulation by three days and a significant amount by 65–75 days (Fig. 3.31). Thus, this species can slowly acclimate to an extremely high salinity.

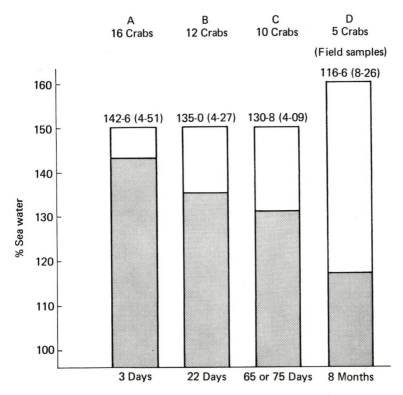

Fig. 3-31. Hypoosmotic regulation in *Hemigrapsus* following different periods of exposure to high salinity. Height of entire bar represents osmotic concentration of the medium at termination of experiment; shaded area of bar represents mean osmotic concentration of blood. Mean values for blood are indicated above shaded areas and respective standard deviations are in parentheses. Only group A was subjected to terminal salinity for entire indicated period of exposure. (From Gross, 1963. Pergamon Publishing Company.)

In addition to Dehnel's study (1962) showing both a temperature and seasonal influence on osmoregulation, Lindquist (1970) reported an inverse relationship between blood osmotic pressure and acclimation temperature (3°, 23°, and 30°C) in the isopod *Porcellio scaber*.

The blood osmotic pressures of animals may depend on size. An inverse relationship between body weight and the freezing-point depression of blood was found in male isopods, *Porcellio scaber,* but no change with body size was reported for another isopod, *Oniscus asellus* (Lindquist, 1970). In *Carcinus maenas* blood conductivity was found to be highest in males and females approaching the onset of sexual maturity and weighing about 35 g, and to be lowest in both smaller and larger animals (Gilbert, 1959). Although no significance was given to these findings in the isopods, it is important to recognize that body size may have to be considered if this response is compared in animals subjected to various environmental stresses. Other variables also must be considered; for example, molting animals and ovigerous females have lowered osmotic pressures (Lindquist, 1970).

The antennal gland (the kidney) of most terrestrial and semiterrestrial crabs regulates the principal ions, but this gland apparently has little to do with osmoregulation. Table 3.9 represents a compilation of various studies involving ionic regulation in semiterrestrial and terrestrial crabs. In contrast, the gills and the posterior diverticulum of the alimentary canal appear to be important in osmoregulation. For example, Mantel (1968) demonstrated that the foregut of the land crab *Gecarcinus lateralis* is permeable in both directions to water and salts. Apparently permeability is under neuroendocrine control, since an extract of thoracic ganglionic mass added to the "hemolymph side" of the foregut in vitro resulted in a significant increase in permeability to water and salts. Similar results were found both in crabs with eyestalks during intermolt and in eyestalkless crabs after ecdysis. This ability to move water and salts between the foregut and hemolymph could help this terrestrial species conserve water, especially during the critical time of ecdysis.

Although various physiological mechanisms have evolved to accomplish osmoregulation in many crabs, a few species which are highly adapted to land have used behavioral responses. The coconut crab *Birgus latro* and the land hermit crab *Coenobita perlatus,* for example, will select from a number of options the proper salinity to maintain normal blood concentrations (Gross, 1957a; Gross and Holland, 1960). It was found that animals kept in 100% sea water had blood sodium levels of 477 to 532 meq/liter. After 20 days in 200% sea water, this value rose to 1022 meq/liter, but then when the animals were given a choice between freshwater and 100% sea water they chose freshwater 45% of the time, sea water 0.1% of the time, and remained out of the water the rest of the time: Their blood sodium dropped to a value of 406 meq/liter. In contrast, an animal kept in freshwater for 14 days chose sea water 56% of the time and freshwater 0.4% of the time and spent 43.6% of the test period out of water. The blood sodium value rose from 398 to 468 meq/liter. Not all ions behaved in the same manner; K levels appeared to be regulated by the antennary gland but not Na, Ca, or Mg.

The cells of those intertidal zone species which do not regulate completely the osmotic or ionic composition of their circulatory fluids, must be able to regulate these properties if cellular metabolism is to continue without wide fluctuation. Even in *Carcinus maenas,* which shows the ability to osmoregulate at low salinities, cellular changes are observed when this species is exposed to a reduced salinity since the water content of muscle fibers rises and some ions are lost. Because chloride loss is greater than accounted for by the increase in muscle water, chloride may be in electrochemical equilibrium across the fiber membrane. Potassium appears to also be in equilibrium, but sodium exclusion from the cell must involve an active process (Shaw, 1955). Much of the osmotic adjustment within *Carcinus* cells is due to variation in amino acid concentration (see Chapter IV for a more detailed discussion).

b. Excretory products

With the approach of a terrestrial mode of existence, the principal excretory product has shifted in certain groups of animals. The classic study of Needham (1938) demonstrated that in gastropods a shift from ammonotelism to uricotelism occurred when snails invaded fresh water or land from the sea. However, within a species the ratio of end products may change at different stages of development or under environmental stress. In contrast, crustaceans are commonly ammonotelic irrespective of their habitats. An earlier study involving a series of amphipods and isopods from marine to terrestrial habitats reported that the percentage of nitrogen excreted as ammonia was 80–90% in aquatic species, but was reduced to 50–60% in terrestrial organisms (Dresel and Moyle, 1950). Recently Hartenstein (1968) reported that ammonia is the chief nitrogen end product of a terrestrial isopod, *Oniscus asellus*. Urea was not measurable, but uric acid levels were high, the greatest concentration (93%) being in the body walls. Molting thus is one method for excretion of the end products of purine catabolism. Hartenstein therefore suggested that the ability to excrete ammonia as a gas offers thermodynamic advantages to this species since it normally does not face dehydration problems. The land crab *Cardisoma guanhumi* excretes ammonia as its principal nitrogen waste product. However, when maintained on a high nitrogen diet, the level of blood urea was markedly increased. In general, the total nitrogen excreted is low in this species, a condition possibly correlated with its herbivorous nature. The picture of nitrogen excretion is not clear, since large deposits of uric acid were found in the hemocoel (Horne, 1968). Apparently this species can tolerate higher blood ammonia concentrations (3 to 8 times higher) than those of aquatic species. In addition, ammonia levels were higher in the stomach fluid. Gifford (1968) proposed that the amount of ammonia stored in the blood and stomach fluid is equivalent to that normally excreted in one or two days, so that when this animal enters water the ammonia load can be easily reduced. Terrestrial crabs have little dependence on the antennal gland as a nitrogen excreting organ. The small amount of urine produced by this species contains little N, and it drains chiefly into the gill chamber.

6. Reproduction and Development

Although many intertidal zone animals tend to lead a semiterrestrial exis-

TABLE 3.9 *Ionic Regulation in Semi-terrestrial and Terrestrial Crabs**

Species	Antennal glands		Bladder	Gills	Gut
	Conserving	Excreting			
Gecarcinus lateralis	Ca (slight)	Na (slight) K (slight) Mg (slight)			Na and Cl (L → H H → L)
Coenobita perlatus		K (slight)			
Cardisoma armatum	Na (high salinity) K Ca Cl (slight)	Na (low salinity) Mg SO$_4$			Probably K and Ca (L → H)
Cardisoma carnifex	Ca	Mg			
Cardisoma guanhumi		K Mg SO$_4$		Na K	

Species				
Ocypode quadrata	Na (high salinity) Ca Cl (low salinity)	Na (low salinity) K Mg Cl SO₄ (slight)		Na (L → H) K Ca (H → L) Cl Mg SO₄
Uca crenulata	Na (high salinity) Ca	Na (low salinity) K Mg		
Uca pugilator	Na (high salinity)	Na (low salinity) K Ca Mg NH₄ Cl SO₄	Na? Mg Cl	Ca NH₄ (H → L) Mg SO₄
Uca pugnax	Na (high salinity)	Na (low salinity) K Ca Mg NH₄ Cl SO₄	Na? Mg Cl	Ca NH₄ (H → L) Mg SO₄
Pachygrapsus crassipes	Na (high salinity) K (high salinity)	Na (low salinity) Mg	Mg (H → Bl) Na (Bl → H) Cl (H → Bl)	
Hemigrapsus nudus		Cl		
Hemigrapsus oregonensis	Na (high salinity) K (high salinity)	Na (low salinity) Mg		

Abbreviations: Bl, Bladder; H, Hemolymph; L, Lumen.
* From Bliss, 1968.
American Zoologist.

tence as adults, most have pelagic larval stages. Thus, the reproductive cycle must be geared to insure release of gametes into a favorable watery environment, and to this end animals have evolved many strategies. The more mobile intertidal animals can migrate to a favorable environment to release their larvae. Gravid females of a Danish population of *Carcinus maenas,* for example, leave the intertidal zone during the breeding season and move out into deeper water, where the water is more saline and more conducive to larval growth than inshore waters. Interestingly, this same migratory behavior pattern has been observed in both male and female *C. maenas* which have undergone parasitic castration by *Sacculina* after the parasite itself has developed gonads. By inducing migration in its host, the parasite insures that its larvae will be released in waters favorable for development (Rasmussen, 1958). Some sessile organisms, such as the barnacle *Balanus balanoides,* have an annual breeding cycle that seems to be largely endogenously controlled, since under optimal conditions of temperature, salinity, and light, it is impossible to shift the date at which egg masses are laid down by more than a few weeks (Patel and Crisp, 1960). On the other hand, the oyster *Crassostrea virginica* depends largely on exogenous factors for timing larval release. This species, which breeds in nature during the warmer months of the year, can be conditioned for spawning throughout the year by manipulation of environmental factors (Loosanoff and Davis, 1963).

Reproductive cycles generally consist of a vegetative or resting stage, which is a period of little or no gonadal activity, and a reproductive stage, beginning with gametogenesis followed by the maturation and release of gametes. Environmental factors, including food, salinity, temperature, oxygen, photoperiod, and lunar cycles, interact with endogenous ones to initiate and influence not only various phases of the reproductive cycle but also larval development.

a. Food

Although the breeding season of many invertebrates coincides with spring algal blooms, there is no experimental evidence to show that the seasonal appearance of any one particular type of food acts as a breeding stimulus. Gonad development and activity, however, may depend on adequate food supply. Experiments on the boreo-Arctic barnacle *Balanus balanoides* showed that adequate food supply was necessary for ovarian development, for if the food supply was interrupted, ovarian tissue regressed. In nature, these barnacles store food when it is abundant, and subsequently utilize it for gonad development. Future development depends on current food supplies once ovarian development is initiated (Barnes and Barnes, 1967, 1968).

Synchronization of naupliar release in *B. balanoides* may be linked to adult nutrition (Crisp, 1964). Fertilization and embryonic development occur in the fall, but the exact date of fertilization depends on any environmental factor that slows down metabolism, such as decreasing temperature and reduced nutrition. In this manner the whole population breeds more or less at once, and the nauplii are released in the spring. Synchronization of naupliar release cannot be linked directly to rising temperatures in the spring; if it could, larvae would be released prematurely during unusually mild winters or delayed if conditions were reversed. Instead, Crisp suggests that the breaking of a period of diapause at the end of embryonic development and the date of release are mediated through a "hatching factor" that is secreted by the adult in response to a feeding stimulus. Such a factor would ensure synchronization of naupliar release with the onset of the spring plankton bloom.

Populations of the polychaete *Cirratulus cirratus* in England have been found to breed throughout the year (Olive, 1971), in contrast to most polychaetes which have been found to breed annually during a restricted time of the year. Olive has suggested that *C. cirratus* is able to reproduce throughout the year because of its feeding habits. Being a nonselective deposit feeder, it finds ample food in the region of its habitat. Just as important, it can exploit this available food even at the low environmental temperatures to which it is exposed. In many animals low temperatures block the feeding response.

b. Temperature

Reproductive processes generally occur over a narrower thermal range than many other physiological functions. The boring gastropod *Urosalpinx cinerea* is capable of locomotor activity at 10°C and will drill oysters at 15°C, but oviposition does not take place until temperatures reach 20°C (Stauber, 1950). The American oyster *Crassostrea virginica* can feed and grow at temperatures ranging from 8–31°C, but breeding does not occur until temperatures are between 16 and 20°C, depending on geographical location (Loosanoff and Nomejko, 1951).

In geographically separated populations the breeding season may vary over different parts of their range. Orton (1920) suggested that since the breeding season of each species occurs at the same temperature throughout the entire distributional range of the animal, it would be longer in warmer parts of the geographical range. The reproductive season of *C. virginica* from southern waters, for example, may last for 7 months of the year, but in Long Island Sound, the reproductive season is limited to only 2 months in mid-summer (Loosanoff and Nomejko, 1951). However, since the temperature necessary for breeding may vary in some species with the typical habitat-temperature range, not all species follow Orton's dictum. More southerly distributed populations may breed at higher temperatures than the more northern ones, as illustrated by *C. virginica*. After 68 days of conditioning, 67% of Long Island Sound oysters at 12°C contained mature eggs or spermatozoa, but oysters from South Carolina and Florida maintained under the same conditions had such poorly developed gonads that the sexes could not be distinguished even when gonadal material was examined microscopically (Loosanoff and Davis, 1963).

Temperature is the primary factor governing the breeding condition in some organisms, and when an animal is exposed to different temperature regimes it can be conditioned to breed throughout the year. The American oyster *C. virginica* is a good example of an animal which can be induced to spawn during the cold months when subjected to gradually increasing water temperatures in the laboratory. The conditioning period at 20°C is approximately 3 to 4 weeks; at 25°C, spawning was induced by day 7, and at 30°C in only 3 days (Loosanoff and Davis, 1963). Spawning and gonad development can also be delayed by low temperatures. *C. virginica* were taken from Long Island Sound early in the season before beginning their natural spawning and transplanted to Maine waters where the temperature was about 7°C lower than that in Long Island Sound. These temperatures permitted slow gonadal development but prevented spawning. This procedure caused spawning to be postponed 6–8 weeks after oysters in Long Island Sound had completely finished spawning. However, if oysters were kept in cold waters for an indefinite period of time, undischarged gonad material was resorbed (Loosanoff and Davis, 1963). The clam *Mercenaria mercenaria,* treated in the same manner as *C. virginica,* does not resorb undischarged gonad material and can be induced to spawn after several months of cold exposure.

In some animals, gamete release is induced by cold temperatures as in two species of Arctic barnacles, *Balanus balanoides* and *B. balanus,* which normally are fertilized once a year, in November and February respectively. Barnacles taken into the laboratory and maintained at temperatures of 14–18°C for long periods of time did not become fertilized or show any copulatory activity during the normal breeding season, but those maintained in the laboratory at temperatures of 3–10°C did become fertilized and produce larvae. *B. balanoides* left under field conditions until just a few days before breeding activity began, and then brought into the laboratory, produced normally fertilized eggs at 17°C; *B. balanus* produced eggs at 11°C. Both of these temperatures are much higher than either species would encounter in nature. Thus, apparently prolonged periods of high temperature prevent these animals from reaching breeding conditions (Crisp, 1957).

Sudden changes in temperature can also initiate spawning in some animals. Rapid temperature changes of 3°C or more and sudden water movements caused the intertidal polychaete *Cirriformia tentaculata* to quickly discharge mature gametes (George, 1964).

Tolerances of larval stages to temperature can be correlated with breeding seasons, as illustrated by a study on different species of barnacles (Patel and Crisp, 1960). Embryos of the winter breeding species *B. balanus* and *B. balanoides* cannot withstand temperatures above 15°C, but embryos of species that breed in the spring and early summer, such as *Verruca stroemia* and *B. crenatus,* can tolerate temperatures as high as 23°C. The embryos of mid-summer breeders, *B. perforatus, B. amphitrite,* and *Chthamalus stellatus,* survive temperatures as high as 30°C, while embryos of *Eliminus modestus,* a species that breeds throughout the year in temperate waters, tolerated temperatures ranging from 3–30°C (Table 3.10).

Rates of development can be directly linked to temperature among some intertidal zone animals, such as the mud-flat snail *Nassarius obsoleta,* studied by Scheltema (1967). This gastropod completes gametogenesis in late fall sev-

TABLE 3.10 *Upper Limits of Temperature Range During Breeding Season and the Range within Which Embryonic Development Proceeded at Maximum Rate in vitro in Different Species of Barnacles***

Species and Geographical Range in Europe	Breeding Season	Range of Monthly Mean Sea Temperatures at Southern Limit of Range during Breeding Season (°C)	Upper Temperature Limit (°C)
B. balanoides Arctic–N. Spain	Nov.–Feb.	11–13	14–16
B. balanus Arctic–English Channel	Feb.–Apr.	9–10	13–14
V. stroemia Iceland–Mediterranean	Jan.–Apr.	13–15	21–23
B. crenatus Arctic–W. France	Jan.–June	11–16	22–24
B. perforatus S. Wales–W. Africa*	Apr.–Aug.*	21–27*	25–27
E. modestus S.W. Scotland– S. Portugal†	Throughout year	14–19	23–25
C. stellatus N. Scotland–W. Africa	Probably through-out year	19–27	29–31
B. amphitrite W. France–Equator	June–Aug.	27–29	>32

* Limits of distribution and breeding habits at its limit not certainly known.
† Still spreading south.
** From Patel and Crisp, 1960.
University of Chicago Press.

eral months before spawning, which occurs as soon as temperatures begin to warm. The exact date varies with latitude. Southern species begin spawning in February, but snails from Cape Cod Bay and Maine do not spawn until early June. Once the eggs have been laid, rates of development depend on environmental temperatures. At 28°C, 90% of the larvae had emerged by day 6, but at 11.5°C, none had emerged by day 16. The time in days required between spawning and veliger emergence is shown in Fig. 3.32. There were no significant differences in development time as correlated with temperature in the two geographically separated populations of *N. obsoleta,* but differences in developmental rates have been reported in other latitudinally separated populations of intertidal zone animals. Consider, for example, the study of Bayne (1965) on growth of the bivalve *Mytilus edulis.* Adult *Mytilus* collected from Talyfoel in

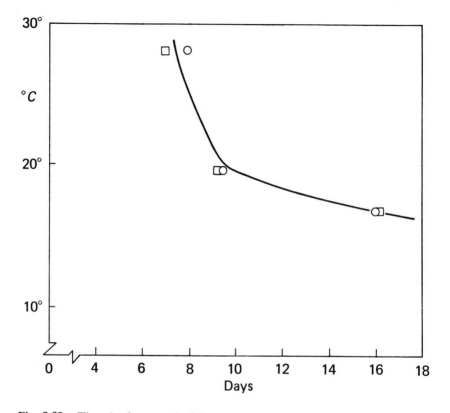

Fig. 3-32. Time in days required between spawning and emergence of *Nassarius obsoletus* larvae from egg capsules as a function of temperature (°C). The points indicate the number of days necessary for emergence of 50 percent of the larvae. Results are from two geographically isolated regions, Beaufort, North Carolina (O) and Cape Cod, Massachusetts (□). No significant difference is discernible in the results between egg capsules obtained from the two populations. (After Scheltema, 1967.)

Anglesey, Wales, and from Helsingør, Denmark, were induced to spawn in the laboratory. The Welsh population of *M. edulis* occurs intertidally in water temperatures ranging from 8° to 18°C and salinities between 29 and 30 0/00 in spring and early summer. In Danish waters *M. edulis* is found subtidally at depths of 8–12 meters, where temperatures range from 8° to 17°C and salinities from 13–23 0/00 in the summer. Between 11° and 17°C the rate of growth of larvae from both populations was similar. Above 17°C the growth rate of larvae from the Talyfoel population remained essentially unchanged, but larvae growth rate from the Helsingør population slowed markedly at temperatures above 17°C (Fig. 3.33). The difference in response is not correlated with differences in habitat temperature of the larvae since exposure to temperatures of 19–20°C would occur only rarely in either population. However, larvae settling at Wales do experience higher intertidal temperature than the subtidal

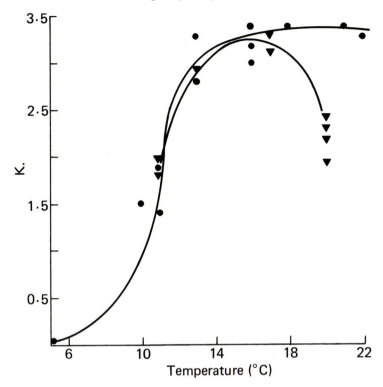

Fig. 3-33. The effect of temperature on the rate of growth of *Mytilus* larvae: •———•, Talyfoel, Wales larvae; ▲———▲, Helsingør larvae. (From Bayne, 1965.)

populations in Denmark. The value K, the instantaneous relative growth rate used to compare larval growth under different conditions, is calculated by the formula:

$$k = \frac{\log e\, L_2 - \log e\, L_1,}{t}$$

where L_1 is the initial mean length of the larvae and L_2 is the mean length at at time t. In Fig. 3.33, K = 100 k. Thus, maximum growth of each population occurred at or near maximum temperatures of their respective habitats.

Even in areas where animals are subject to varying temperatures within their biokinetic zone, periods of maximum growth are not necessarily correlated with maximum temperatures. The peak growth rate in a southern population of *C. virginica*, for example, occurred after the reproductive seasons and the periods of maximum temperature (Dame, 1971; Fig. 3.34). Since food supplies in southern estuaries reach maximal quantities in October and November (de la Cruz, 1965), Dame speculated that the growth pattern of these oysters is a function of food supply, reproduction, and temperature.

Early cleavage stages tend to be limited to a narrower temperature range than eggs or larvae. Cleavage stages of *M. edulis* from Wales required a more

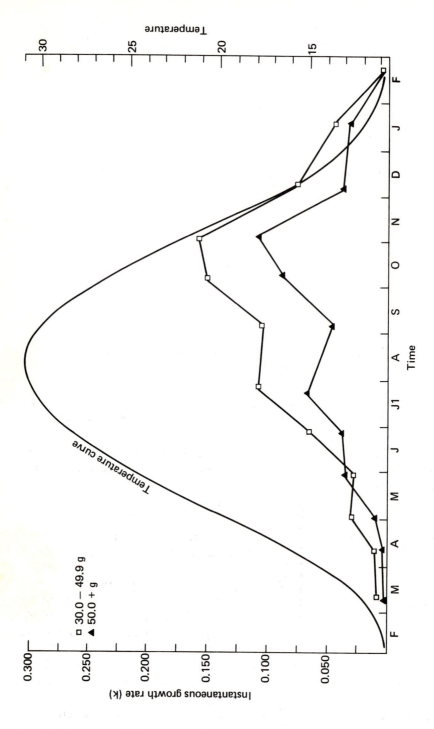

Fig. 3-34. Seasonal growth rate in the southern intertidal oyster, *Crassostrea virginicus.* (After Dame, 1971.)

restricted temperature range than did developing larvae (Bayne, 1965). Mussels were induced to spawn at 17–18°C, then eggs and sperm were pipetted directly from the spawning trays into small dishes of filtered sea water at a temperature ranging from 5° to 22°C. Although fertilization occurred at all temperatures, normal cleavage occurred only at temperatures from 8° to 18°C. At 20–22°C, cleavage was abnormal, although larvae were not adversely affected. Over the temperature range of 13–18°C, the growth rate was relatively temperature independent.

Some intertidal zone animals have larvae that are more tolerant of temperature extremes than are the adults. The LD_{50} of the adult stages of tropical zone species of fiddler crabs, genus *Uca,* was less than one hour at 7°C, while larvae of these same species were able to survive for four hours at 5°C. Larvae of two temperature zone species, *U. pugnax* and *U. pugilator,* molted when maintained at 15°C, but this temperature inhibits molting in adults (Vernberg F., 1969).

After comparing growth rates of latitudinally separated populations of gastropods, Dehnel (1955) concluded that embryos and larvae from a northern population (Alaska) grew two to nine times as fast as a southern population (California) of the same species at a comparable temperature. Laboratory rearing studies demonstrated that growth was apparently independent of the physiological temperature range. The growth curve of the northern form was displaced to the left of the curve of the southern form when the rate of relative growth was plotted against temperature, as illustrated in Fig. 3.35 for one species studied, *Thais emarginata.*

It has been suggested that one of the ways in which low temperature affects larval growth is through inactivation of certain digestive enzymes (Loosanoff and Davis, 1963). Larvae of the clam, *Mercenaria mercenaria,* can ingest food organisms at 10°C, but apparently cannot digest them. At 15°C, the larvae can digest and assimiliate naked flagellates and can grow slowly, although at this temperature they cannot utilize *Chlorella* sp. However, at 20°C, the larvae can utilize both naked flagellates and *Chlorella* sp.

c. Salinity

Many intertidal animals, particularly in estuarine areas, can and do live in low salinity waters. Often, however, they cannot reproduce there and their repopulation depends on the invasion of larvae or adult migratory populations from higher salinity waters. The amphipod *Corophium volutator,* for example, was found in salinities as low as 2 0/00, but breeding occurred only in salinities greater than 7.5 0/00 (McLusky, 1968).

There is some evidence that salinity tolerances of progeny may depend upon the salinity to which the parent stock is adapted. Offspring of the amphipod *Gammarus duebeni,* reared to a salinity of 2 0/00, tolerate transfer to fresh water better than do offspring from a different stock grown at either 30 0/00 or 40 0/00 (Kinne, 1953). *C. virginica* collected in areas where the salinity was 8.7 0/00 and spawned in salinities of 7.5, 10, and 15 0/00 produced larvae in

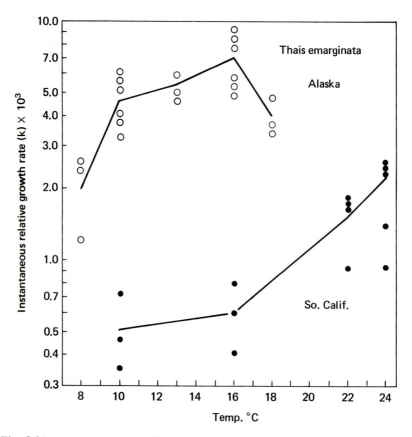

Fig. 3-35. A comparison of the instantaneous relative growth rates (k) of length of shell of veliger larvae of *T. emarginata* from Mount Edgecumbe, Alaska, and Big Rock, California, as a function of temperature. Each point represents the mean rate (k) of a group of larvae, reared at one temperature for varying periods of time. Lines connect means. (From Dehnel, 1955. University of Chicago Press.)

which optimal salinity for normal development was between 12 and 15 0/00. In contrast, the optimal salinity for growth of oysters that had developed gonads at a salinity of 27 0/00 was 17.5 0/00 (Loosanoff and Davis, 1963).

Salinity differentially influences various stages of the reproductive process of *Mytilus edulis* from Wales. Fertilization, for example, may occur over a wider range of salinities than subsequent larval growth and development. Fertilization occurred in salinities from 15–40 0/00, but successful development to the trochophore stage occurred only in salinities of 30–40 0/00 (Bayne, 1965). Similar observations have been made by *Mytilus californianus* (Young, 1941) and for *Adula californiensis* (Lough and Gonor, 1971).

Larval development of some estuarine intertidal zone crabs is limited to the lower salinities of estuarine waters, as it is for the wharf crab *Sesarma cinereum*. Not all of the larval stages had the same salinity tolerance. This crab has four zoeal stages and a megalops stage before it undergoes development to the crab stage, and it is the fourth zoeal stage which is most sensitive. Maximum survival and molting of this stage to the megalops occurs in salinities of 26.7 0/00, and if the crabs are carried out to the high salinities of the sea, they may be killed. Once it has reached the megalops stage, however, then *Sesarma* can withstand a wide range of salinities and temperatures (Costlow et al., 1960).

The influence of salinity on early development has been studied on two populations of the polychaete worm *Nereis diversicolor* (Smith, 1964), one population from Finland, the other from Sweden. The resulting data (Fig. 3.36) demonstrate certain trends in development. The population from the low salinity waters of Finland are better adapted for the low salinity environment in terms of reproduction and development than is the Swedish population. The cleavage stage is the most critical developmental stage in response to salinity extremes, and it differs in the two populations; for Swedish populations the salinity range for cleavage is 9–27 0/00, for Finnish animals, 5–14 0/00.

d. Photoperiod

The reproductive cycle of many animals is influenced by their photoperiod, although the role of photoperiod in reproductive processes in intertidal organisms is as yet poorly understood. It is known, for example in the crabs *Hemigrapsus nudus* and *Lophopanopeus bellus* that copulation begins during and continues after the shortest day in the year as the photoperiod lengthens (Knudsen, 1964). Furthermore, in three tropical species of *Littorina* that normally spawn throughout the year, spawning times could be altered by imposing different light-dark cycles on the females (Struhsaker, 1966).

The correlation between laboratory experiments and field observations are not always obvious. Consider, for example, a study on the effects of light and temperature on the reproductive cycle of the barnacle, *Balanus balanoides* (Crisp and Patel, 1969), which breed only once annually. Gonadal enlargement begins in the spring soon after the last year's brood has been liberated, and the gonads appear to be fully developed by late summer. Fertilization occurs in late autumn, hatching the following spring. In animals collected from the field at monthly intervals from May to early October and kept at low temperatures in the dark, fertilization occurred only slightly later than it did in the field animals, but under conditions of low temperature and ambient daylight, onset of breeding was delayed as much as three months. However, in the field the well-illuminated, high water species are known to breed about a month before low water species, which are exposed less to lengthening photoperiods. Thus, although light exercises an important and prolonged effect in delaying breeding in *B. balanoides* in the laboratory, the role of photoperiod in field animals is unclear. Another interesting aspect of the laboratory study was the finding that light and temperature were effective in delaying breeding only during certain

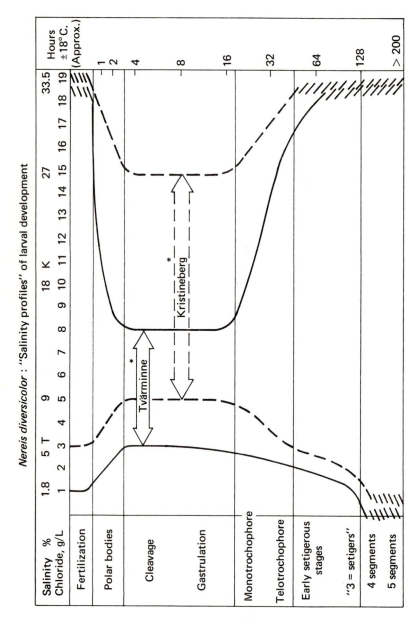

Fig. 3-36. "Salinity profiles" of larval development of *N. diversicolor* at Kristineberg and Tvärminne. "T" and "K" in line of salinity values represent approximate surface salinity of these areas. Asterisks (*) above double arrows mark the optimal salinities for larval development as observed at the respective stations. Stages and times are approximations only at ± 18°C. (From Smith, 1964. Wistar Press.)

phases of the reproductive cycle. Animals collected after early October when their gonads were mature were only slightly influenced by these environmental conditions.

e. Lunar phases

The reproductive cycle of many marine animals is timed to coincide precisely with certain phases of the moon. On the basis of behavioral studies on species of *Uca* in the intertidal zone, Hagen (1970) concluded that for species restricted to the lower levels of the intertidal zone most phases of sexual behavior were confined to the days around full and new moon.

Epidemic swarming is one of the most spectacular phenomena that has been linked to lunar periodicity in intertidal animals, and it has been reported in many species. The crab *Cardisoma guanhumi* will serve as an illustration (Gifford, 1962). Although this crab lives intertidally and along drainage ditches on the south Florida coast and sometimes is found as much as 5 kilometers from the coast, it must return to the sea to spawn. The spawning season extends off and on from late June to early December and spawning occurs in sharp peaks near the time of full moon and in lesser peaks at new moon periods. Ovigerous females appear simultaneously, moving toward the sea, reaching the highest concentration between one night before the full moon and one night after; then the spawning period stops abruptly.

Lunar periodicity has also been described in an intertidal insect, the chironomid *Clunio marinus,* which lives in tangles of algae in the intertidal zone. The imagos are short-lived and must copulate and deposit eggs within a few hours after leaving the pupa. The eggs, which are deposited in strings, are glued to the substratum (Korringa, 1957). Because all of these activities must take place when the habitat area is relatively dry, the semilunar periodicity of hatching and emergence during spring and neap tides is not surprising (Neumann, 1966). Under laboratory conditions of LD 16:8 and a tidal cycle of 12.5 hours, a hatching rhythm was observed every 12.6 days. When the light-dark cycle was shifted there was a corresponding shift in hatching period (Neumann, 1970). Animals exposed to a LD cycle, but without a tidal cycle, exhibited a free-running semilunar cycle, so that both diurnal and semilunar rhythms appeared to be endogenous, with the diurnal rhythm synchronized by the LD cycle, the semilunar one by tidal conditions.

REFERENCES

Ansell, A. D. (1969). Defensive Adaptations to Predation in the Mollusca. Proc. Sympo. on Mollusca, Part II. Pp. 487–512.

———, and A. Trevallion (1969). Behavioural adaptations of intertidal molluscs from a tropical sandy beach. *J. exp. mar. Biol. Ecol.,* 4:9–35.

Arudpragasan, K. D., and E. Naylor (1964). Gill ventilation volumes, oxygen consumption and respiratory rhythms in *Carcinus maenas* L. *J. exp. Biol.,* **41:**309–321.

Bannister, W. H., J. V. Bannister, and H. Micallef (1966). A biochemical factor in the zonation of marine molluscs. *Nature, London,* **211:**747.

Barnes, H., and M. Barnes (1967). The effect of starvation and feeding on the time of production of egg masses in the boreo-Arctic cirripede *Balanus balanoides* (L.). *J. exp. mar. Biol. Ecol.,* **1:**1–6.

——, —— (1968). Egg numbers, metabolic efficiency of egg production and fecundity; local and regional variations in a number of common cirripedes. *J. exp. mar. Biol. Ecol.,* **2:**135–153.

——, ——, and D. M. Finlayson (1963a). The seasonal changes in body weight, biochemical composition and oxygen uptake of two common boreo-Arctic cirripedes, *Balanus balanoides* (L.) and *Balanus balanus* (L.). *J. mar. biol. Ass. U.K.,* **43:**185–211.

——, ——, —— (1963b). The metabolism during starvation of *Balanus balanoides. J. mar. biol. Ass. U.K.,* **43:**213–233.

Barnwell, F. H. (1963). Observations on daily and tidal rhythms in some fiddler crabs from equatorial Brazil. *Biol. Bull.,* **125:**399–415.

—— (1966). Daily and tidal patterns of activity in individual fiddler crabs (Genus *Uca*) from the Woods Hole region. *Biol. Bull.,* **130:**1–7.

—— (1968). The role of rhythmic systems in the adaptation of fiddler crabs to the intertidal zone. *Am. Zool.,* **8:**569–584.

Bayne, B. L. (1963). Responses of *Mytilus edulis* larvae to increases in hydrostatic pressure. *Nature,* **198:**406–407.

—— (1964). The responses of the larvae of *Mytilus edulis* (L.) to light and to gravity. *Oikos,* **15:**162–174.

—— (1965). Growth and the delay of metamorphosis of larvae of *Mytilus edulis* L. *Ophelia,* **2:**1–47.

—— (1969). The gregarious behavior of the larvae of *Ostrea edulis* (L.) at settlement. *J. Mar. Biol.,* **49:**327–356.

Berguist, P. R., and M. E. Sinclair (1968). The morphology and behavior of larvae of some intertidal sponges. *New Zealand Journal of Marine and Freshwater Research,* **2:**426–437.

Bliss, D. E. (1968). Transition from water to land in decapod crustaceans. *Am. Zool.,* **8:**355–392.

——, and D. M. Skinner (1963). Tissue Respiration in Invertebrates. Am. Museum of Natural History, New York. 139 pp.

Brett, J. R. (1954). The sense organs: the eye. *In* The Physiology of Fishes. M. Brown, ed. Academic Press, Inc., New York. **2:**121–154.

Brown, A. C. (1960). Desiccation as a factor influencing the vertical distribution of some South African Gastropoda from intertidal rocky shores. *Portugaliae Acta Biologica* (B), 7n. 1:11–23.

Brown, F. A., Jr. (1968). Endogenous biorhythmicity reviewed with new evidence. *Scientia,* **103:**1–16.

——, M. F. Bennett, and H. M. Webb (1954). Persistent daily and tidal rhythms of O_2 consumption in fiddler crabs. *J. Cell. Comp. Physiol.,* **44:**477–506.

——, M. Fingerman, M. I. Sandeen, and H. M. Webb (1953). Persistent diurnal and tidal rhythms of color change in the fiddler crab, *Uca pugnax. J. Exp. Zool.*, **123**:29–60.

Bruce, J. R. (1928). Physical factors on the sandy beach. Part II. Chemical changes—carbon dioxide concentration and sulphides. *J. mar. biol. Ass. U.K.*, **15**:553–565.

Bullock, T. H. (1955). Compensation for temperature in the metabolism and activity of poikilotherms. *Biol. Rev.*, **30**:311–342.

Carr, W. E. S. (1967a). Chemoreception in the mud snail *Nassarius obsoletus*. I. Properties of stimulatory substances extracted from shrimp. *Biol. Bull.*, **132**:90–105.

———— (1967b). Chemoreception in the mud snail, *Nassarius obsoletus*. II. Identification of stimulatory substances. *Biol. Bull.*, **132**:106–127.

Carriker, M. R. (1967). Ecology of estuarine benthic invertebrates: A perspective. *In* Estuaries. G. H. Lauff, ed. AAAS Publ. No. 83, Washington, D.C. Pp. 442–487.

————, D. V. Zandt, and G, Charlton (1967). Gastropod *Urosalpinx:* pH of accessory boring organ while boring. *Science*, **158**:920–922.

Clarke, G. L., and R. H. Oster (1934). The penetration of the blue and red components of daylight into Atlantic coastal waters and its relation to phytoplankton metabolism. *Biol. Bull.*, **67**:59–75.

Cole, H. A., and E. W. Knight-Jones (1949). The setting behavior of larvae of the European flat oyster, *Ostrea edulis* (L.) and its influence on methods of cultivation and spat collection. *Fishery Invest., Lond. Ser. II*, **17**:1–39.

Costlow, J. D., C. G. Bookhout, and R. Monroe (1960). The effect of salinity and temperature on larval development of *Sesarma cinereum* (Bosc) reared in the laboratory. *Biol. Bull.*, **118**:183–202.

Crane, J. (1941). Eastern Pacific expeditions of the New York Zoological Society XXVI. Crabs of the genus *Uca* from the west coast of Central America. *Zoologica*, **26**:145–208.

———— (1943). Crabs of the genus *Uca* from Venezuela. *Zoologica*, **28**:33–44.

———— (1957). Basic patterns of display in fiddler crabs (Ocypodidae, Genus *Uca*). *Zoologica*, **42**:69–82.

Crisp, D. J. (1957). Effect of low temperature on the breeding of marine animals. *Nature*, **179**:1138–1139.

———— (1964). The effects of the severe winter of 1962–1963 on marine life in Britain. *J. Anim. Ecol.*, **33**:165–210.

———— (1965a). Surface chemistry, a factor in the settlement of marine invertebrate larvae. Botanica Gothoburgensia III. Proceedings of the 5th Marine Biological Symposium, Göteborg. Pp. 51–65.

———— (1965b). The ecology of marine fouling, pp. 99–117. *In* G. Goodman, ed. Ecology and the Industrial Society. *Fifth Symp. Brit. Ecol. Soc.* Blackwell, Oxford.

———— (1967a). Chemical factors inducing settlement in *Crassostrea virginica* (Gmelin). *J. Anim. Ecol.*, **36**:329–335.

———— (1967b). Chemoreception in cirripedes. *Biol. Bull.*, **133**:128–140.

————, and P. S. Meadows (1962). The chemical basis of gregariousness in cirripedes. *Proc. Roy. Soc. Lond. B*, **156**:500–520.

————, and B. Patel (1969). Environmental control of the breeding of three boreo-Arctic Cirripedes. *Mar. Biol,* **2**:283–295.

————, and H. G. Stubbings (1957). The orientation of barnacles to water currents. *J. Anim. Ecol.,* **26:**179–196.

Croghan, P. C. (1958). The osmotic and ionic regulation of *Artemia salina. J. exp. Biol.,* **35:**219–233.

Dales, R. P. (1958). Survival of anaerobic periods by two intertidal polychaetes, *Arenicola marina* (L.) and *Ovenia fusiformis* Delle Chiaje. *J. mar. biol. Ass. U.K.,* **37:**521–529.

Dame, R. F. (1971). Growth, Respiratiton and Energy Flow in the Intertidal Oyster, *Crassostrea virginica.* Unpublished Ph.D. dissertation, University of South Carolina. 75 pp.

Daniel, A. (1957). Illumination and its effect on the settlement of barnacle cyprids. *Proc. Zool. Soc. Lond.,* **129:**305–313.

Davies, P. S. (1969). Physiological ecology of *Patella.* III. Desiccation effects. *J. mar. biol. Ass. U.K.,* **49:**291–304.

Dean, J. M., and F. J. Vernberg (1965a). Variations in the blood glucose level of Crustacea. *Comp. Biochem. Physiol.,* **14:**29–34.

————, ———— (1965b). Effects of temperature acclimation on some aspects of carbohydrate metabolism in decapod Crustacea. *Biol. Bull.,* **129:**87–94.

————, ———— (1966). Hypothermia and blood of crabs. *Comp. Biochem. Physiol.,* **17:**19–22.

DeCoursey, P. J. (1961). Effect of light on the circadian activity rhythm of the flying squirrel *Glaucomys volans. Zeit. vergl. Physiol.,* **44:**311–354.

Dehnel, P. A. (1955). Rates of growth of gastropods as a function of latitude. *Physiol. Zool.,* **28:**115–144.

———— (1958). Effect of photoperiod on the oxygen consumption of two species of intertidal crabs. *Nature,* **181:**1415–1417.

———— (1960). Effect of temperature and salinity on the oxygen consumption of two intertidal crabs. *Biol. Bull.,* **118:**215–249.

———— (1962). Aspects of osmoregulation in two species of intertidal crabs. *Biol. Bull.,* **122:**208–227.

De la Cruz, A. A. (1965). A Study of Particulate Organic Detritus in a Georgia Salt Marsh–Estuarine Ecosystem. Ph.D. Thesis, University of Georgia. 110 pp.

Dill, D. B., E. F. Adolph, and C. G. Wilber, eds. (1964). Adaptation to the environment. *In* Handbook of Physiology. Section 4. Am. Physiol. Soc., Washington, D.C.

Doty, M. S. (1957). Rocky intertidal surfaces. *In* Treatise on Marine Ecology and Paleoecology. J. W. Hedgpeth, ed. Geol. Soc. Am. Mem. No. 67, Washington, D.C. **1:**535–585.

Dresel, E. I. B., and V. Moyle (1950). Nitrogenous excretion of amphipods and isopods. *J. exp. Biol.,* **27:**210–225.

Duncan, A. (1966). The oxygen consumption of *Potamopyrgus jenkinsi* (Smith) (Prosobranchiata) in different temperatures and salinities. *Verh. int. Ver. Limnol.,* **16:**1739–1751.

Edney, E. B. (1961). The water and heat relationships of fiddler crabs (*Uca* spp.). *Trans. roy Soc. S. Afri.,* **36:**71–91.

———— (1962). Some aspects of the temperature relations of fiddler crabs (*Uca* spp.). *Biometeorology:* 79–85.

Edwards, G. A., and L. Irving (1943a). The influence of temperature and season upon the oxygen consumption of the sand crab, *Emerita talpoida* Lay. *J. Cell. Comp. Physiol.,* **21**:169–182.

———, ——— (1943b). The influence of season and temperature upon the oxygen consumption of the beach flea, *Talorchestia megalophthalma*. *J. Cell. Comp. Physiol.,* **21**:183–189.

Eliassen, E. (1955). The oxygen supply during ebb of *Arenicola marina* in the Danish Waddensea. *Naturv. rekke, Nr.,* **12**:1–9.

Englemann, M. D. (1966). Energetics, terrestrial field studies, and animal productivity. *Adv. Ecol. Res.,* **3**:73–115.

Enright, J. T. (1965). Entrainment of a tidal rhythm. *Science,* **147**:864–867.

——— (1968). Differences and similarities of tidal and circadian rhythms. *Proc. 5th Int. Cong. Photobiol.,* **1**:44.

Farmanfarmaian, A., and A. G. Giese (1963). Thermal tolerance and acclimation in the western purple sea urchin, *Strongylocentrotus purpuratus*. *Physiol. Zool.,* **36**:237–243.

Fingerman, M. (1955). Persistent daily and tidal rhythms of color change in *Callinectes sapidus*. *Biol. Bull.,* **109**:255–264.

Flemister, L. J., and S. C. Flemister (1951). Chloride ion regulation and oxygen consumption in the crab *Ocypode albicans* (Bosc). *Biol. Bull.,* **101**:259–273.

Foster, B. A. (1971). Desiccation as a factor in the intertidal zonation of barnacles. *Mar. Biol.,* **8**:12–29.

Fox, H. M. (1939). The activity and metabolism of poikilothermal animals in different latitudes. V. *Proc. Zool. Soc. Lond. A,* **109**:141–156.

——— (1955). The effect of oxygen on concentration of haem in invertebrates. *Proc. Roy. Soc. Lond. B.* **143**:203–214.

———, and A. E. R. Taylor (1955). The tolerance of oxygen by aquatic invertebrates. *Proc. Roy. Soc. Lond. B.* **143**:214–225.

Friedrich, H. (1961). Physiological significance of light in marine ecosystems. *In* Oceanography. AAAS Publ. No. 67, Washington, D.C. Pp. 257–270.

Fraenkel, G. (1966). The heat resistance of intertidal snails at Shirahama, Wakayamaken, Japan. *Publ. of the Seto Marine Biological Laboratory,* **14**:185–195.

——— (1968). The heat resistance of intertidal snails at Bimini, Bahamas; Ocean Spring, Miss.; and Woods Hole, Mass. *Physiol. Zool.,* **41**:1–13.

Fretter, V., and M. C. Montgomery (1968). The treatment of food by prosobranch veligers. *J. mar. biol. Ass. U.K.,* **48**:499–520.

Gamble, J. C. (1970). Effect of low dissolved oxygen concentrations on the ventilation rhythm of three tubicolous crustaceans, with special reference to the phenomenon of intermittent ventilation. *Mar. Biol.,* **6**:121–127.

Galtsoff, P. S. (1964). The American oyster: *Crassostrea virginica* Gmelin. *U. S. Fish Wildlife Serv. Fish Bull.,* **64**:1–480.

George, J. D. (1964). The life history of the cirratulid worm, *Cirriformia tentaculata*, on an intertidal mudflat. *J. mar. biol. Ass. U.K.,* **44**:47–65.

Gibson, R. N. (1969). The biology and behavior of littoral fish. *Oceanogr. Mar. Biol. Ann. Rev.,* **1**:367–410.

Gifford, C. A. (1962). Some observations on the general biology of the land crab, *Cardisoma guanhumi* (Latreille) in South Florida. *Biol. Bull.,* **123**:207–223.

—————— (1968). Accumulation of uric acid in the land crab, *Cardisoma guanhumi*. *Am. Zool.*, **8**:521–528.

Gilbert, A. B. (1959). The composition of the blood of the shore crab *Carcinus maenas* Pennant, in relation to sex and body size. II. Blood chloride and sulphate. *J. exp. Biol.*, **36**:356–362.

Gray, I. E. (1957). A comparative study of the gill area of crabs. *Biol. Bull.*, **112**: 34–42.

Gray, J. S. (1966). The response of *Protodrilus symbioticus* (Giard) (*Archiannelida*) to light. *J. Anim. Ecol.*, **35**:55–64.

Gross, W. J. (1957a). An analysis of response to osmotic stress in selected decapod crustacea. *Biol. Bull.*, **112**:43–62.

—————— (1957b). A behavioral mechanism for osmotic regulation in a semiterrestrial crab. *Biol. Bull.*, **113**:268–274.

—————— (1961). Osmotic tolerance and regulation in crabs from a hypersaline lagoon. *Biol. Bull.*, **121**:290–301.

—————— (1963). Acclimation to hypersaline water in a crab. *Comp. Biochem. Physiol.*, **9**:181–188.

—————— (1964). Trends in water and salt regulation among aquatic and amphibious crabs. *Biol. Bull.*, **127**:447–466.

——————, and P. V. Holland (1960). Water and ionic regulation in a terrestrial hermit crab. *Physiol. Zool.*, **33**:21–28.

Gunter, G. (1967). Vertebrates in hypersaline waters. *Contributions in Marine Science*, **12**:230–241.

Hagen, H. O. (1970). Zur Deutung langstieliger und gehörnter Augen bei Ocypodiden (Decapoda, Brachyura). [On the significance of elongated and horned eyes in ocypodid crabs (Decapoda, Branchyura).] *Forma Functio*, **1**:13–57.

Halcrow, K., and C. M. Boyd (1967). The oxygen consumption and swimming activity of the amphipod *Gammarus oceanicus* at different temperatures. *Comp. Biochem. Physiol.*, **23**:233–242.

Hammen, C. S. (1969). Lactate and succinate oxidoreductases in marine invertebrates. *Mar. Biol.*, **4**:233–238.

Hartenstein, R. (1968). Nitrogen metabolism in the terrestrial isopod, *Oniscus asellus*. *Am. Zool.*, **8**:507–519.

Hazlett, B. A. (1968). Stimuli involved in the feeding behavior of the hermit crab *Clibanarius vittatus* (Decapoda, Paguridea). *Crustaceana*, **15**:305–311.

Hedgpeth, J. W. (1957). Classification of marine environments. *In* Treatise on Marine Ecology and Paleoecology. J. W. Hedgpeth, ed. Geol. Soc. Am. Mem. 67, Washington, D.C. **1**:17–281.

Herrnkind, W. F. (1968). Adaptive visually-directed orientation in *Uca pugilator*. *Am. Zool.*, **8**:585–598.

Hidu, H. (1969). Gregarious setting in the American oyster *Crassostrea virginica* Gmelin. *Chesapeake Sci.*, **10**:85–92.

Hoffmann, R. J., and C. P. Mangum (1970). The function of coelomic cell hemoglobin in the polychaete *Glycera dibranchiata*. *Comp. Biochem. Physiol.*, **36**: 211–228.

Hopkins, A. E. (1936). Adaptation of the feeding mechanism of the oyster (*O. gigas*) to changes in salinity. *Bull. U. S. Bur. Fish.*, **48**:345–363.

Horne, F. R. (1968). Nitrogen excretion in crustacea. I. The herbivorous land crab *Cardisoma guanhumi* Latreille. *Comp. Biochem, Physiol.,* **26**:687–695.

Huggett, A. St. G., and D. A. Nixon (1957). Enzymatic determination of blood glucose. *Biochem. J.,* **66**:12.

Hughes, R. N. (1970). An energy budget for a tidal-flat population of the bivalve *Scrobicularia plana* (Da Costa). *J. Anim. Ecol.,* **39**:357–380.

Jacubowa, L., and E. Malm (1931). Die Beziehungen einiger Benthosformen des Schwarzen Meeres zum Medium. *Biol. Zbl.,* **51**:105–116.

Jansson, B-O. (1967a). The availability of oxygen for the interstitial fauna of sandy beaches. *J. exp. mar. Biol. Ecol.,* **1**:123–143.

——— (1967b). Diurnal and annual variations of temperature and salinity of interstitial water in sandy beaches. *Ophelia,* **4**:173–201.

Johnson, R. G. (1965). Temperature variation in the infaunal environment of a sand flat. *Limnol. Oceanog.,* **10**:114–120.

——— (1967). Salinity of interstitial water in a sandy beach. *Limnol. Oceanogr.,* **12**:1–7.

Jones, D. A., and E. Naylor (1970). The swimming rhythm of the sand beach isopod *Eurydice pulchra. J. exp. mar. Biol. Ecol.,* **4**:188–199.

Jones, J. D. (1955). Observations on the respiratory physiology and on the haemoglobin of the polychaete genus *Nephthys*, with special reference to *N. hombergii* (Aud. et M. Edw.). *J. exp. Biol.,* **32**:110–125.

Jørgensen, C. B. (1952). Efficiency of growth in *Mytilus edulis* and two gastropod veligers. *Nature,* Lond., **170**:714.

——— (1966). Biology of Suspension Feeding. Pergamon Press, New York.

Kanwisher, J. W. (1955). Freezing in intertidal animals. *Biol. Bull.,* **109**:56–63.

——— (1959). Histology and metabolism of frozen intertidal animals. *Biol. Bull.,* **116**:258–264.

Kenny, R. (1969). The effects of temperature, salinity and substrate on distribution of *Clymenella torquata* (Leidy) Polychorta. *Ecology,* **50**:624–631.

Kinne, O. (1953). Zur Biologie and Physiologie von *Gammarus duebeni Lillj. Z. wiss. Zool.,* **157**:427–491.

——— (1964). The effects of temperature and salinity on marine and brackish water animals. II. Salinity and temperature salinity combinations. *Oceanogr. Mar. Biol. Ann. Rev.,* **2**:281–339.

Kleinholz, L., V. J. Havel, and R. Reichart (1950). Studies in regulation of blood sugar concentration in crustaceans. II. Experimental hyperglycemia and the regulatory mechanisms. *Biol. Bull.,* **99**:454–468.

Knight-Jones, E. W. (1952). Gregariousness and some other aspects of the setting behaviour of *Spirorbis. J. mar. Ass. U.K.,* **30**:201–222.

Knudsen, J. W. (1964). Observations of the reproductive cycles and ecology of the common Brachyura and crablike Anomura of Puget Sound, Washington. *Pacif. Sci.,* **18**:3–33.

Korringa, P. (1952). Recent advances in oyster biology. *Quart. Rev. Biol.,* **27**:266–308, 339–365.

——— (1957). Water temperature and breeding throughout the geographical range of *Ostrea edulis. Ann. Biol.,* **33**:1–17.

Kuenzler, E. J. (1961). Structure and energy flow of a mussel population in a Georgia salt marsh. *Limnol. Oceanogr.,* **6**:191–204.

Laverack, M. S. (1968). On the receptors of marine invertebrates. *Oceanogr. Mar. Biol. Ann. Rev.,* **6**:249–324.

Lent, C. M. (1968). Air-gaping by the ribbed mussel, *Modiolus demissus* (Dillwyn): Effects and adaptive significance. *Biol. Bull.,* **134**:60–73.

Lewis, J. B. (1963). Environmental and tissue temperature of some tropical intertidal marine animals. *Biol. Bull.,* **124**:277–284.

Lewis, J. R. (1964). The Ecology of Rocky Shores. English Univ. Press, London, 323 pp.

——— (1965). The littoral fringe on rocky coasts of Southern Norway and Western Sweden. *Botanica gothoburg,* **3**:129–143.

Lilly, S. J., J. F. Sloane, R. Bassingdale, F. J. Ebling and J. A. Kitching (1953). The ecology of Lough Ine with special reference to water currents. IV. The sedentary fauna of sublittoral boulders. *J. Anim. Ecol.,* **22**:87–122.

Lindquist, O. V. (1970). The blood osmotic pressure of the terrestrial isopods *Porcellio scaber* Latr. and *Oniscus asellus* L., with reference to the effect of temperature and body size. *Comp. Biochem. Physiol.,* **37**:503–510.

Lockwood, A. P. M. (1962). The osmoregulation of Crustacea. *Biol. Rev.,* **37**:257–305.

Loosanoff, V. L., H. C. Davis, and P. E. Chanley (1953). Behavior of clam larvae in different concentrations of food organisms. *Anat. Rec.,* **117**:586–587.

———, ——— (1963). Rearing of bivalve mollusks. *In* Advances in Marine Biology. F. S. Russell, ed. **1**:1–136.

———, and C. A. Nomejko (1946). Feeding of oysters in relation to tidal stages and its periods of light and darkness. *Biol. Bull.,* **90**:244–264.

———, ——— (1951). Existence of physiologically-different races of oysters, *Crassostrea virginica. Biol. Bull.,* **101**:151–156.

———, and F. D. Tommers (1947). Effect of low pH upon rate of water pumping of oysters, *Ostrea virginica. Anat. Rec.,* **99**:112–113.

———, ——— (1948). Effect of suspended silt and other substances on rate of feeding of oysters. *Science,* **107**:69–70.

Lough, R. C., and J. J. Gonor (1971). Early embryonic stages of *Adula californiensis* (Pelecypoda: Mytilidae) and the effect of temperature and salinity on developmental rate. *Mar. Biol.,* **8**:118–125.

Lund, E. J. (1957). Self-silting, survival of the oyster as a closed system, and reducing tendencies of the environment of the oyster. *Publ. Inst. Marine Sci.,* **4**:313–319.

McLusky, D. S. (1968). Some effects of salinity on the distribution and abundance of *Corophium volutator* in the Ythan Estuary. *J. Mar. Biol.,* **48**:443–454.

Madanmohanrao, G., and K. P. Rao (1962). Oxygen consumption in a brackish water crustacean, *Sesarma plicatum* (Latreille) and a marine crustacean, *Lepas anserifera* L. *Crustaceana,* **4**:75–81.

Mangum, C. (1962). Source of dichromatism in two maldanid polychaetes. *Nature,* **195**:198–199.

——— (1969). Low temperature blockage of the feeding response in boreal and temperate zone polychaetes. *Chesapeake Sci.,* **10**:64–65.

——— (1970). Respiratory physiology in annelids. *Am. Scientist,* **58**:641–647.

———, and C. Sassaman (1969). Temperature sensitivity of active and resting metabolism in a polychaetous annelid. *Comp. Biochem. Physiol.,* **30**:111–116.

Mantel, L. H. (1968). The foregut of *Gecarcinus lateralis* as an organ of salt and water balance. *Am. Zool.,* **8:**433–442.

Manwell, C. J. (1960). Comparative physiology: Blood pigments. *Ann. Rev. Physiol.,* **22:**191–244.

—— (1963). The chemistry and biology of hemoglobin in some marine clams. I. Distribution of the pigment and properties of the oxygen equilibrium. *Comp. Biochem. Physiol.,* **8:**209–218.

——, E. C. Southward, and A. J. Southward (1966). Preliminary studies on haemoglobin and other proteins of the Pogonophora. *J. mar. biol. Ass. U.K.,* **46:**115–124.

Miller, D. C. (1962). The feeding mechanism of fiddler crabs, with ecological considerations of feeding adaptations. *Zoologica,* **46:**89–101.

Morrison, P. R., and K. C. Morrison (1952). Bleeding and coagulation in some Bermudan crustacea. *Biol. Bull.,* **103:**395–406.

Moulton, J. M. (1962). Intertidal clustering of an Australian gastropod. *Biol. Bull.,* **123:**170–178.

Needham, J. (1938). Contributions of chemical physiology to the problem of reversibility in evolution. *Biol. Rev.,* **13:**225–251.

Neumann, D. (1966). Die lunare und tägliche Schlupfperiodik der Mücke *Clunio.* *Z. vergl. Physiol.,* **53:**1–61.

—— (1970). Die Kombination verschiedener endogener Rhythmen bei der zeitlichen Programmierung von Entwicklung und Verhalten. *Oecologia,* **3:**166–183.

Newell, R. C. (1970). The Biology of Intertidal Animals. American Elsevier Publishing Co., Inc., New York. 555 pp.

——, and V. I. Pye (1971). The influence of thermal acclimation on the relation between oxygen consumption and temperature in *Littorina littorea* (L.) and *Mytilus edulis* L. *Comp. Biochem. Physiol.,* **34:**385–397.

Nicolaisen, W., and E. Kanneworff (1969). On the burrowing and feeding habits of the amphipods *Bathyporeia pilosa* Lindström and *Bathyporeia sarsi* Watkin. *Ophelia,* **6:**231–250.

North, W. J., and C. F. A. Pantin (1958). Sensitivity to light in the sea-anemone *Metridium senile* (L.): Adaptation and action spectra. *Proc. Roy. Soc. Lond. B.,* **148:**385–396.

Odum, E. P., and A. E. Smalley (1959). Comparison of population energy flow of a herbivorous and a deposit-feeding invertebrate in a salt marsh ecosystem. *Proc. natn. Acad. Sci., U.S.A.* **45:**617–622.

Oglesby, L. C. (1969). Salinity-stress and desiccation in intertidal worms. Symp. Adaptations of Intertidal Organisms. *Am. Zool.,* **9:**319–331.

Olive, P. J. W. (1971). Ovary structure and oogenesis in *Cirratulus cirratus* (Polychaeta: Cirratulidae). *Mar. Biol.,* **8:**243–259.

Orton, J. H. (1920). Sea-temperature, breeding and distribution in marine animals. *J. mar. biol. Ass. U.K.,* **12:**339–366.

Palmer, J. D. (1962). A persistent diurnal phototactic rhythm in the fiddler crab, *Uca pugnax. Biol. Bull.,* **123:**507–508.

—— (1967). Daily and tidal components in the persistent rhythmic activity of the crab, *Sesarma. Nature,* **215:**64–66.

Patel, B., and D. J. Crisp (1960). Rates of development of the embryos of several species of barnacles. *Physiol. Zool.,* **33:**104–119.

Paulson, T. C., and R. S. Scheltema (1968). Selective feeding on algal cells by the veliger larvae of *Nassarius obsoletus* (Gastropoda, Prosobranchea). *Biol. Bull.,* **134**:481–489.

Pearse, A. S. (1929). Observations on certain littoral and terrestrial animals at Tortugas, Fla., with special reference to migrations from marine to terrestrial habitats. Papers Tortugas Sta. Reprinted from Carnegie Inst., Washington. Publ. No. 391. Pp. 205–223.

———, and G. Gunter (1957). Salinity. *In* Treatise on Marine Ecology and Paleoecology. J. Hedgpeth, ed. Geol. Soc. America. Pp. 129–158.

Postma, H. (1967). Sediment transport and sedimentation in the estuarine environment. *In* Estuaries. G. H. Lauff, ed. AAAS Publ. No. 83, Washington, D.C. Pp. 180–184.

Potts, W. T. W., and G. Parry (1964). Osmotic and Ionic Regulation in Animals. Pergamon Press, London and Oxford.

Prosser, C. L. (1955). Physiological variation in animals. *Biol. Rev.,* **30**:229–262.

——— (1958). General summary: The nature of physiological adaptation. *In* Physiological Adaptation. C. L. Prosser, ed. Am. Physiol. Soc., Washington, D.C. Pp. 167–180.

———, ed. (1967). Molecular mechanisms of temperature adaptation. AAAS Publ. No. 84, Washington, D.C. 390 pp.

Raben, K. van (1934). Veränderungen in dem Kiemendeckel und in den Kiemen einiger Brachyuren (Decapoden) im Verlauf der Anpassung an die Feuchtluftatmung. *Z. wiss, Zool.,* **145**:425–461.

Rao, K. P. (1953). Rate of water propulsion in *Mytilus californianus* as a function of latitude. *Biol. Bull.,* **104**:171–181.

——— (1954). Tidal rhythmicity in the rate of water propulsion in *Mytilus* and its modifiability by transplantation. *Biol. Bull.,* **106**:353–359.

Rasmussen, E. (1958). Behavior of sacculinized stone crabs (*Carcinus maenas* Pennant). *Nature,* **183**:479–480.

Read, K. R. H. (1962). The hemoglobin of the bivalved mollusc, *Phacoides pectinatus* Gmelin. *Biol. Bull.,* **123**:605–617.

——— (1963). Thermal inactivation of preparations of aspartic/glutamic transaminase from species of bivalved molluscs from the sublittoral and intertidal zones. *Comp. Physiol.,* **46**:209–247.

Redmond, J. R. (1955). The respiratory function of hemocyanin in Crustacea. *J. Cell. Comp. Physiol.,* **46**(2):209–247.

——— (1962a). Oxygen-hemocyanin relationships in the land crab, *Cardisoma guanhumi. Biol. Bull.,* **122**:252–262.

——— (1962b). The respiratory characteristics of chiton hemocyanins. *Physiol. Zool.,* **35**:304–313.

——— (1968). Transport of oxygen by the blood of the land crab, *Gecarcinus lateralis. Am. Zool.,* **8**:471–479.

Reish, D. J., and J. L. Barnard (1960). Field toxicity tests in marine waters utilizing the polychaetous annelid *Capitella capitata* (Fabricius). *Pacific Nat.,* **1**:1–8.

Ritz, D. A., and D. J. Crisp (1970). Seasonal changes in feeding rate in *Balanus balanoides. J. mar. biol. Ass. U.K.,* **50**:223–240.

Ross, D. M. (1960). The association between the hermit crab *Eupagurus bernhardus*

(L.) and the sea anemone *Calliactus parasitica* (Couch). *Proc. Zool. Soc. London,* **134**:43–57.

Salmon, M. (1965). Waving display and sound production in the courtship behavior of *Uca pugilator,* with comparisons to *U. minax* and *U. pugnax. Zoologica,* **50:** 123–149.

———, and S. P. Atsaides (1968). Visual and acoustical signalling during courtship by fiddler crabs (Genus *Uca*). Am. Zool., **8**:623–639.

———, ——— (1969). Sensitivity to substrate vibration in the fiddler crab *Uca pugilator* Bosc. *Anim. Behav.,* **71**:68–76.

Sameoto, D. D. (1969). Comparative ecology, life histories, and behavior of intertidal sand-burrowing amphipods (Crustacea: Haustoriidae) at Cape Cod. *J. Fish. Res. Bd. Can.,* **26**:361–388.

Sandeen, M. I., G. C. Stephens, and F. A. Brown, Jr. (1954). Persistent daily and tidal rhythms of oxygen consumption in two species of marine snails. *Physiol. Zool.,* **27**:350–356.

Scheltema, R. S. (1961). Metamorphosis of the veliger larvae of *Nassarius obsoletus* (Gastropoda) in response to bottom sediment. *Biol. Bull.,* **120**:92–109.

——— (1967). The relationship of temperature to the larval development of *Nassarius obsoletus* (Gastropoda). *Biol. Bull.,* **132**:253–265.

Schlieper, C. (1929). Über die Einwirkung niederer Salzkonzentrationen auf marine Organismen. *Zeit. Vergl. Physiol.,* **9**:478–514.

——— (1935). Neuere Ergebnisse und Probleme aus dem Gebiet der Osmoregulation wasserlebender Tiere. *Biol. Rev.,* **10**:334–360.

——— (1955). Über die physiologischen Wirkungen des Brackwassers (Nach Versuchen an der Miesmuschel *Mytilus edulis*). *Kieler Meeresf.,* **11**:22–33.

——— (1957). Comparative study of *Asterias rubens* and *Mytilus edulis* from the North Sea (30 per 1000 S) and the western Baltic Sea (15 per 1000 S). Anèe. *biol.,* **33**:117–127.

Schöne, H. (1961). Complex behavior. *In* The Physiology of Crustacea. T. H. Waterman, ed. Academic Press, New York. **2**:465–515.

Schwabe, E. (1933). Über die Osmoregulation verschiedener Krebse (Malacostracan). *Zeit. vergl. Physiol.,* **19**:183–234.

Segal, E., K. P. Rao, and T. W. James (1953). Rate of activity as a function of intertidal height within populations of some littoral molluscs. *Nature,* **172**:1108–1109.

Shaw, J. (1955). Ionic regulation in the muscle fibers of *Carcinus maenas.* II. The effect of reduced blood concentration. *J. exp. Biol.,* **32**:664–680.

——— (1961). Studies on ionic regulation in *Carcinus maenas* (L.). *J. exp. Biol.,* **38**:135–152.

Shelford, V. E., and E. B. Powers (1915). An experimental study of the movements of herring and other marine fishes. *Biol. Bull.,* **28**:315–334.

Simpson, J. W., and J. Awapara (1966). The pathway of glucose degradation in some invertebrates (*Rangia cuneata, Crassostrea virginica, Valsella demissus, Ascaris lumbricoides, Hymenolepis diminuta*). *Comp. Biochem. Physiol.,* **18**:537–548.

Smalley, A. E. (1960). Energy flow of a salt marsh grasshopper population. *Ecology,* **41**:672–677.

Smith, F. G. W. (1948). Surface illumination and barnacle attachment. *Biol. Bull.,* **94**:33–39.

Smith, R. I. (1964). On the early development of *Nereis diversicolor* in different salinities. *J. Morphology,* **114**:437–463.

——— (1970). The apparent water-permeability of *Carcinus maenas* (Crustacea, Brachyura, Portunidae) as a function of salinity. *Biol. Bull.,* **139**:351–362.

Southward, A. J. (1955). On the behavior of barnacles. II. The influence of habitat and tide-level on cirral activity. *J. mar. biol. Ass. U.K.,* **34**:423–433.

——— (1958). Note on the temperature tolerance of some intertidal animals in relation to environmental temperatures and geographic distribution. *J. mar. biol. Ass. U.K.,* **37**:49–66.

———, and D. J. Crisp (1965). Activity rhythms of barnacles in relation to respiration and feeding. *J. mar. biol. Ass. U.K.,* **45**:161–185.

Stauber, L. A. (1950). The problem of physiological species with special reference to oysters and oyster drills. *Ecology,* **31**:107–118.

Stephenson, T. A., and A. Stephenson (1949). The universal features of zonation between tide-marks on rocky coasts. *Jour. Ecol.,* **37**:289–305.

Steven, D. M. (1963). The dermal light sense. *Biol. Rev.,* **38**:204–240.

Struhsaker, J. W. (1966). Breeding, spawning, spawning periodicity and early development in the Hawaiian *Littorina: L. pintado* (Wood), *L. picta* (Philippi) and *L. scabra* (Linne). *Proc. Malac. Soc. Lond.,* **37**:137–166.

Stutman, L. J., and M. Dolliver (1968). Mechanism of coagulation in *Gecarcinus lateralis. Am. Zool.,* **8**:481–489.

Swedmark, B. (1964). The interstitial fauna of marine sand. *Biol. Rev.,* **39**:1–42.

Tavolga, W. N. (1958). Underwater sounds produced by males of the blennid fish *Chasmodes bosquianus. Ecology,* **39**:759–760.

Teal, J. M. (1962). Energy flow in the salt marsh ecosystem of Georgia. *Ecology,* **43**: 614–624.

———, and F. G. Carey (1967). The metabolism of marsh crabs under conditions of reduced oxygen pressure. *Physiol. Zool.,* **40**:83–91.

Theede, H., A. Ponat, K. Kioroki, and C. Schlieper (1969). Studies on the resistance of marine bottom invertebrates to oxygen-deficiency and hydrogen sulphide. *Mar. Biol.,* **2**:325–327.

Thorne, M. J. (1968). Studies on homing in the chiton *Acanthozostera gemmata. Aust. J. Mar. & Freshwater Res.,* **19**:151–160.

Thorson, G. (1964). Light as an ecological factor in the dispersal and settlement of larvae of marine bottom invertebrates. *Ophelia,* **1**:167–208.

Trueman, E. R. (1967). Activity and heart rate of bivalve molluscs in their natural habitat. *Nature, London,* **214**:832–833.

———, and A. D. Ansell (1969). The mechanisms of burrowing into soft substrata by marine animals. *Oceanogr. Mar. Biol. Ann. Rev.,* **7**:315–366.

———, and G. A. Lowe (1971). The effect of temperature and littoral exposure on the heart rate of a bivalve mollusc. *Isognomum alatus,* in tropical conditions. *Comp. Biochem. Physiol.,* **38A**:555–564.

Ushakov, B. (1964). Thermostability of cells and proteins of poikilotherms and its significance in speciation. *Physiological Reviews,* **44**:518–560.

Van Weel, P. B., J. E. Randall, and M. Tukata (1954). Observations on the oxygen consumption of certain marine Crustacea. *Pacif. Sci.,* **8**:209–218.

Vernberg, F. J. (1956). Study of the oxygen consumption of excised tissues of certain marine decapod crustacea in relation to habitat. *Physiol. Zool.,* **29**:227–234.

——— (1959a). Studies on the physiological variation between tropical and temperate zone fiddler crabs of the genus *Uca*. II. Oxygen consumption of whole organisms. *Biol. Bull.*, **117**:163–184.

——— (1959b). Studies on the physiological variation between tropical and temperate zone fiddler crabs of the genus *Uca*. III. The influence of temperature acclimation on oxygen consumption of whole organisms. *Biol. Bull.*, **117**:582–593.

——— (1962). Latitudinal effects on physiological properties of animal populations. *Ann. Rev. Physiol.*, **24**:517–546.

——— (1963). Temperature effects on invertebrate animals. Temperature—Its measurement and control in science and industry. Reinhold Publ. Co., New York. **3**(pt. 3):135–141.

——— (1969). Acclimation of intertidal crabs. *Am. Zool.*, **9**:333–341.

——— (1972). Dissolved gases—animals. *In* Marine Ecology—Environmental Factors. Interscience Publishers-John Wiley and Sons, New York. **1**, pt. 3. (In press.)

———, and R. E. Tashian (1959). Studies on the physiological variation between tropical and temperate zone fiddler crabs of the genus *Uca*. I. Thermal death limits. *Ecology*, **40**:589–593.

———, and W. B. Vernberg (1966a). Studies on physiological variation between tropical and temperate zone fiddler crabs of the genus *Uca*. VII. Metabolic-temperature acclimation responses in southern hemisphere crabs. *Comp. Biochem. Physiol.*, **19**:489–524.

———, ——— (1966b). Studies on the physiological variation between tropical and temperate zone fiddler crabs of the genus *Uca*. VI. The rate of metabolic adaptation to temperature in tissues. *Journ. of the Elisha Mitchell Sci. Soc.*, **82**: 30–34.

———, ——— (1967). Thermal lethal limits of southern hemisphere *Uca* crabs. Studies on the physiological variation between tropical and temperate zone fiddler crabs of the genus *Uca* IX. *Oikos*, **18**:118–123.

———, ——— (1970). The Animal and the Environment. Holt, Rinehart and Winston, New York. 398 pp.

Vernberg, W. B. (1963). Respiration of digenetic trematodes. *Ann. New York Acad. Sci.*, **113**:(Art. 1):261–271.

——— (1969). Adaptations of host and symbionts in the intertidal zone. *Am. Zool.*, **9**:357–365.

———, and F. J. Vernberg (1966). Studies on the physiological variation between tropical and temperate zone fiddler crabs of the genus *Uca*. V. Effect of temperature on tissue respiration. *Comp. Biochem. Physiol.*, **17**:363–374.

———, ——— (1968a). Studies on the physiological variation between tropical and temperate zone fiddler crabs of the genus *Uca*. X. The influence of temperature on cytochrome-*c* oxidase activity. *Comp. Biochem. Physiol.*, **26**:499–508.

———, ——— (1968b). Physiological diversity of metabolism in marine and terrestrial crustacea. *Am. Zool.*, **8**:449–458.

Walshe-Metz, B. M. (1956). Controle respiratoire et metabolisme chez les Crustaces. *Vie Milieu*, **7**:523–543.

Wells, H. W. (1958). Feeding habits of *Murex fulvescens*. *Ecology*, **39**:556–558.

Wieser, W., and J. Kanwisher (1959). Respiration and anaerobic survival in some seaweed-inhabiting invertebrates. *Biol. Bull.*, **117**:594–600.

Wildish, D. J. (1970). Locomotory activity rhthyms in some littoral *Orchestia* (Crustacea: Amphipoda). *J. mar. biol. Ass. U.K.,* **50**:241–252.

Williams, B. G. (1969). The rhythmic activity of *Hemigrapsus edwardsi* (Hilgendorf). *J. exp. mar. Biol. Ecol.,* **3**:215–223.

———, and E. Naylor (1969). Spontaneously induced rhythm of tidal periodicity in laboratory-reared *Carcinus. J. exp. Biol.,* **47**:229–234.

Wood, L. (1968). Physiological and ecological aspects of prey selection by the marine gastropod *Urosalpinx cinerea* (Prosobranchia: Muricidae). *Malacologia,* **6**:267–320.

Yonge, C. M. (1928). Feeding mechanisms in the invertebrates. *Biol. Rev.,* **3**:21–76.

Young, R. T. (1941). The distribution of the mussel (*Mytilus californianus*) in relation to the salinity of its environment. *Ecology,* **22**:379–386.

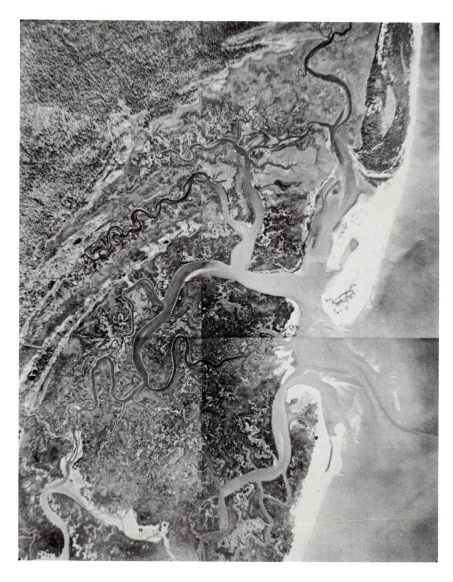

Aerial view of North Inlet Estuary at the Belle W. Baruch Coastal Research Institute field station near Georgetown, South Carolina. This representation is a composite of aerial photographs which were photographed at an altitude of approximately 24,000 feet. (Photographs from the United States Department of Agriculture).

CHAPTER IV

ESTUARIES

Estuaries represent a region of dynamic environmental encounter where water from a fresh-water system pours into the sea. From the viewpoint of a physical oceanographer, an estuary is "a semi-enclosed coastal body of water which has a full connection with the open sea and within which sea water is measurably diluted with fresh water derived from land drainage" (Pritchard, 1967). Although other definitions have been proposed to describe the many different types of estuarine areas throughout the world, one feature is common to all: estuaries are restricted to the mouths of rivers in an area of tidal oceanic fluctuations.

One of the chief characteristics of an estuary is fluctuation in salinity, which depends on fresh-water inflow. Typically, water will be fresher at the surface, more saline on the bottom, and the saline water will penetrate further up the estuary near the bottom than near the surface. Tidal changes, which vary temporally and geographically, also affect salinity by increasing salinities on flood tide, decreasing it on an ebb tide. These fluctuations will be greater in the middle than at either the head or mouth of the estuary. Salinity may vary seasonally, ranging from near fresh water during winter and spring rains to full strength sea water in summer. Since the inflowing river water tends to be colder than the sea water in winter and warmer in summer, temperature fluctuations are greater in estuaries than those of shore waters or the open ocean. The amount of temperature and salinity fluctuation is also influenced in part by estuarine circulation patterns, but since many physical factors influence circulation patterns, no one pattern is typical. Consult Bowden (1967) for a characterization of the principal types of estuaries.

Estuaries are generally turbid because of the amounts of silt in the water. Thus, light cannot penetrate to as great a depth as it does in coastal and open ocean waters. The substratum in an estuary is nearly always covered with silt, and mud is the most common type of bottom. The bottom muds are rich in organic detritus, which is derived primarily from the vegetation along the upper tidal levels. This vegetation consists mainly of swamp grasses and reeds in temperate regions and of mangroves in the tropics.

Estuarine animals are recruited mainly from the sea. Many neritic species (species that live in the shallow-water zone on the continental shelf) use the estuary as a nursery ground before migrating to the open sea. Actual numbers of estuarine species are few, but those that have become adapted to this fluctuating environment are present in large numbers. Euryhalinity is the trademark of these animals.

A.   RESISTANCE ADAPTATIONS

1. Temperature

Animals from colder geographical areas are generally less heat tolerant and more cold tolerant than animals from warmer regions. Thus, specimens of the jellyfish *Aurelia* from Nova Scotia die at a water temperature of 29–30°C, while *Aurelia* from Florida tolerate temperatures up to 38.5°C. Similarly, *Limulus,* the horseshoe crab, from Woods Hole will die at 38.5°C, but specimens from Florida survive up to 46.2°C (Mayer, 1914). It is not known whether these reflect genetic differences between the two populations or simply physiological adaptation.

Within an estuary, temperature tolerance of a species may be correlated with its habitat in that animals from the upper reaches of the intertidal zone are more thermally resistant than organisms inhabiting the deeper channels of estuaries. See section on resistance adaptations of intertidal zone for specific comparisons. The species composition of an estuary, especially in the temperate zone, may change seasonally, apparently in response to temperature. For example, during the winter, northern-affinity crustaceans from offshore waters immigrate into the Newport Estuary, North Carolina, but leave when the waters warmed up; other species then occupy their vacated niches.

Knowledge of the heat tolerances of estuarine animals (including intertidal species) has become of special interest in recent years because of the increasing use of these waters as a discharge site for heated effluents from power plants. Conceivably, this thermal addition could alter the species composition of an estuarine system so drastically that a new ecosystem would be established. After the construction of a power plant on a Maryland estuary, an extensive multidisciplinary investigation under Dr. J. Mihursky clearly showed the relative heat resistance of six estuarine crustacean species (Fig. 4.1). The two least heat-tolerant species, *Neomysis americana* and *Crangon septemspinosa,* are cold-water species and Maryland is near their southern limit. The amphipod *Monoculodes* sp. lives in deep water and is less heat tolerant than the other two amphipods, *Gammarus fasciatus* and *G. macronatus.* The most tolerant

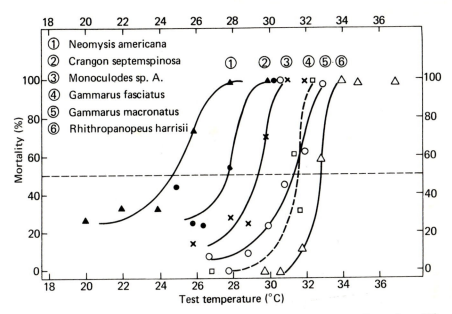

Fig. 4-1. Upper lethal thermal limits of adult invertebrates acclimated to 15°–16°C. (Adapted from Mihursky and Kennedy, 1967. American Fishery Society.)

species is *Rhithropanopeus harrisii,* a eurythermic species with very broad zoo-geographical limits.

Although the temperature tolerances of a number of larval estuarine crustaceans have been determined, differences in experimental procedures and methods of expressing results make detailed comparisons difficult. In one study the larvae of five species of fiddler crabs (genus *Uca*) from the tropical and temperate zones had similar upper thermal lethal points (Fig. 4.2), and the larvae survived longer than adults submerged in water (Vernberg, F., 1969). The zoeae of tropical fiddler crabs survived low temperature better than adults, while the adults withstood a greater range in salinities. These responses indicate that the larvae are not a priori more sensitive than adults. Another method of presenting data on larval mortality in various combinations of salinity and temperature is seen in Fig. 4.3 (Costlow et al., 1960). Although this technique lacks a time component which would be helpful in predicting the success or failure of an organism in fluctuating tidal conditions, data from this study indicate that the larvae of *Sesarma cinereum* are more thermally sensitive than the adult which lives intertidally.

The resistance to thermal extremes has been shown to be influenced both by salinity and by previous thermal history. Acclimation to high salinity increases the resistance of the oligochaete *Enchythaeus albidus* to temperature extremes. Acclimation to temperature not only increases thermal resistance in a predictable manner but also it influences the effect of salinity on heat resis-

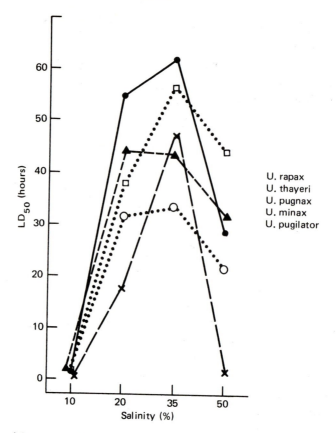

Fig. 4-2. Survival of larval tropical and temperate zone fiddler crabs exposed to different salinities at 38°C. LD_{50} indicates time required for 50 percent mortality. (From Vernberg, 1969.)

tance. For example, the beneficial effect of high salinity is reduced when animals are acclimated to higher temperatures. Furthermore, specific ions influence thermal resistance; sodium, potassium, calcium, and magnesium reduce cold resistance in cold-acclimated animals (Kähler, 1970). The importance of ions on the cellular resistance of isolated sections of gills from bivalves has been demonstrated by Schlieper (1966) for increased calcium levels resulted in an increase in thermal cellular resistance time.

Thermal acclimation can influence the temperature limits of isolated ciliated gill sections from certain molluscs. For example, the bay scallop *Aequipecten irradians,* which lives subtidally, is less resistant to high temperature than two intertidal species, *Modiolus demissus* and *Crassostrea virginica.* The bay scallop is mobile and can escape high water temperatures more easily than the sessile bivalves which must endure both high air and water temperatures. A further indication of differential environmental adaptations by these species is that the

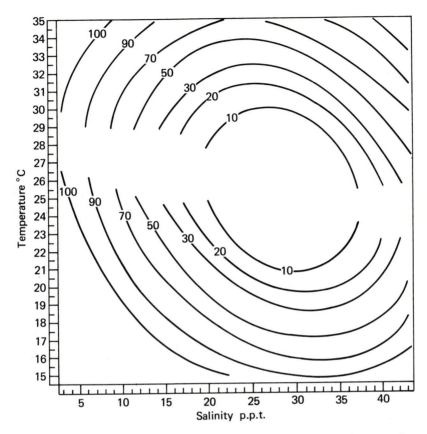

Fig. 4-3. Estimate of percent mortality of first stage zoeae of *Sesarma cinereum* based on fitted response surface to observed mortality determined at 12 temperature-salinity combinations. (From Costlow *et al.*, 1960.)

upper thermal limits of tissues of the subtidal species is not altered when animals are acclimated to different temperatures (Fig. 4.4). In sharp contrast, the responses of the intertidal species were significantly altered for the gill tissue of warm-acclimated animals (25°C) was more resistant to high temperature than cold-acclimated tissues (10°C) (Vernberg et al., 1963). Cellular adaptation to low temperature is seen in North Sea bivalves where isolated gill pieces from the subtidal *Spisula solida* could not survive low temperature as well as those from the intertidal *Mytilus edulis* (Theede, 1965). This ability to acclimate to new environmental stresses seems to be characteristic of eurythermal organisms.

2. Salinity

Salinity is one of the most conspicuously fluctuating environmental factors in the estuary. It is no wonder that many marine scientists have studied this

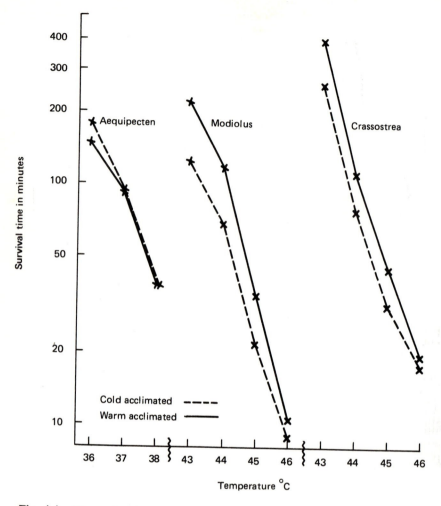

Fig. 4-4. Thermal resistance time (in minutes) of isolated gill pieces from three lamellibranchs as influenced by thermal acclimation. (From Vernberg *et al.,* 1963. Pergamon Publishing Company.)

parameter from various viewpoints. This section will stress the importance of salinity in limiting the distribution of estuarine animals; the reader should consult the section on chemical regulation for a discussion of the ionic and osmoregulatory adaptations necessary for an organism to live in different salinities (p. 197).

The number of marine species present in an estuary decreases with a decrease in salinity (Gunter, 1945; Remane and Schlieper, 1958; Wells, 1961). Not always is the cause-and-effect relationship understood, since other factors, such as temperature, oxygen, and type of substrate and food, may also vary along with salinity changes. An excellent investigation that considered many variables is that of Wells (1961), who studied animals associated with an oyster-bed community distributed in different parts of an estuary. As graphically represented in Fig. 4.5, the mean number of species associated with this oyster-bed community decreased from a high of 67 in high salinity water to only 16 in low salinity water, while the total number of species from these two extremes was 220 and 56, respectively. The results of the salinity death point determination of 20 species agreed well with the observed field

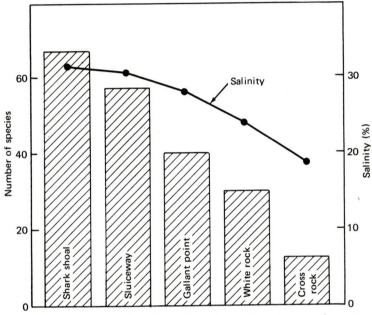

Fig. 4-5. Mean member of species collected at different areas of an estuary and the mean salinity for each station. (From Wells, Duke University Press. Copyright 1961.)

distribution of these species since echinoderms are restrictd to higher salinity waters, crustaceans are tolerant of low salinity, and some molluscs are eury-halinic while others are stenohalinic (Table 4.1). Of particular interest in understanding the distribution of these animals is the observation that the predators of oysters, including the oyster drill *Urosalpinx cinerea*, are less toler-ant of reduced salinity than their prey, the oyster *Crassostrea*, and the popula-tions of oysters in turn are most dense in low salinity waters.

Not only do benthic estuarine animals differ in their ability to survive salinity stress, but also planktonic crustaceans living in estuaries vary in their tolerance to salinity (Table 4.2). The salinity limits of various copepod species, as determined in the laboratory, varied as follows: *Acartia tonsa* > *A. bifilosa* > *A. discaudata* > *A. clausi* > *Centropages hamatus* > *Temora longicornis*, a

TABLE 4.1 *Ranking of Twenty Species of Estuarine Animals on the Basis of Their Distribution Limits and Tolerance to Low Salinities. (The most tolerant species are listed first.)* *

Distribution		Tolerance	
1 *Mercenaria*	(M)	1 *Mercenaria*	(M)
2 *Panopeus*	(C)	2 *Panopeus*	(C)
3 *Eurypanopeus*	(C)	3 *Eurypanopeus*	(C)
4 *Modiolus*	(M)	4 *Modiolus*	(M)
5 *Crassostrea*	(M)	8 *Clibanarius*	(C)
6 *Nassarius*	(M)	5 *Crassostrea*	(M)
7 *Pagurus*	(C)	6 *Nassarius*	(M)
8 *Clibanarius*	(C)	7 *Pagurus*	(C)
9 *Odostomia*	(M)	17 *Thais*	(M)
10 *Urosalpinx*	(M)	9 *Odostomia*	(M)
11 *Brachidontes*	(M)	10 *Urosalpinx*	(M)
12 *Busycon*	(M)	12 *Busycon*	(M)
13 *Fasciolaria*	(M)	11 *Brachidontes*	(M)
14 *Chione*	(M)	13 *Fasciolaria*	(M)
15 *Arbacia*	(E)	19 *Asterias*	(E)
16 *Cantharus*	(M)	16 *Cantharus*	(M)
17 *Thais*	(M)	14 *Chione*	(M)
18 *Cerithium*	(M)	15 *Arbacia*	(E)
19 *Asterias*	(E)	18 *Cerithium*	(M)
20 *Lytechinus*	(E)	20 *Lytechinus*	(E)

C = Crustacean
E = Echinoderm
M = Mollusc
* Wells, 1961.

ranking which corresponds well with their distribution in the field (Lance, 1964). Survival also varied with stage of the life history, for the adult copepod *Acartia tonsa* is more resistant to low salinity than the copepodite stages; in contrast, zoeal *Porcella longicornis* are more resistant than the post-larval form.

A similar study on copepods inhabiting a mangrove region of an estuary in Brazil also demonstrated marked species differences which could be correlated with distribution (Tundisi and Tundisi, 1968). Acclimation to reduced salinity increased the salinity resistance of *Acartia lilljeborgi*. A salinity of 11 0/00 was lethal to animals acclimated to high salinities, but those acclimated to low salinity water could survive dilutions to 6.3 0/00. Other data on estuarine plankton demonstrated that the larvae of fiddler crabs from tropical and temperate zones had similar salinity tolerances, and all were less resistant to low salinity than the adult forms (Vernberg, F., 1969).

A critical salinity boundary of 5–8 0/00 separates fresh-water and marine faunas. This concept was first suggested by Remane (1934) and later docu-

TABLE 4.2 *Salinity Tolerance of* Acartia *Species at Different Temperatures (The duration of each experiment, in days, is indicated in parentheses)* *

Species	Range of salinities (% seawater) causing mortality at various temperatures (C)				
	4.5°	10°	16°/17°†	20°	24°
A. tonsa	0–60 (7½)	0–40 (8½)	0–25 (8)	0–15 (8)	0–30 (7)
A. discaudata	0–50 (5½)	0–50 (5½)	0–35 (5½)		
A. bifilosa	0–50 (10½)	0–40** (9)	0–40 (6½)	0–60 (5)	

Species	Range of lethal salinities (% seawater) at various temperatures (C)				
	4.5°	10°	16°/17°†	20°	24°
A. tonsa	0–25	0–15	0–10	0–10	0–15
A. discaudata	0–25	0–25	0–25		
A. bifilosa	0–25	0–20	0–25	0–40	

** 0–35% seawater after 5 days.
† A. tonsa at 17C, other species at 16C.
* Lance, 1964.

mented by Khlebovich (1969) (Fig. 4.6). The critical salinity principle is based on a number of physiological parameters including survival, growth, locomotion, and osmoregulation. For example, the body fluid concentration of brackish water animals may be reduced to a salinity of about 5 0/00 before serious damage results. Below 5 0/00, distortion of cellular electrochemical properties occurs, and the tissue albumin fraction undergoes a marked change. Above a 5–8 0/00 salinity, poikilosomotic organisms can survive, but below this concentration hyperosmotic regulation is required.

The salinity acclimation capacity in some organisms indicates that animals acclimated to low salinity generally have greater resistance to low salinity than animals acclimated to higher salinities (see review by Kinne, 1964). The mud crab *Hemigrapsus oregonensis*, for example, can be acclimated to live in hypersaline water with detectable acclimation after 5 days and strong acclimation by the 22nd day (Gross, 1963).

In addition to observed organismic effects, it has been shown that altered salinity differentially influenced the lethal limits of isolated gill sections from various molluscs occupying different estuarine habitats. The intact stenohalinic bay scallop does not survive well in salinities below about 14 0/00, and the gill pieces also die at this salinity. In contrast, two euryhaline species (the oyster

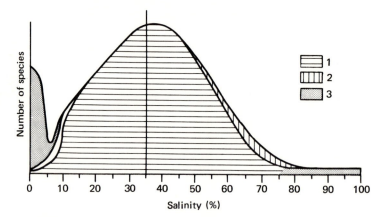

Fig. 4-6. Number of marine animal species (1), brackish water and euryhaline estuarine-living species of marine origin (2), and freshwater-living species (3). (From Khlebovich, 1969.)

Crassostrea virginica and the ribbed mussel *Modiolus demissus*) can withstand low salinity; the adult intact oyster dies at 7 0/00 and its gill tissue at 4 0/00, while these values for the ribbed mussel are 5 0/00 and 2 to 3 0/00. Temperature acclimation influenced salinity resistance (Table 4.3). Tissue from warm-

TABLE 4.3 *Osmotic Resistance of Isolated Gill Tissue from Lamellibranchs Subjected to Various Salinities for Twenty-four Hours (Determined March 1961)**

Salinity (p.p.t.)	Modiolus W-A†	Modiolus C-A‡	Crassostrea W-A	Crassostrea C-A	Aequipecten W-A	Aequipecten C-A
30	3	3	3	3	3	3
27	3	3	3	3	3	3
24	3	3	3	3	3	3
21	3	3	3	3	3	3
18	3	3	3	3	2.6	2.7
15	3	3	3	3	1.7	2.4
12	3	3	3	3	0	1.0
9	3	3	2.8	2.5		
6	3	3	2.1	1.6		
5	3	3	1.9	1.4		
4	2.6	2.4	1.2	0.4		
3	1.6	0.7	0	0		
2	0.2	0	0	0		

† W-A = Warm-acclimated animals.
‡ C-A = Cold-acclimated animals.
* Vernberg et al., 1963.
Pergamon Publishing Company.

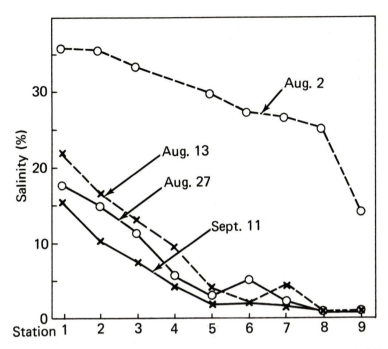

Fig. 4-7. Salinity profiles at different areas of an estuary before and following Hurricanes Connie (August 12) and Diane (August 17). Station 1 is near the mouth of the estuary and Station 9 is farthest from the mouth. (From Wells, 1961. Duke University Press. Copyright 1961.)

acclimated oysters and ribbed mussels survived low salinity better than tissues from cold-acclimated animals. However, the cold-acclimated tissue of the bay scallop withstood reduced salinities better than that of warm-acclimated animals (Vernberg et al., 1963).

Unpredictable natural occurrences, such as the extended heavy rains commonly associated with hurricanes, may drastically reduce the salinity of an estuary and thereby influence faunal distribution. Wells (1961) published one of the best documented papers dealing with the influence of markedly dilute estuarine water on animals. Three hurricanes and associated rains dumped approximately 43 inches of rain into one estuarine system in North Carolina within 30 days in late August and early September, and the effects of this fresh water on the salinity of the estuary are depicted in Fig. 4.7. As might be expected, a marked decline in the number of species resulted. The reappearance of species could be correlated with their reproductive period, for those species normally reproducing in the early fall were not found until the following year, while many early winter spawners quickly reappeared.

Although the preceding paragraphs emphasize the effect of a single environmental factor on the resistance capacity of an organism, there are some examples of the synergistic interaction of two or more factors. Of utmost importance to understanding environmental biology is the need for more detailed analyses of multiple factor interactions on organisms and to model the variation of environmental factors in these studies to match those conditions the animal experiences in the field. One of the often cited studies dealing with multiple factor-organismic interaction is that of McLeese (1956). In general, he found that the thermal lethal limits of lobsters were reduced if these animals were subjected to extreme salinities and low oxygen concentrations. A more detailed discussion of this important paper may be found in Chapter V.

3. Oxygen

The brackish-water copepod, *Eurytemora hirundoides,* is very tolerant of low oxygen levels for it has survived in polluted areas of the River Tyne (U.K.), except in regions of complete oxygen deprivation (Bull, 1931). Certain amphipod species with high resistance to anoxia are also able to inhabit estuarine regions influenced by pollution (Waldichuk and Bousfield, 1962). The bivalve *Mytilus edulis,* which may live in polluted waters with low oxygen content, can survive oxygen lack for weeks. The resistance capacity of other bivalves depends on their burrowing habits since those burrowing in mud are more resistant to anoxia than those from well-aerated waters. *Nucula tenuis,* which does not maintain contact with the overlying water as it burrows, survived oxygen lack for 5 to 17 days, but *Syndosmya alba,* which maintains contact with the water by means of extensible siphons, is less resistant and dies within 3–5 days (Dodgson, 1928).

Anoxic tolerance changes with stage of the life cycle. Larvae of fiddler crabs are sensitive to reduced oxygen levels, while the adults can survive anoxia for long periods of time. The adaptive nature of these responses is obvious in that the larvae live in estuarine waters, while the adults normally experience low oxygen conditions in their burrows in marshes and sand beaches. Free-swimming larval stages of trematodes cannot survive anaerobiosis long, while the adults, which are endoparasites, can survive anoxia for long periods (Vernberg, W., 1969). In certain salmonid fish, the young (1 to 4 weeks after hatching) are tolerant of low oxygen tensions, but the fry (5 to 16 weeks after hatching) avoid water which has a concentration of oxygen less than 4.6 mg/liter, and the older fish are less sensitive to oxygen deprivation (Bishai, 1962).

Field observations by Reish (1955) demonstrated that species of polychaetes were common only in a given type of bottom which had a specific oxygen regime. Data from laboratory studies substantiate Reish's observation. A mud-dwelling polychaete, *Scoloplos armiger,* survived better in poorly aerated water (4% oxygen) than in well-aerated water (21% oxygen). *Arenicola marina,* which lives in a sand-mud habitat, survived equally well at 4% and 21% oxygen. In contrast, a tube-dwelling worm, *Sabella pavonina,* could not withstand water which was only 10% oxygen saturated, but survived exposure to higher oxygen levels (Fox and Taylor, 1955).

Two burrowing crustaceans, *Callianassa californiensis* and *Upogebia pugettensis,* living in an Oregon estuary, were resistant to anoxia for an average time of 138 hours and 81 hours respectively. This prolonged resistance time would permit these species to survive for a number of tidal changes during which time the oxygen levels could be increased. This interspecific difference in resistance capacity can be correlated with habitat differences: *Upogebia* burrows are found at or slightly below the low tide mark, while *Callianassa* are higher in the intertidal zone where they are exposed longer to anoxic conditions in the subsurface (Thompson and Pritchard, 1969).

Temperature influences the anoxic survival of the clam *Mya arenaria.* This species can survive oxygen lack for weeks at low temperature, 8 days at 14°C, and approximately 24 hours at 31°C (Collip, 1921).

Other variables influence anoxic resistance of animals:

Sex: Male fish (*Fundulus parvipinnis*) resist anoxia longer than females during the breeding season (Keys, 1931).

Body size: Large prawns (*Penaeus indicus*) and large *Fundulus* are more tolerant of oxygen lack than small individuals (Subrahmanyam, 1962; Keys, 1931). However, in other species of fish body size is not a variable (Job, 1957).

Parasitism: *Fundulus* which are parasitized or diseased die more quickly when exposed to anoxic waters than "healthy" individuals (Keys, 1931).

Molting: Molting crabs are less resistant to oxygen lack than intermolt crabs (McLeese, 1956).

Acclimation: Animals acclimated to reduced, but sublethal, oxygen concentrations may be more tolerant to hypoxia, as demonstrated by Tagatz (1961) with shad. However, acclimation to various oxygen concentrations did not significantly change the lethal oxygen level (McLeese, 1956).

Other factors probably influence the lethal limits of animals to oxygen, but research on this problem has not received as much attention as thermal and salinity changes.

B. CAPACITY ADAPTATIONS

1. Perception of the Environment

a. Orientation and environment selection

As in other environments, animals in estuaries rely on light perception for orientation and exploration. Among free-swimming animals placement in a favorable position in the water column, which is of utmost importance for survival, does not occur by chance, for each species occupies defined areas in the water column. Some prefer surface waters, others the water near the substrate, and still others mid-water areas. Preference for one area over another often changes throughout the day, and the diurnal vertical migration of zooplankton and fish through the water column is one of the most conspicuous and widespread phenomena in aquatic environments. Cushing (1951) proposed a generalized pattern of diurnal vertical migration in which the organisms ascend to surface or near-surface water from the deeper depths as evening approaches; they stay at this level until about midnight, then descend to a lower depth. During the dawn hours, organisms again ascend to the surface, then descend sharply as light intensity increases. Many factors, however, influence migratory patterns, such as stage of life cycle, sex, and phase of the reproductive cycle, along with environmental variables. There are a multitude of changing environmental parameters in the estuary, including salinity, tidal currents, water turbulence and wave action, temperature, food supply, and turbidity, as well as light intensity. In a stratified estuary the two-layered drift strongly influences migratory behavior, for the upstream flow on the floor and downstream flow of the surface are superimposed on the tidal oscillations (Carriker, 1967). It is not surprising, then, that the behavior patterns of planktonic forms appear highly adaptive and geared toward keeping the animals within the bounds of a compatible environment.

What sensory modalities do these animals use in migrating through the water column? Response to light is generally considered an important factor. One animal, the freshwater cladoceran *Daphnia,* depends upon both photokinetic and phototactic responses in its migrations (Harris and Mason, 1956). The migratory movement is independent of the direction of light and is entirely photokinetic. That is, light stimulates locomotion at a certain intensity, irrespective of the direction of the light. At high intensities, however, the response is phototactic, and the light source governs the direction of locomotion. When light intensities are reduced, *Daphnia* moves toward the light source, and when light intensity is increased beyond a certain point, then the animal moves away from it. Blinded animals will respond photokinetically, but not phototactically. For directional responses compound eyes are necessary; dermal responses are necessary for the kinetic responses.

Changes in hydrostatic pressure are also thought to play a role in vertical migratory behavior. Knight-Jones and Morgan (1966) have postulated that since an organism in the water column can adapt to gradual changes in pressure up to a certain point, when pressure builds up beyond this point, the animal responds by compensatory swimming. Thus, pressure could aid in setting the limits of vertical migration. Indeed some pelagic animals are extraordinarily sensitive to pressure changes. The pinfish *Lagodon rhomboides* and the black sea bass *Centropristus striatus,* for example, proved sensitive to changes in water pressure of less than 0.5 cm with a reaction time of about 0.1 second (McCutcheon, 1966).

Control mechanisms, however, are not fully understood, and they are probably not the same for all migrating animals. Consider, for example, the ingenious studies by Enright and Hamner (1967) on 13 species of zooplankton. The animals were kept in a large concrete tank 2.5 meters deep, under a controlled light-dark cycle. Light at the water surface decreased from a daytime maximum of 100 lux to a nighttime minimum of 0.02 lux. Animals in the top 30 cm were sampled every four hours by net tows throughout the length of the tank. In the presence of a light-dark cycle most organisms studied showed significant vertical migratory behavior. Several species were found in greater numbers at the surface of the water during the hours of light than during dark periods. When the zooplankton were maintained under constant dim light, the vertical migration of the amphipod *Nototrotis* sp. differed only in amplitude from migrations during the light cycle. The persistence of the rhythm under constant conditions suggests control by an endogenous circadian timer. Furthermore, since the timing of these migrations under constant light conditions corresponded to the immediately prior light cycle rather than to outdoor daytime, light was the entraining mechanism. A less pronounced endogenous rhythm was also observed in peltidiad copepods. Other species of zooplankton did not show any rhythmicity of behavior under constant conditions. One species of amphipod, *Tiron* sp., showed a nonrhythmic migratory pattern; the other species responded directly to the experimental light regimes. This wide range of responses led Enright and Hamner (1967) to conclude: "Although the vertical migration of zooplankters may appear to be a single phenomenon in an ecological context, conceivably evolved on the basis of a common selective pressure, the physiological mechanisms underlying the field behavior are by no means uniform."

Regardless of what controlling mechanisms are involved in vertical migrations, environmental factors can markedly modify behavior patterns in migration. In a series of experiments on the effects of reduced salinity on vertical migration in estuarine zooplankton, Lance (1962) found that the behavior of copepods was changed in all dilutions, and distribution in the water column partly depended on the degree of the dilution of the sea water. In full-strength sea water the animals generally swam toward a light, although some were photonegative. Copepods placed in 90% and 80% sea water initially reacted by irregular swimming movements and sinking, but generally recovered rapidly so that distribution then differed only slightly from that of control animals in full strength sea water. Salinities lower than 80% sea water caused the copepods to swim violently, and many sank to the bottom of the experimental vessel. There was a wide variation in response of different copepods, with the behavior varying according to the salinity tolerances of the species. Thus, *Acartia tonsa,* which is extremely tolerant to lower salinities, did not move from the upper half of the column until salinities were lower than 60% sea water, while other less euryhaline species disappeared from the top of the water column at salinities of 70% and 60% sea water. Even at these salinities, however, the

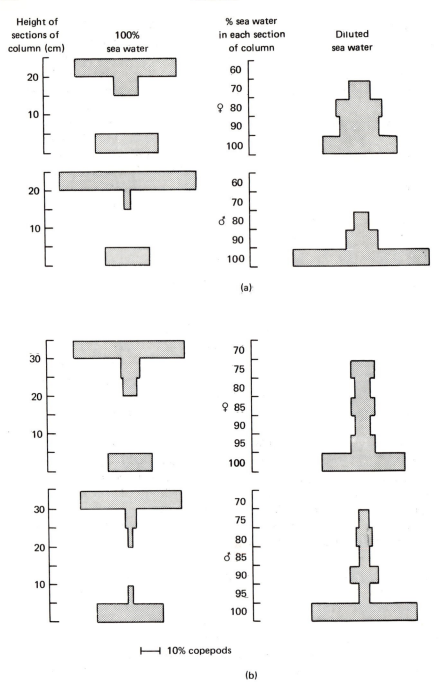

Fig. 4-8. Vertical distribution of *Centropages hamatus* in the presence of a series of discontinuity layers. (a) Salinity differences of 10 percent S.W.; (b) salinity differences of 5 percent S.W. (From Lance, 1962. Cambridge University Press.)

copepods could recover and eventually would respond to the light. Copepods placed in a column of water where they had to pass through one or more salinity discontinuity layers to reach the upper layers of the column were not able to enter the reduced salinity zone when an extreme dilution was used; the reduced salinity acted as a barrier in keeping the animals from migrating upward (Fig. 4.8). These results confirm field studies indicating that low surface salinities will cause zooplankton that ordinarily migrate vertically to concentrate in the more saline bottom water (Lance, 1962; Grindley, 1964).

Laboratory studies have shown that temperature also can limit the extent of vertical migration, since some vertically migrating species have definite preferences for certain temperatures. Parthenogenetic females of the cladoceran *Podon polyphemoides*, for example, were acclimated to 15°C and then presented with a number of temperature choices. Where lower temperatures were about 8°C, no significant differences in choices were observed. However, when one of the two temperatures was 8°C or less, they showed a very strong preference for the lower temperature (Fig. 4.9) (Ackerfors and Rosen, 1970). This

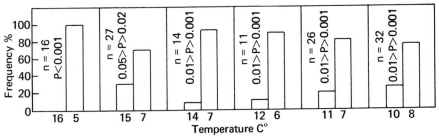

Fig. 4-9. Temperature preferences of the parthenogenetic female cladoceran, *Podon polyphemoides*, after acclimation to 15°C and presented with a number of temperature choices. (From Ackefors and Rosen, 1970. North-Holland Publishing Company.)

preference correlates well with field observations. When the thermocline was above the 15-m level, the animals accumulated at the level in the thermocline where temperatures were about 7–8°C (Ackefors, 1969).

Responses of the pelagic larvae of intertidal zone animals to the fluctuating estuarine environment have been discussed in Chapter III. As noted there, phototactic response may be reversed in response to thermal or salinity stress. The introduction of chemicals into the environment can also alter vertical migratory patterns. Exposure to a xanthene dye, rhodamine B, has been shown to increase photopositive responses in *Acartia lilljeborgi* and abolish normal differences between day and night responses (Bjornberg and Wilbur, 1968). Such changes in phototactic responses, of course, result in abnormal behavior patterns, which could be of considerable consequence in estuaries subject to industrial pollution.

There has been much speculation about the adaptive value of vertical migrations. Some workers have suggested that the migrations are essentially a feeding response, allowing animals to feed in plankton-rich surface waters; some animals, however, that make diurnal migrations do not feed in the upper waters. Migratory behavior could have a significant role in interspecific competition, placing competing species at different positions in the water column throughout the 24-hour period. Other workers have suggested that migrations to surface waters where currents are stronger aid in horizontal dispersion. McLaren (1963) has suggested yet another adaptive value. He postulated that the zooplankton feed in surface waters, then descend to cooler, deeper water to conserve metabolic energy which in turn could be put into fecundity. However, as Woodhead (1966) has pointed out, adaptive significance might be expected to differ for each species and probably also change for a single species during different stages of its life history.

Estuarine animals also rely on other sensory modalities in their perception of the environment. Some are able to utilize tidal salinity changes to their advantage in seeking a suitable environment using rheotactic and chemoreceptive modalities. While reduced salinities can prevent animals from carrying out normal vertical migrations, changing salinities associated with the ebb and flow of the tide can be utilized to advantage by others, such as the European eel *Anguilla vulgaris* (Creutzberg, 1961). The eels hatch in the tropical waters of the Sargasso Sea and slowly migrate to inland waters along the coast of Europe and Africa, reaching the area as far east as Sicily in the Mediterranean in the autumn of their third year. Then they metamorphose from willow leaf–shaped larvae into transparent elvers and become distributed along the European and North African coast. The elvers are able to move into the estuaries in these areas with the tides by staying in the surface waters during incoming flood tides and then descending to bottom waters during ebb tides; thus, they avoid being washed out to sea. This migration pattern continues both day and night, although the elvers do not reach the surface waters during the daytime. Once they are in the estuary with the ebb tide flowing over them, the elvers can discriminate between the overhead water masses. When a water mass containing inland water passes over them, they orient themselves by heading into the current and then make their way to inland waters to complete their development. The animals apparently are able to distinguish the specific odor of inland water by chemoreception, for they showed no response when exposed to tap water. Inland water, however, proved to be a very strong attractant.

There is some evidence that salinity also influences the migration of some species of shrimp in and out of estuaries. In laboratory studies on the pink shrimp *Penaeus duorarum*, Hughes (1969) indicated that inshore movement of post-larval shrimp and subsequent offshore movements of juveniles were aided by flood and ebb tides, respectively. The usual positive rheotactic response of the juveniles was reversed in decreased salinities, and as a result they swam actively downstream but gave way to passive drifting when currents became too strong. Thus, if the same responses are elicited in nature during flood tides, juveniles would orient and swim in an offshore direction during flood tides. During ebb tide when salinities decrease, juveniles would swim or be passively displaced with the current, still in an offshore direction. It was demonstrated that post-larvae were active in the water column until salinity decreased, then they sank to the substrate or remained low in the water column. During flood tide, then, when salinities are high, they would be active in the water column and displaced shoreward. On an ebb tide when salinities are lowered, the post-larvae would sink to the substrate and avoid being carried out to sea. It is not known whether these responses are related to salinity per se, or if the shrimp are able to detect differences in water quality as the elvers do.

b. Communication

Because visual display would be of limited value in murky waters, audition may provide the best means of communication. The male toadfish *Opsanus tau,* for example, has a well-defined ritual before spawning occurs. He first seeks a suitable nesting site, such as a tin can or a large empty shell, and stands guard; when he is ready to spawn, he emits a characteristic boat whistle call, which in turn acts as a stimulus to attract females that are ready to lay eggs. After the female lays a clutch of eggs, she leaves the nest, but the male remains until the young attain free-swimming stage (Gray and Winn, 1961). During the time that the male toadfish is guarding the nest, the approach of toadfishes and other objects elicits aggressive grunt sounds from the guarding male. Winn (1964) has hypothesized that these grunts cause other toadfish to leave the area of the nest.

2. Feeding

a. Detritus feeders

Zooplankton are often assumed to be the primary herbivores in pelagic systems, serving to transfer energy from phytoplankton to secondary consumers. In estuaries, however, especially in shallow, warm-water ones, efficiency of utilization of plant food by estuarine zooplankton decreases. The most important primary producers are the marsh grasses, sea grasses, reeds, and other marsh vegetation. Much of this plant material becomes a part of the sediment, and in estuaries, food chains are based on organic detritus derived from these various plants, or on the benthic and epiphytic microflora. Thus benthic herbivores rather than zooplankton represent the critical herbivore link in the estuarine food web (Darnell, 1961; Odum, 1970; Qasim, 1970).

Zooplankton in estuaries characteristically are large in volume, but small in number of species. Species composition varies markedly with seasonal changes, and during the warmer months of the year there are large numbers of larvae of benthic and intertidal zone animals (Riley, 1967). In Long Island Sound, New York, the zooplankton consume approximately 50–60% of the net phytoplankton produced (Riley, 1967). Zooplankton become less important as

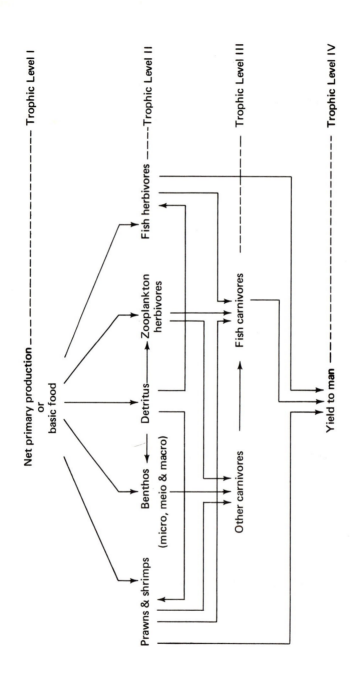

Fig. 4-10. Generalized representation of a food web in the Cochin Backwater. (From Qasim, 1970. The Regents of the University of California.)

herbivores in the vicinity of Cape Hatteras, North Carolina, especially in areas shallower than 100 meters. Zooplankton prove to be relatively unimportant in food chains, grazing only 2 to 9% of the available phytoplankton (Williams et al., 1968). Further south in Lake Pontchartrain, Louisiana, Darnell (1961) found that phytoplankton was utilized directly by only a few of the consumer species, and in the Cochin Backwater, a tropical estuary in India, phytoplankton production far exceeded the rate of consumption by the zooplankton herbivores (Qasim, 1970). A generalized representation of a food web in a tropical estuary is given in Fig. 4.10.

Feeding habits of the benthic estuarine organisms have been correlated with the hydrodynamic characteristics of their environment. In a study of the relationship between distribution of megabenthos and sediment types in an estuary in North Carolina, Brett (1963, cited in Carriker, 1967) found that most of the animals were detritus feeders in areas of slow currents where the mean diameter of sediments is less than 0.09 mm. In areas where mean grain size exceeded 0.09 mm, filter feeding was the dominant form of feeding behavior, while predation was common where mean sediment size was greater than 0.15 mm and clay and organic matter were low.

Most of the larger estuarine consumers are omnivorous, and any given size group of a species can utilize several sources of food (Darnell, 1961, 1967). For example, in Lake Pontchartrain, Darnell (1961) found that the diet of adult blue catfish consisted of 12% molluscs, 17% macro-crustacea, 30% fishes, 26% vegetation, and 15% miscellaneous material. Generally, alternate foods may be utilized by a species from time to time and from place to place, depending upon the availability of food, and these food substitutions are not necessarily confined to a single trophic level. The trophic spectra for the most important consumer species of the Lake Pontchartrain community show that these species rely heavily upon detritus, bottom animals, and fishes. Phytoplankton, vascular plant material, or zooplankton are little utilized (Darnell, 1961) (Fig. 4.11). Since an estuary has large quantities of dead protoplasm, shallow bottoms, and moving water, Darnell (1967) has characterized the entire estuary as "a thin mud containing nutritious opportunities for the consumer species." In the Lake Pontchartrain community, even the most carnivorous species took in a certain amount of organic detritus, and most consumer species utilized detritus at some stage in their life cycle (Darnell, 1961). Food preferences of the Atlantic croaker *Micropogon undulatus,* change from small zooplankton to small bottom animals to detritus to larger crustaceans and fish, while the pinfish *Lagodon rhomboides* prefers the change from small crustaceans to detritus to vegetation (Darnell, 1961).

One of the most successful estuarine fish, the striped mullet *Mugil cephalus,* illustrates well some of these aspects of estuarine nutrition. This species feeds on mosquito larvae, copepods, and other zooplankton until it attains a length of about 30 mm. Then it either feeds on detritus by sucking up the surface layers of mud, or browses on micro-algae attached to submerged surfaces. Sedi-

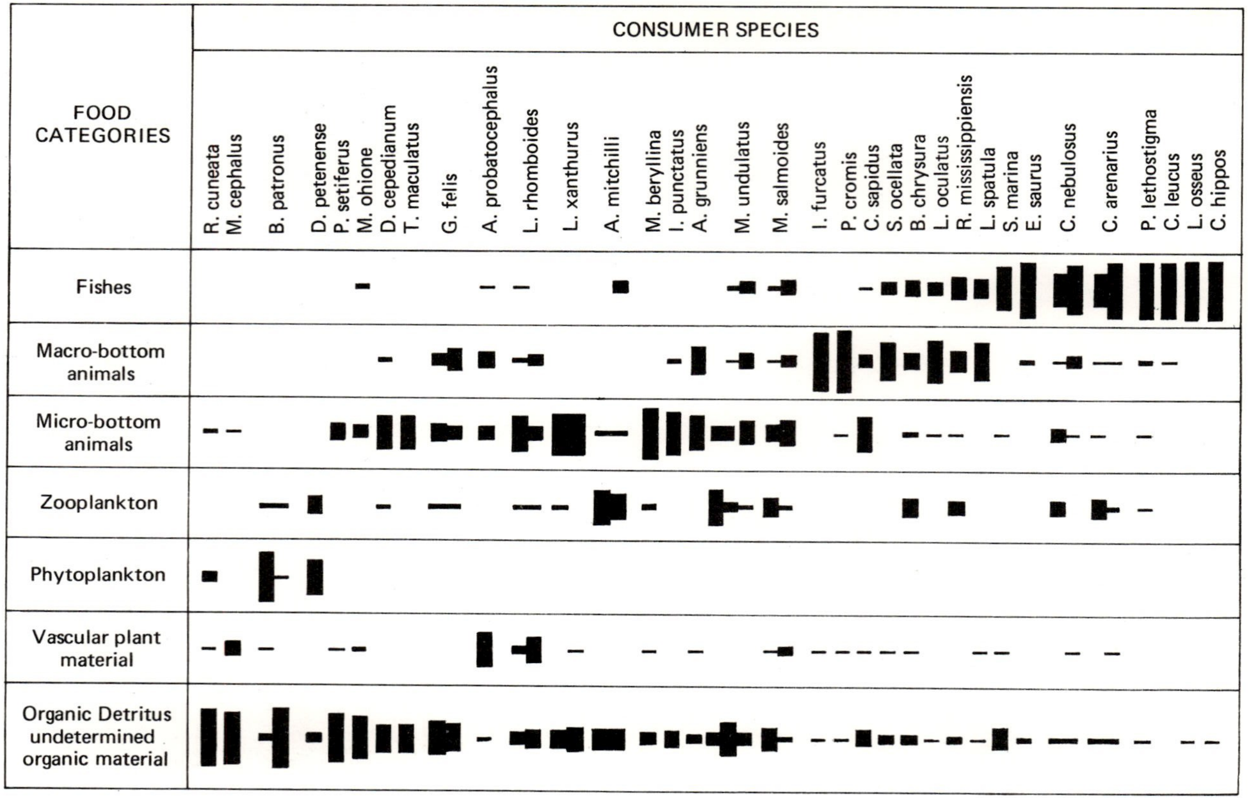

Fig. 4-11. Trophic spectra for the most important consumer species of the Lake Pontchartrain community. Note relatively little utilization of phytoplankton, vascular plant material, or zooplankton, but heavy emphasis upon detritus, bottom animals, and fishes. From Darnell, 1961. (Duke University Press. Copyright 1961.)

ment-feeding fish may take up small mouthfuls at random or may skim along the bottom, lips barely touching the sediment, and suck up the top layer. Both types of feeders strain a small quantity of the sediment by the pharyngeal filter and expel undesired material. Browsing animals nibble the attached algae, digest suitable food, and expel the remainder. When offered a choice, the mullet showed a definite preference for live material over plant detritus. Studies indicated that the algal diet has a higher caloric content than a detritus one for the mullet (Odum, 1970).

The considerable amounts of organic material in solution in estuaries are also thought to be a part of the estuarine detritus (Darnell, 1967). Whether this dissolved organic material, such as amino acids and sugars, can contribute to the nutrition of aquatic animals has been the subject of debate for some time. Pütter (1909) originally proposed that dissolved organic material could be used in nutrition, but Krogh (1931) concluded that there was no substantial evidence to support this idea. The question is still being debated today. Stephens and his coworkers (see Stephens, 1967, 1968 for reviews) suggested that dissolved organic material, particularly amino acids, can serve as a supplementary nutritional source for marine and estuarine invertebrates, while others (Johannes et al., 1969) argued that the evidence for the use of dissolved amino acids as an energy source is not substantial.

There are, however, some areas of general agreement. Amino acids are found in measurable amounts in sea water, and they can be taken up by soft-bodied marine animals. Stephens (1963) found 12 neutral and acidic amino acids in interstitial sea water of a mud flat, ranging in concentration from 2.5×10^{-5} moles per liter to trace amounts. Total amino acid concentrations ranged from 6×10^{-5} to 10×10^{-4} moles per liter. Thus, he assumed that the total concentration of amino acids in inshore waters ranges between 10^{-5} and 10^{-4} M. The uptake seems to be linearly related to concentration at low ambient concentration. Then, as the concentration of the material increases, eventually further increases no longer modify the rate at which material is removed from solution. Interestingly, amino acid uptake values are reduced in low salinity water.

When specimens of the polychaete *Darnillea articulata* were exposed to labeled amino acids, Stephens (1968) found that the rates of uptake of individual amino acids exceeded by a large margin the apparent rates of leakage. That is, more amino acids were taken up by the worms than were lost, and thus amino acids represented an accumulation of these compounds into the free amino acid pool of the animal. Could this be of value as a nutritive supplement? By comparing caloric value with heat production per unit time to assess nutritive value, Stephens (1963) contended that if material obtained directly from solution is approximately equal to the reduced carbon necessary to support the oxygen consumption of an organism, then it would probably be significant as a nutritive supplement. After measuring the oxygen consumption rate of the polychaete *Clymenella*, Stephens calculated that the mixture of amino acids observed in its habitat could provide approximately 150% of the reduced carbon necessary to support its observed oxygen consumption. Thus, the free amino acids absorbed from the ambient medium could have supplementary nutritional value to *Clymenella*.

Johannes et al. (1969) interpreted the observations made using ^{14}C-labeled free amino acids very differently from Stephens. Since marine invertebrates have very high concentrations of free amino acids, the concentration of these acids in the tissues of marine invertebrates is generally higher than are the levels in sea water. Therefore, Johannes et al. argued that the labeled free amino acids moving out of the animal represented the flux of a greater quantity of unlabeled free amino acids than a unit of ^{14}C entering them. Thus free amino acids from the medium could not represent an accumulation in the animal's endogenous free amino acid pool. Although these workers conceded that dissolved organic compounds could possibly satisfy some micronutrient or vitamin requirements, they did not believe these compounds could have significant nutritional value.

b. Predators

Some estuarine animals are not omnivorous and do not feed on detritus. Consider, for example, the planktonic chaetognath *Sagitta hispida*, which is a voracious carnivore. Reeve (1966) found that feeding habits could be correlated with food availability. Copepods over 1.0 mm in length were the preferred prey, although the chaetognaths would accept smaller organisms. Only live, swimming material was eaten; dead plankton, eggs, or particulate detritus being moved by water currents were unacceptable. Potential prey probably are recognized through nonmotile cilia that lie between fine groups of setae projecting from the general body surface. It was found that temperature greatly influenced feeding behavior, but salinity variations did not. The lowest temperature at which these animals would feed was about 11°C and feeding rates continued increasing up to 25°C, but at 30°C feeding was somewhat reduced and the animals died when temperatures reached 33°C. Feeding was not appreciably affected over a salinity range of 25 to 50 0/00.

c. Rate and conversion of food

While many animals utilize a variety of foods, several factors affect the rate and efficiency of conversion, which can be determined by relating the amount of absorbed food to the amount converted into the growth of the animals. Kinne (1960) has suggested that "conversion of food into body substance and into biologically useful energy" is a sensitive parameter for assessing rates and efficiencies of metabolic processes. The euryhaline fish *Megalops cyprinoides* will feed on both the mosquito fish *Gambusia affinis* and the prawn *Metapenaeus monoceros*, but feeding and digestive rates of *Gambusia*-fed fish as well as efficiency of conversion are higher than for fish fed with the prawn (Pandian, 1967a). The amount of food eaten also affects the efficiency of conversion. The maximum amount of food *M. cyprinoides* consumed in a day proved to be approximately 5% of its body wet weight; if force-fed beyond this level the fish would disgorge the entire meal. Moreover, an increased rate of intake did not result in increased efficiency, and an individual fed the equivalent of 5% of its body weight per day absorbed at about the same efficiency as a fish fed an equivalent of 2% of its body weight per day (Pandian, 1967b). Certain other trends have been noted for efficiency of conversion and the conversion rate. The efficiency of food and nitrogen conversion decreases with increasing body weight; and interspecific differences may be correlated with growth pattern. For example, fish that rapidly grow to a large size have a faster rate of food conversion than species that grow more slowly.

Environmental factors may also affect rate and conversion of food. In a detailed study on food intake and conversion in the flatfish *Limanda limanda,* Pandian (1970) found that besides food quality and body size, temperature also affected food conversion. He also noted that optimum temperatures for maximum food intake shifts downward with increasing body weight. While small individuals will feed very little at 8°C, large females consume maximum amounts at this temperature. The optimum temperature for food conversion efficiency also shifts downward with size, shifting from between 13° and 18°C for immature *L. limanda* to 8°C for mature individuals.

3. Respiration

The success of an animal living in the fluctuating estuarine environment depends on proper functioning of its metabolic machinery. Although in this section we will emphasize the influence of abiotic factors on metabolism, it should be borne in mind that biotic factors also influence estuarine animals. These factors, including body size, acclimation phenomena, sex, starvation, stage of life cycle, and reproductive stage, have been discussed in detail in Chapters I, III, and V.

a. Temperature

Estuarine animals living at different latitudes may experience dissimilar thermal ranges which may be reflected in their metabolic response to temperature. It is not always certain, however, whether observed differences in metabolic response are genetically or environmentally induced. Consider, for example, the study on metabolic adaptation in the widely distributed xanthid crab *Rhithropanopeus harrisii.* This crab is distributed along the east coast of the Americas from New Brunswick, Canada, to northeast Brazil. Field-collected populations from Maine, North Carolina, and Florida all had distinctive metabolic-temperature curves. When metabolic-temperature patterns were determined on laboratory-reared animals from the three populations, however, most of the interpopulation differences disappeared. Therefore, environmentally induced phenotypic variation, rather than genetic factors, accounted for differences in metabolic-temperature patterns between populations (Schneider, 1967).

Metabolic-temperature responses of first-stage zoeae of five species of fiddler crabs from temperate and tropical zones were not distinctively different over the thermal range of 15–35°C. Distinct intraspecific differences between temperate zone larval populations were noted, however (Vernberg and Costlow, 1966) (Fig. 4.12). First-stage zoeae of *Uca pugnax* demonstrated translation to the left, a common response. In contrast, *U. pugilator* larvae from Florida had a

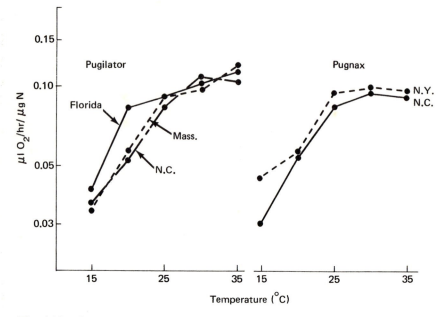

Fig. 4-12. Oxygen consumption of first-stage zoeae of two species of temperate-zone fiddler crabs from latitudinally separated populations. Determinations were made over a wide temperature gradient. (From Vernberg and Costlow, 1966. University of Chicago Press.)

higher metabolic rate at low temperatures than did the population either from North Carolina or Massachusetts. These distinctive differences between populations and species were also noted for the megalops of *U. pugnax* and *U. pugilator* (Fig. 4.13). Again, larval *U. pugilator* did not show the same type of response as noted for *U. pugnax,* but by the time *U. pugilator* had developed to the crab stage, the typical metabolic–temperature response was reported. Therefore, in this species marked ontogenetic differences in temperature-metabolic rates occur.

Rate of metabolism may also vary on a temporal basis, as illustrated in a study on the estuarine amphipod *Bathyporeia pilosa* and *B. pelagica* (Fish and Preece, 1970). *B. pilosa,* which occurs intertidally, had higher metabolic rates over the thermal range of 15–25°C during the winter than in the summer. In contrast, metabolic peaks in the subtidal *B. pelagica* occurred in May and September. These differences were correlated with reproductive behavior since

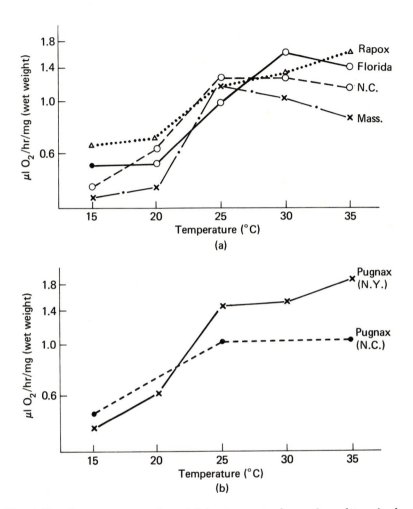

Fig. 4-13. Oxygen consumption of laboratory-reared megalops determined over a wide temperature range. (a) *Uca rapax* from Puerto Rico and various populations of *U. Pugilator.* (b) *Uca pugnax* from New York and North Carolina. (From Vernberg and Costlow, 1966.)

B. pilosa reproduced throughout the winter, but *B. pelagica* did not. However, since there was no observed metabolic adaptation at low temperature, the significance of this response is not clear. Of more demonstrable biological meaning is the marked reduction in metabolic rate observed at higher temperatures during the summer in *B. pilosa*. This response would tend to conserve energy and indicates metabolic regulations (see p. 114 for seasonal adjustments by intertidal animals).

Within the estuary both activity and habitat influence metabolic patterns (Fig. 4.14). In three species of harpacticoid copepods, the lethargic species,

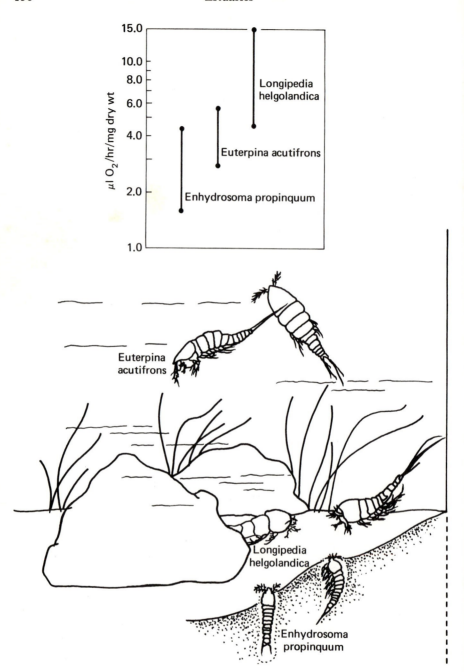

Fig. 4-14. Range of respiratory rates of harpacticoid copepods from different habitats: *Euterpina* is a pelagic species, the other two species are meiobenthic with *Enhydrosoma* being more lethargic than *Longipedia*. (From Coull and Vernberg, 1970.)

Enhydrosoma, had the lowest rate of oxygen uptake while the active species, *Longipedia,* had the highest metabolic rate. The pelagic species, *Euterpina,* was intermediate between these two meiobenthic copepods (Coull and Vernberg, 1970).

Distinctive differences in thermal-metabolic acclimation patterns have been reported within one species. In the copepod *Euterpina acutifrons,* dimorphic male forms occur. One is distinctly smaller than the other, and there are also morphological differences in the antennules, antennae, and second pair of legs. Haq (1965) reported that the dimorphism was genetically controlled and independent of environmental temperature. Moreira and Vernberg (1968) found that cold enhanced metabolic activity at low temperatures (15°C), but not at high temperatures (25°C). In large males the reverse was observed; cold enhanced metabolic activity at high temperatures, but not at low temperatures (Fig. 4.15). Female *E. acutifrons* show metabolic adaptation to both high and

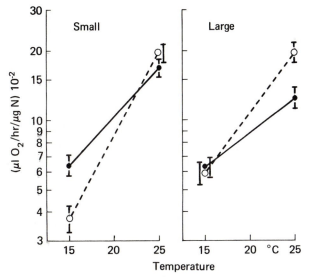

Fig. 4-15. Thermal metabolic acclimation patterns in dimorphic males of *Euterpina acutifrons.*The copepods were cold-acclimated at 15°C (-----), warm-acclimated at 25°C (——). (From Moreira and Vernberg, 1968.)

low temperatures. Thus, the female would appear to be more labile metabolically than either large or small males (Vernberg, 1971). These differences in metabolic responses in the female and dimorphic males could be of value to the species in invading new areas with different thermal regimes.

The metabolic rate of an animal at a heightened state of activity is called *activity metabolism,* while the metabolic rate at a prescribed low rate of locomotor activity is the *standard metabolism.* The difference between these two values is the *scope for activity* (Fig. 4.16) and represents the energy available

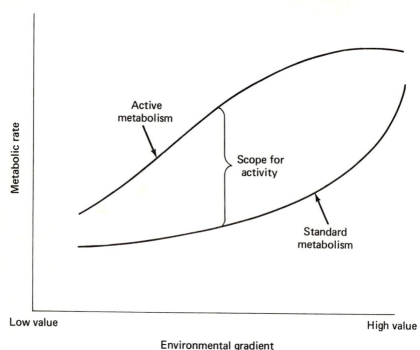

Fig. 4-16. Generalized representation of the influence of an environmental factor on the active and standard metabolic rates of animals. The difference between these two values is the scope for activity. (Based on terminology of Fry, 1964.)

to the organism to meet environmental stress. For example, when the scope of activity value is small, the organism is utilizing most of its energy for maintenance activities and has little left to mobilize for meeting additional demands. At increased temperatures the scope for activity is then typically decreased (Fig. 4.17), as demonstrated by fish (Brett, 1964) and invertebrates (Halcrow and Boyd, 1967). Reduced oxygen content and increased CO_2 tensions also decrease the scope for activity value in fish (Basu, 1959).

  b. Salinity

  Since changing salinities are characteristic of estuaries, it is not surprising that estuarine animals exhibit a wide diversity of metabolic responses to variable salinities. Four general types of metabolic response have been described (Fig. 4.18): (1) The rate of oxygen consumption is not influenced by salinity changes. The crab *Eriocheir* and fish *Fundulus heteroclitus* respond in this manner. (2) The metabolic rate increases in reduced salinity and decreases in high salinities. Examples of this type of response can be found in the polychaete *Nereis diversicolor* and the crab *Carcinus maenas*. (3) In some animals, such as the ghost crab *Ocypode quadrata* and the shrimp *Metapenaeus*, metabolic rates increased in both low and high salinities. (4) In still other species, oxygen

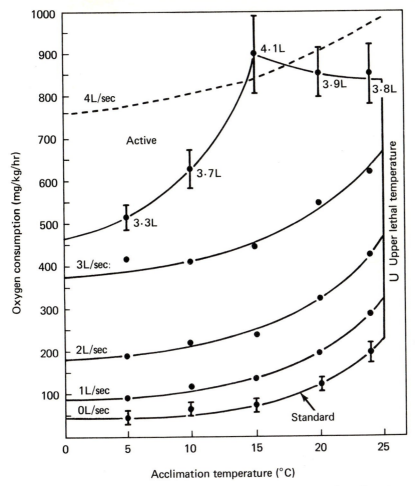

Fig. 4-17. Relation between rate of oxygen consumption and acclimation temperature at various swimming speeds (L/sec = total length per second) for *Oncorhynchus nerka*. The broken line for 4 L/sec is drawn in an area where rapid fatigue would occur since the speed at these temperatures demands a metabolic rate in excess of the active rate. (After Brett, 1964.)

uptake rates decrease whenever the animals are moved to a new salinity. The anthozoan *Metridium marginatum* and the bivalve *Mytilus edulis* are examples of this type of response. These categories are based on the new steady metabolic state the organisms reach when exposed to a new salinity (Kinne, 1967). A species initially shows a transitory metabolic response (an over- and under-shoot response) upon exposure to a new salinity; varying periods of time are required to reach the steady state level. The rate of acclimation to salinity change may be of significance in habitat selection by a species.

Salinity does not have a uniform effect on tissue respiration of estuarine

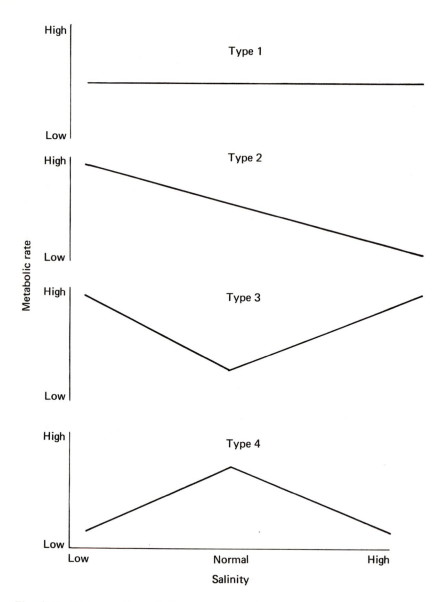

Fig. 4-18. Patterns of metabolic response to salinity change (adapted from scheme proposed by Kinne, 1967. American Association For the Advancement of Science.)

animals. For example, the metabolism of gill tissue of two bivalve species, *Crassostrea* and *Mytilus,* was unchanged over the salinity range of 5 to 30 0/00. In contrast, gill tissue from two other species, *Mercenaria* and *Modiolus,* had higher metabolic rates at low salinities. The reason for interspecific differences is unknown. The results cannot be correlated with the lower salinity boundary of these animals in the estuary. Furthermore, when considering the multiple effects of season, temperature, and salinity on respiration, there was no statistically significant difference in the three-factor interaction in these four species. Seasonal differences in metabolism were extremely variable; winter rates were higher, unchanged, or lower (Van Winkle, 1968).

c. Oxygen tension

Estuarine animals may be faced with fluctuations in the ambient oxygen content which greatly influence their metabolic machinery. Not all species have solved this problem and hence are restricted to highly oxygenated waters. Others, however, have evolved adaptive mechanisms that enable them to live under hypoxic conditions. Some fishes, such as menhaden and mackerel, which invade estuaries during their life cycle, require high oxygen levels; in contrast, bottom-dwelling species—for example, the toadfish *Opsanus tau*—can survive prolonged periods of anoxia. In fact, metabolic characteristics of estuarine species frequently can be correlated with habitat preferences, as illustrated by studies on species of crabs (Van Weel et al., 1954). The oxygen consumption rates of species living in well-oxygenated waters or in sand flats were proportional to the ambient oxygen tension. In mud-dwelling species on the other hand, metabolism was oxygen independent.

Burrowing animals are particularly well adapted to low oxygen levels. Two species of burrowing crabs, *Callianassa californiensis* and *Upogebia pugettensis,* offer a case in point. Both species regulate their metabolic rates over a wide range of oxygen concentrations and have a P_c value below 50 mm Hg. But *Callianassa,* which normally lives under more hypoxic conditions than *Upogebia,* has the lower metabolic rate and the lower P_c value, and can survive anoxia longer. Although anaerobic metabolic pathways are presumed to be active in these animals during anoxic stress, no data are available (Thompson and Pritchard, 1969).

Animals faced with increasing hypoxic conditions need not necessarily appreciably alter their metabolic rate. Instead they might seek out regions that are oxygen-rich in the estuary. Sessile animals may adjust their oxygen uptake rate and/or withdraw within protective devices and wait for more favorable conditions. During such stressful periods, anaerobic metabolism can supply the energy needed to survive for varying periods of time.

4. Circulation

When estuarine animals are exposed to fluctuating environmental variables, the resulting changes in some of the various blood constituents reflect only one aspect of a general environmental adaptive response. For example, blood analyses of the pinfish *Lagodon rhomboides,* which had been exposed to high temperatures in the field, indicated increases in both hemoglobin concentration and red blood cell numbers, but decreases in hematocrit and mean erythrocyte volume (MEV). Thermal acclimation studies in the laboratory gave different results in that the hematocrit value did not change. All estuarine fish do not exhibit the same hematological response since data collected on the field population of the striped mullet *Mugil cephalus* were different from those of the pinfish. The hemoglobin, hematocrit, and number of red blood cells increased, and the MEV was unchanged. Cameron and Cech (1970) suggested that the increased hemoglobin concentrations at high temperature would aid in transport of oxygen to meet needs of increased respiration at elevated temperatures. Greater oxygen-carrying capacity of blood resulted when pinfish were exposed to low ambient oxygen levels, increased exercise, and increased salinity.

Other factors may influence the composition of blood. The concentration of blood proteins in the crustacean *Crangon vulgaris* varied with the stage of the molting cycle and with season. Starvation resulted in a decreased concentration of blood proteins, especially glycoproteins, but this response was reversed after feeding. Further, it was observed that copper accumulated in the hepatopancreas during starvation and the copper concentration in the blood varied with the molt stage (Djangmah, 1970).

The alternate exposure to low oxygen and "normal" oxygen content water greatly influenced the serum protein patterns of some fish but not all (Bouck and Ball, 1965).

The pH of the vascular fluid of the estuarine worm *Amphitrite cirrata* was 7.06 for animals acclimated to 10°C and 6.90 for animals maintained at 15°C, while previous measurements on this species gave values of 7.4 (Mangum, 1970). This difference is important since the "normal" oxygen equilibrium curve for respiratory pigment *of Amphitrite,* as well as many other species, has been determined at pH 7.4. Conceivably, the pH of blood of other species varies with environmental conditions. If so, this would cast doubt on the functional significance of previous studies on respiratory pigments.

When the ambient oxygen tension decreases, the frequency of contraction of the dorsal blood vessel increases in *Nereis diversicolor* (Jürgens, 1935); in contrast, there was no observed pulsation rate change in *Nereis virens* over a wide range of oxygen tensions (Lindroth, 1938). The rate of beating of isolated heart ventricle preparations of the clam *Tivela stultorum* was independent over a wide range of oxygen tensions (Baskin and Allen, 1963).

Various ions differentially influence the heart rate, but the effects of specific ions are not easily determined. In some cases the ratio of ions present may be of more significance than absolute concentrations, since some ions are mutually antagonistic. For example, excess sodium stimulates the pacemaker neurons of crustacean hearts, but this effect is antagonized by calcium, potassium, and magnesium in that order (Bullock and Terzuola, 1957). Although a sizable literature on the effects of ions on heart rate exists, few papers emphasize the environmental effects, especially the multiple interaction of salinity, temperature and oxygen.

As discussed in the section on osmoregulation, the volume of circulating fluids may change with the external salinity, depending in great part on the ability of the organism to regulate water balance.

The reader is referred to the excellent treatises of Prosser and Brown (1961) and Altman and Dittmer (1971) for extensive details on various aspects of circulation, including blood volume, blood pressure, electrocardiograms, and types of hearts.

5. Chemical Regulation and Excretion

a. Ionic and osmoregulation

By its nature, the estuary represents a salinity gradient ranging from fresh to hypersaline waters. Some organisms occupy restricted portions of this gradient while others range over wide sectors. Because of the multiplicity of microenvironments found in estuaries, interaction between environmental factors and functional capacity of organisms to cope with these factors has resulted in a great degree of physiological diversity in estuarine animals. Although a cause and effect relationship between environment and physiological response has not been demonstrated in all species, a trend for the reduction in the number of marine species with decreasing salinity has been reported for invertebrates and vertebrates from different estuarine systems found throughout the world (Ekman, 1953; Gunter, 1945; Wells, 1961). The following discussion will emphasize the functional strategies of chemical regulation and excretion which have evolved to enable organisms to live where they do in estuaries.

Estuarine organisms may escape the stress of a fluctuating salinity by actively seeking a more favorable environment, either by seaward movement or vertical migration. However, species having limited locomotor ability must remain to endure cyclic salinity changes. In general, this latter group of organisms attempt to escape from the harmful effects of their external environment by using a variety of ploys. Some animals simply penetrate the bottom substrate either to occupy existing burrows or to create temporary microhabitats. Others retreat within resistant outer body coverings, such as the hard-shell structures found in barnacles, or the mucus secreted by some fish.

Although the need for the maintenance of an osmoconcentration necessary for sustaining protoplasmic integrity is obvious, the relative concentration of the various osmotically active substances within an organism may vary in response to environmental change. Not only may one species respond differently under different environmental stresses, but different species may also vary greatly in their ability to regulate these substances. In one species, the NaCl level change may largely account for variation in the osmotic concentration of the body fluid, while an equivalent osmoconcentration change may occur in another species for other reasons, such as concentration changes in amino acids, or urea, or other ionic species.

Many estuarine animals are confronted with an external milieu which may be hypotonic to their body fluid. Under these conditions the body fluids of organisms would tend to lose ions and/or gain water. That many organisms are capable of osmo- and ionic-regulation has been known for many years, but, since the development of microchemical and isotopic techniques, an enormous amount of literature about this subject has appeared.

In general, estuarine species exposed to low salinities exhibit several mechanisms for maintaining hyperosmotic body fluids. These include reduced permeability of body surface to salts and/or water, active uptake of salts by the body surface from the external environment into the body fluids, and production of a hypoosmotic urine.

Differences in the permeability of the integument of animals which occupy fresh, brackish, and marine waters have been correlated with the differential salinity regimes they encounter. It has been demonstrated that crabs and polychaetes from fresh water lose less salt and less water through their body covering than estuarine species, which in turn are more resistant than marine species (Herreid, 1969; Lockwood, 1964; Oglesby, 1965; Rudy, 1967; Smith, 1963 and 1970a, b). Typical results are represented in Fig. 4.19 (Herreid, 1969). These adaptive differences undoubtedly reflect genotypic differences between species. Some experimental studies indicate that environmentally induced (phenotypic) changes are observed in some animals as a function of exposure to different salinities. Two examples will illustrate this line of research. First, the euryhaline crabs *Rhithropanopeus harrisii* are able to alter the rate of water loss in response to ambient salinity levels (Smith, 1967, 1970b)—i.e., at low salinity this rate is lower than at higher salinities. Although the adaptive mechanism is not known, several possibilities exist. The permeability of the exoskeleton may be altered, and circulation of the blood may be reduced or there may be a reduction in the flow of the medium bathing the gills or gut. Second, the body shape of an organism may vary with salinity, as seen in hydroids of the same species grown at different salinities. For example, *Cordylophora* from fresh water are shorter and have long and narrow cells compared with colonies maintained at 15 0/00. As a result much less surface area is available to interact with the ambient environment (Kinne, 1958).

In addition to the reduction in exchange between the organism and the ambient environment by means of changes in body wall or body shape, many estuarine animals have met the challenge of salinity change by regulating either their volume or their osmotic or ionic composition. One of the classic examples of volume regulation by an estuarine animal is that reported for the flatworm *Procerodes ulvae* (=*Gunda*). When exposed to distilled water, these animals increase in volume until they eventually burst. In contrast, an animal exposed to natural hard water or calcium-enriched water increases about 70% and then gradually decreases with time because of possible changes in the membrane permeability. These results may be related to calcium present in the ambient water. Also, the epithelial cells lining the gut remove and store

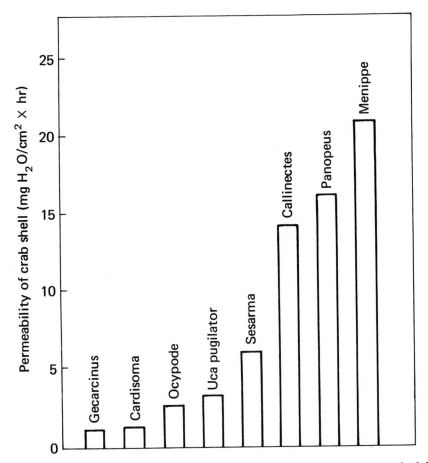

Fig. 4-19. Permeability of crab shells to water in air. Land crabs are on the left, aquatic species on the right. An average of six determinations was made for each species. The average standard deviation among the measurements for each species was ± 30 percent of the mean value. (From Herreid, 1969. Pergamon Publishing Company.)

the excess water in vacuoles until the tide changes (Beadle, 1934; Weil and Pantin, 1931).

Interspecific differences in response to salinity between closely related nereid worms further illustrate the importance of physiological adaptation to environmental stresses. Marine species or those restricted to high salinity water rapidly gain weight when subjected to low salinity water, while estuarine species demonstrate some degree of volume control as well as some osmoregulating ability. Marine species have more permeable body walls than either estuarine or freshwater species (sodium, Fretter, 1955; chloride, Oglesby, 1965, and Smith, 1970a). The species *Nereis limnicola,* which can survive fresh water, is not only the most impermeable to salt loss of the nereid worms, but it also has the ability to acclimate within a few days to low salinities by altering permeability (Oglesby, 1968a). Smith (1964) reported that this species is successful in low salinity water because it produces more urine and has a less permeable integument than marine species. Other soft-bodied animals have shown a similar response. However, organisms such as the spider crab *Libinia emarginata* with a hard body covering are prevented from expanding their volume in a dilute medium, although they may increase in weight.

Some estuarine animals are able to maintain the osmotic concentration of their body fluids at a relatively constant level over a wide range of lowered salinity (hyperosmotic regulation). Some can hyperosmotically regulate and also maintain the osmoconcentration below that of the external medium. For example, estuarine calanoid copepods from Australian estuaries have been shown to regulate hypo- and hyperosmotically with the point of isosmoticity being between 13 and 18 0/00 (Brand and Bayly, 1971).

The brackish-water prawn *Macrobrachium equidens* is capable of both hypo- and hyperosmotic regulation while the freshwater *M. australiense* is a hyperosmotic regulator. Both species can produce urine which is hypoosmotic to blood and both can regulate Na and Cl blood concentration (Denne, 1968).

Other euryhalinic species do not osmoregulate their body fluids to any extent, but instead appear to tolerate tissue dilution and may depend more on cellular osmoregulating abilities. Numerous examples of species which can hyperosmoregulate can be found in various review papers and books (Lockwood, 1964; Kinne, 1964; Prosser and Brown, 1961; Vernberg and Vernberg, 1970). A generalized representation of the comparative osmoregulatory responses of marine, estuarine, freshwater, and terrestrial animals is presented in Fig. 4.20.

Adaptive mechanisms to low salinity have been demonstrated in the Chinese wool-handed crab *Eriocheir sinensis,* which ranges from fresh to sea water (see review of Remane and Schlieper, 1958). The blood of this species is hyperosmotic to dilute sea water. Salts are absorbed from dilute solutions, as demonstrated by sodium being absorbed by isolated gills from a medium with an 8 mM sodium concentration even though the blood within the gills has a 300 mM concentration. The general permeability of this crab to ions and water relative to other crabs is low as measured by the iodine tracer technique. The kidneys are operative in ionic regulation but not osmoregulation. Urine output is reduced, and the urine may be hyperosmotic to blood. Not only has it been demonstrated experimentally that the chief route of salt absorption for animals in reduced salinities is through the gills, but also the fine structural basis for this function has been suggested by Copeland (1968) for both the blue crab *Callinectes sapidus* (an estuarine crab) and the land crab *Gecarcinus lateralis* (Fig. 4.21). The salt-absorbing tissue is found as a highly interdigitated epithelium within the respiratory lamellae of the gills. In the osmoregulatory tissue are found "mitochondrial pumps" which are defined as a close, parallel arrangement of plasma membranes and mitochondrial envelope-membranes.

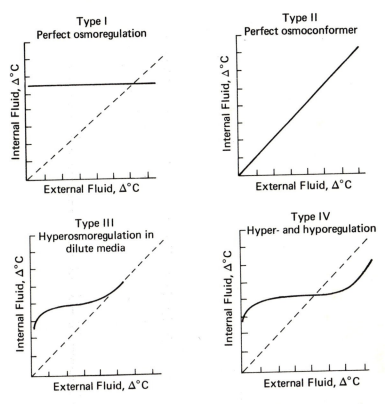

Fig. 4-20. A generalized representation of the comparative osmoregulatory response of marine, estuarine, fresh water, and terrestrial animals.

In another prominent estuarine group of animals, the polychaetes, Oglesby (1969) has reported active uptake of Na^+ and Cl^- by *Nereis diversicolor* at ambient concentrations of 2% and 22% sea water. In addition, a marked decrease in body wall permeability was reported in various species of nereid worms living at low salinities. Smith (1970a) has demonstrated that hyposomotic urine is produced by *N. diversicolor* when exposed to low salinity. The active uptake of various ions to replace those lost in the urine or through the body surface has been demonstrated in various animals including crustaceans (Lockwood, 1962), annelids (Oglesby, 1969), and fish (Hickman and Trump, 1969).

In general, there is a critical concentration of a given ion in the external medium termed the critical value below which the blood concentration cannot be maintained while above this value the transport mechanism is fully saturated. Lockwood (1960) has proposed that in *Asellus aquaticus* the critical value for sodium is 90 μmoles/liter. This value for sodium uptake by animals not as well adapted for freshwater life is higher: *Eriocheir*, 4 mmoles/liter; *Gammarus duebeni*, 10 mmoles/liter; and *Carcinus maenas*, 70 mmoles/liter

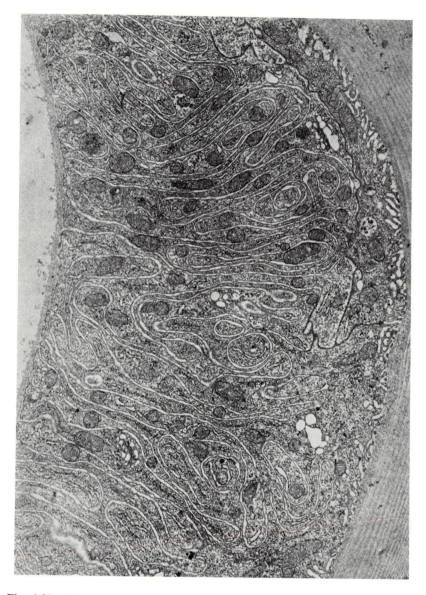

Fig. 4-21. Fine structure of osmoregulatory tissue in gill tissue. Cuticle at right and at left are plasma membrane and granular basement membrane X 12,500. (From Copeland, 1968. American Zoologist.)

(Shaw, 1961a, b; Shaw and Sutcliffe, 1961). The critical value for other ions may be different, and any attempt to relate distribution of an organism to ionic transport mechanisms must consider various ions, not only sodium and chloride (Lockwood, 1962).

The uptake of ions is influenced by temperature as well as by variations in the chemical composition. A temperature decrease of $10°C$ reduces the rate of sodium uptake by half, but apparently does not affect the rate of sodium loss. With time a new steady state is reached, but the blood concentration is below that at the higher temperatures (Lockwood, 1960).

The extremely important question concerning the complex mechanisms of ionic transport is not within the scope of this book, but the reader is referred to a number of excellent reviews (Kirschner, 1970; Potts and Parry, 1964; Rothstein, 1968). Although many significant studies have been published which deal with transport mechanisms, the importance of environmental fluctuations has received minor consideration.

Many estuarine animals, especially molluscs and worms, are classified as osmoconformers. However, it should be kept in mind that all organisms are not perfect osmoconformers or osmoregulators, but demonstrate a response somewhere in between, which is probably correlated with their habitat. Recent work on estuarine molluscs and worms demonstrates this point. The blood of a number of gastropods is hyperosmotic to that of the isosmotic line, with the slope of the curve for blood osmotic concentration being parallel to the isosmotic line (Remmert, 1968; Seelemann, 1968). Data from other studies indicate that some annelids, sipunculids, and bivalves appear to hyperregulate at least over some part of their normally encountered salinity range (Oglesby, 1968a, b; Pierce, 1970). For example, polychaetes inhabiting fresh water or low salinity water regulate better at low salinity than oceanic species. Species differences can be correlated with habitat preference within an estuary as demonstrated by two closely related species of bivalves, *Modiolus demissus*, which is found intertidally in salt marshes, and *Modiolus squamosus*, which is a subtidal species (Pierce, 1971). Not only do these species exhibit habitat differences, but their lethal salinity points are different. *M. demissus* survived in all salinities between 3 0/00 and 48 0/00, but no animals survived in 1.5 0/00 for more than 16 days, while at least 50% of the *M. squamosus* died at salinities below 18 0/00 or above 45 0/00. Although the osmotic concentrations of body fluids for both species were hyperosmotic over the nonlethal salinity range, it was attributed to passive equilibrium rather than active regulation. However, in dilute salinities, these species became volume regulators since water concentration in the tissues remained constant, while there was an intracellular solute loss. The source of solute appeared to be the intracellular amino acid pool which varied in a controlled manner, while the ionic concentration of the body fluids was not regulated. Taurine, alanine, glycine, and proline decreased in concentration more than the other amino acids in the intertidal species; but in the subtidal species, only taurine, alanine, and glycine decreased significantly, whereas the remainder of the free amino acids increased slightly with decreasing salinity. The differential behavior of the free amino acid pool of these two species appears to account for the greater salinity tolerance of the intertidal species, which has more expendable intracellular solute than *M. squamosus* and can maintain its cellular volume over a wide salinity range (Pierce, 1971). Lange (1970) also reported another example in which the relative capability for volume regulation of a series of bivalves could be correlated with their degree of euryhalinity. Stenohalinic species were unable to regulate the volume of their adductor muscles completely, while the euryhalinic species *Mya arenaria* and *Cardium edule* exhibited complete regulation (Fig. 4.22).

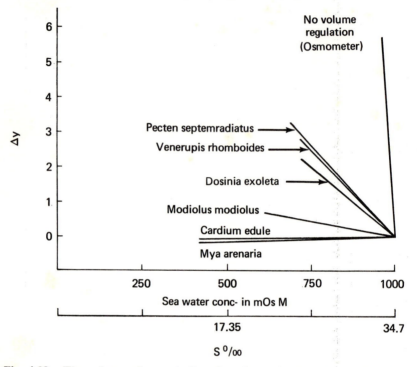

Fig. 4-22. The volume of osmotically adapted muscle tissue in some bivalves in relation to the sea water salinity: the curves are drawn for the salinity range in which the animals have been tested; y gives the increment in the water content which cannot be accounted for by the cellular exchange process between osmotically active solutes and pure water; the muscle volume of animals in sea water of 1000 mOsM is set = 100 percent so that y gives the percent variation of this reference value. (From Lange, 1970. North-Holland Publishing Company.)

Other studies indicate that when the soft-shelled clam *Mya arenaria* is exposed to salinities ranging from 2 to 30 0/00, a linear relationship is noted between the total ninhydrin-positive substances (NPS, an indication of amino acid), in their adductor muscles and the external salinity (Virkar and Webb, 1970). Although immediate changes were observed at 5 0/00 salinity, a period of 48 hours was required for acclimation. The concentration of free amino acids at a salinity of 2 0/00 was 45.85 mmoles/kg of tissue water, 125.26 at 10 0/00, 244.40 at 20 0/00, and 350.72 at 30 0/00. Glycine and alanine were responsible in large part for this linear relationship between NPS and salinity. In nature, this species is not subjected to salinity changes alone but may be confronted with marked temperature changes. DuPaul and Webb (1970) found that the rate of NPS accumulation was faster in warm-acclimated animals (25°C) than those acclimated to 8°C (Fig. 4.23). They suggested this response might be correlated with the need for a faster rate of manufacture of amino acid–

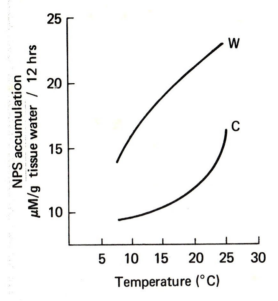

Fig. 4-23. Rate-temperature of NPS (ninhydrin positive substances) accumulation for warm- and cold-acclimated clams, *Mya arenaria*. (From DePaul and Webb, 1970. Pergamon Publishing Company.)

synthesizing enzyme in animals living at higher temperatures, where a more active metabolism occurs.

Intracellular regulation of osmotic pressure has been studied in other euryhalinic species, especially the wool-handed crab *Eriocheir,* in which the free amino acid concentration varies with osmotic concentration of the blood, with marked alterations in alanine and glycine levels. In other crustaceans, alanine and glycine as well as proline are the amino acids most involved in this regulation. Working with isolated nerves of *Eriocheir,* Schoffeniels (1960) demonstrated a similar regulation of free amino acids which suggested that this mechanism is of intracellular origin and that a hormonal mechanism need not be involved. Furthermore, osmotic pressure per se was not responsible for these changes; rather the presence of sodium or potassium is necessary. In stenohalinic species, Florkin and Schoffeniels (1965) suggested that changes in cellular ionic concentration result in a change in the rate of amino acid turn–

over but not in concentration. In euryhalinic forms, differential ionic changes produce an alteration in the equilibrium level.

The influence of salinity stress on the kinetics and substrate-binding property of myosin ATPase from the estuarine crab *Scylla serrata* was demonstrated by Krishnamoorthy and Venkatramiah (1971). They suggested that the kinetic changes imply the existence of a shift in control of an enzymatic reaction due to salinity change. The V_{max} and K_m values gradually drop as adaptation salinity increases, and an Arrhenius plot indicates that the energy of activation for the enzyme increases when the animal is exposed to an abnormal salinity.

Salinity stress influences not only enzyme kinetics, but also nerve conduction. When nerves of the gastropod *Scutus breviculus* were placed in a solution with a cation composition comparable to the minimum concentration of the blood reached during a tidal cycle, the conduction velocity was similar to that observed in normal physiological solution. However, a significant decrease in conduction velocity of all spikes of the compound action potential occurred when the nerve was placed in a medium whose cation concentration was similar to that of blood from animals which had been in 75% sea water for 24 hours. This response could be correlated with the sluggishness and loss of muscle tone of animals exposed to 75% sea water and might explain why this species is not successful in low salinity waters (Tucker, 1970).

Some species have evolved osmoregulating abilities in response to both an external environment and to the symbiont with which it is associated. The pinnotherid crab *Pinnotheres ostreum* is commensal with oysters, and *P. maculatus* is associated with the bay scallop. Unlike *P. maculatus*, which has very limited osmoregulatory ability, *P. ostreum* can hyperregulate at very low salinities. The response of *P. maculatus* is similar to that of the scallop for neither species can cope with salinity below 15 0/00, while *P. ostreum* and its host, the oyster, can tolerate low salinity (Read, 1968).

The ability to osmoregulate may vary with the organism's stage of development. For example, both the larval and adult stages of the euryhaline mud crab *Rhithropanopeus harrisii* osmoregulate in dilute sea water (Kalber and Costlow, 1966). In contrast, the adult fiddler crab can both hyper- and hyporegulate, but the first stage zoeae cannot survive extended exposure to low salinity since a salinity of 20 0/00 will inhibit development (Vernberg, F., 1969).

Sexual differences in osmoregulatory ability have been reported in the blue crab *Callinectes sapidus*, but only under certain environmental conditions. Over the salinity range of 20 to 30 0/00, males and females have similar blood osmoconcentrations, but at the lower salinities of 10 to 20 0/00, the males consistently have lower osmoconcentrations than the females. However, at extremely low (0–1 0/00) or very high (40–50 0/00) salinities, the female, but not the male, showed a breakdown in osmoregulatory ability—a response which might explain the observed preference of males for low salinity water (Ballard and Abbott, 1969).

Environmental factors other than salinity profoundly influence the osmoregulating ability of organisms. Although only a few of these environmental factors

have been studied, there is reason to believe that other environmental parameters are of importance and warrant study. Some work has been reported on
the effects of temperature on osmoregulation. Panikkar (1940) reported that
as temperature increases the minimum blood osmotic pressure is lowered, and
thus osmotic work to maintain a hypertonic blood is reduced at high temperatures. On the basis of this observation, he postulated that in the warm waters
of the tropics it is easier for marine animals to colonize fresh and brackish
waters. However, northern water species which do not migrate from the estuary to offshore waters apparently withstand low salinity and low temperature well, i.e., the osmoregulating capacity of *Rhithropanopeus* and *Gammarus
duebeni* is greater near the lower end of the temperature range (7°C) than at
higher temperatures (20°C) (Kinne, 1952; Kinne and Rotthauwe, 1952).

An inverse relationship between blood osmoconcentration and temperature
and/or season was reported for the blue crab (Ballard and Abbott, 1969) and
for two species of *Hemigrapsus* (Dehnel, 1962). Further, higher blood sodium
levels were reported for blue crabs collected during the winter (Mantel, 1967).
The functional significance of seasonal changes is not entirely known.

When killifish *Fundulus heteroclitus* from sea water were acclimated to
near-freezing temperatures, two principal changes in blood composition occurred: the concentration of inorganic electrolytes increased and the glucose
level increased by as much as sixfold. However, when this species was first
adapted to fresh water and then subjected to a near-freezing temperature, the
serum sodium decreased by 30% and the serum chloride decreased 42%, but
the serum osmolality decreased only 15%. A marked increase in serum glucose
(almost twentyfold) partly offset the reduction in osmolality caused by electrolyte loss. Whether previous exposure of this animal was to fresh water or sea
water, low temperature impaired ionic regulation so that increased glucose appears to play some undefined adaptive role (Umminger, 1971).

Another aspect of osmoregulation important to interaction between physiological adaptation and distribution of estuarine animals is its role in intraspecific competition. Croghan (1961) suggested that as organisms invaded low
salinity waters, mechanisms of osmoregulation evolved, so that energy became
expended by these organisms for osmoregulation. However, if organisms can
increase the efficiency of osmoregulation, they require less energy. Either such
organisms may require less food or this "unemcumbered" energy may be available for other active phases of their life processes. Croghan further suggested
that these organisms that are better adapted to osmoregulate would have a
competitive advantage over other estuarine species. This increased competitive
advantage may be reflected in the distributional differences of animals in an
estuary. For example, some animals have the ability to osmoregulate in low
salinities as determined in the laboratory, and yet they may not be found in
low salinity waters which are occupied by other organisms. It is conceivable that
the species found in these waters is successful because it is a more efficient
osmoregulator, and thus it is more competitive than the other species.

Differences in the osmoregulatory ability of geographically separated populations of one species have been reported. For example, interpopulational differences in the worm *Nereis diversicolor* apparently are a result of acclimation rather than genetic differences (Smith, 1955).

A population sample of the polychaete worm *Nereis limnicola* from fresh water demonstrated a higher degree of hyperosmotic regulation, a lower critical salinity level, and a lower salt permeability of the body surface than animals from an estuary (Oglesby, 1968a). Populations of the isopod *Mesidotea entomon* from the Baltic Sea can be acclimated to low salinities, but not to the fresh water from Scandinavian lakes in which another population of this species lives (Croghan and Lockwood, 1968). Two physiological differences between these two populations were reported. The permeability of the fresh-water animals is greatly reduced over that of the Baltic animals, and secondly, the active transport mechanism in the lake forms has a greater affinity for the uptake of sodium from the surrounding medium than in Baltic Sea animals. The fresh-water crab *Potamon edulis* from Malta differs in response to increased salinity from populations on the European continent (Harris and Micallef, 1971). The Malta animals cannot be acclimated to salinities in excess of 80% sea water, whereas animals from the continent tolerate full-strength sea water. After 12 days of exposure to 80% sea water, the hemolymph concentration reached a steady state and it was hyperosmotic to the medium. However, at higher salinities, the hemolymph was hypoosmotic to the medium and both the sodium and the chloride concentrations were below those of the medium. A short time before the death of these animals, the osmotic pressure and the concentration of sodium and chloride were similar to those of the medium. Thus, it was suggested that the tissues of Malta animals could not tolerate raised hemolymph concentrations as well as those of continental animals.

An example of how organisms differentially adapt to their chemical environment can be seen by comparing the ionic and osmoregulatory ability of three closely related species of fish, genus *Cottus,* which occupy different habitats: *Cottus morio* from fresh water, *C. bubalis* from intertidal and subtidal zones, and *C. scorpius* from deeper oceanic waters (Foster, 1969). The fresh-water species cannot withstand 100% sea water, while the two marine species died in fresh water. The open ocean species *C. scorpius* was the least tolerant of salinity changes, while *C. bubalis* was the most tolerant species. With decreasing external salinity concentration, the total body sodium levels decreased. A marked decrease in plasma osmotic pressure and sodium concentration followed exposure to reduced salinity in the stenohaline *C. scorpius*. Also the body water content of this species increased in reduced salinities, unlike the response of the other two species which apparently regulated water levels. The rate of sodium efflux is less in the fresh-water species when determined at different salinities from the marine species, although all species demonstrated some degree of regulation over sodium loss at low salinity. However, unlike the two marine species, the water flux rate is significantly altered in *C. morio* at different salinities—a response indicating a change in water permeability. Foster suggested that *C. scorpius* cannot live in fresh water because it cannot reduce its salt loss sufficiently to maintain an equilibrium. This reason is probably true for *C. bubalis,* although this fish can partially regulate to lower salinities. In contrast, the fresh-water species is unable to adapt to 100% sea water because it is unable to get rid of the excess salt taken in at higher salinities.

Although present-day reptiles are not a major biotic component of the estuaries, some species are regular inhabitants. One of their major functional problems in this habitat is that of osmotic and ionic regulation. Since the kidneys of reptiles cannot produce urine hyperosmotic to the blood plasma, a different evolutionary strategy has evolved, namely the development of the salt gland, an exocrine gland capable of producing salt solutions more concentrated than sea water (see review by Dunson, 1969). The diamondback terrapin *Malaclemys terrapin,* which lives in estuaries, has a salt gland located in the eye region that is intermediate between terrestrial and marine reptiles in its secretory ability (Dunson, 1970). When this species is acclimated to sea water, sodium concentration of the salt gland secretion increases along with the rate of secretion. Although this gland is operative during osmotic stress periods, Dunson suggested that this species cannot occupy marine waters indefinitely without drinking fresh water, since the extracellular fluid sodium concentration was observed to rise when the turtle was kept in sea water. Preliminary work on the American and "salt water" crocodiles indicates that salt glands are not a major pathway of electrolyte excretion.

Some birds such as the herring gull, which are common along the shores of estuaries and feed on marine organisms, have well-developed functional salt glands located in the nasal region. Thus, excess salt load, especially monovalent ions, is excreted by this gland and the kidneys (Schmidt-Nielsen, 1960). All estuarine-dwelling birds do not have salt glands, as illustrated by the salt marsh Savannah sparrow, which eliminates excess salt loads via the kidneys. Adaptive mechanisms to salt stress in mammals from marine habitats are not well known. The kidneys are probably of paramount importance in eliminating salts, since the urine of marine mammals is strongly hyperosmotic to their blood. It is conceivable that extrarenal pathways for salt elimination may exist, such as the salivary gland.

b. Excretion

Although information is still fragmentary and the influence of environmental stress on excretion is far from being well studied, the nitrogenous waste products of estuarine animals appear to fit into the classical relationship between nitrogen excretion and an animal's habitat described in Chapter I. For example, in seven species of prosobranch snails, all excreted ammonia but not urea. Although these species were from different estuarine habitats and had different dietary requirements, no correlation between these variables and the amount of ammonia excreted could be drawn. However, it was suggested that the uric acid products might serve as a mobile nitrogen depot (Duerr, 1968).

When the crayfish *Orconectes rusticus* is acclimated to sea water, a reversible shift from ammonotelic to ureotelic excretion occurs. This shift toward ureotelism appears to be triggered by an increase in concentration of salts in the animal and not related to the need to denitrify ammonia (Sharma, 1969).

6. Reproduction and Development

In intertidal animals, gamete or larval release must be synchronized with the tidal cycle. The more mobile species move down to the water to release their larvae; sessile organisms must time their release of reproductive products to coincide with periods when the adults are submerged. Subtidal estuarine animals face somewhat different problems since they are always covered with water, and when and where their larvae are released depends in large part on larval tolerance to estuarine conditions. Larvae of sessile estuarine animals and some mobile ones can tolerate a wide range of environmental conditions. Larvae of the Atlantic menhaden *Brevoortia tyrannus,* for example, can survive in salinities ranging from 5 to 30 0/00 at temperatures of 4°C and above (Lewis, 1966). However, many of the estuarine animals migrate long distances offshore at breeding time into the deeper, more stable oceanic waters (Allen, 1966). The principal environmental factors that initiate and influence the reproductive cycles of estuarine animals are temperature, photoperiod, salinity, oxygen tension, and lunar phases.

a. Temperature

As noted previously, the temperature range necessary for reproduction is generally much narrower than the temperature extremes within which other vital bodily functions can occur. Thorson (1950) observed that many aquatic invertebrate animals also do not release their gametes when temperatures are too low to allow early development of the young. An interesting and clear representation of this relationship has been pointed out by Brett (1970) in several species of marine fish, which spawn during temperatures that are about one-fourth to one-third that of the lethal range for these animals. When optimum temperatures for spawning are compared with temperature limits necessary for embryonic survival, spawning temperatures fall into the intermediate position (Table 4.4). Thus, as these fish approach the stage of liberating reproductive cells, they must seek a thermal environment suitable for the embryos to develop.

As is true for animals living in any marine environment where temperatures fluctuate, spawning in estuarine animals can be initiated by changes in the thermal environment. The temperature necessary to trigger spawning may not be the same in geographically separated populations of a species. The Massachusetts population of the bay scallop *Aequipecten irradians,* for example, will spawn when water temperatures reach 14–16°C, but North Carolina and Florida scallops require higher temperatures. Other population differences have been observed in this species, for the Massachusetts scallops will start spawning

TABLE 4.4 *Spawning and Lethal Temperatures (° C) of Some Marine Fishes**

| Species and Locality | Spawning temperature | | Lethal | |
	Range	Optimum	Lower	Upper
Liopsetta pinnifasciata				
Kamchatka Coast	−2 to 4	−1 to 2	—	—
Platessa quadrituberculata				
Kamchatka Coast	−1 to 6	1.5 to 4	—	—
Gadus callarius (morhua)				
Barents Sea to Atlantic	0.4 to 7	1.5 to 4	−2	22 ± 2
Clupea harengus (spring)				
Barents Sea to Atlantic	0 to 12	4 to 9	−1.8	23
Pleuronectes platessa				
North Sea	4 to 7	6 to 7	—	26
Clupea pallasii				
Hokkaido West Coast	4 to 8	5 to 6	—	—
Clupea harengus (autumn)				
North Sea to Atlantic	6 to 15	9 to 13	−1.8	23
Sardinia pilchardus				
English Channel	9 to 16.5	—	—	—
Scomber scombrus				
North Atlantic	10 to 15	—	—	—
Girella nigricans				
California Coast	12 to 19	—	4.6	31
Sardinops melanosticta				
Sea of Japan	13 to 17	14 to 15.5	8	28
Sardinops caerulea				
California Coast	13 to 22	15 to 16	—	—
Plecoglossus altivelis				
Sea of Japan	14 to 19	—	(10)	24
Sardina ocellata				
Southwest Africa Coast	15 to 19.6	—	—	—
Cyprinodon macularius				
Salton Sea (California)	20 to 30 ± 2	—	<9	>40

* Brett, 1970.
John Wiley and Sons, Limited.

when the temperature is increasing, but the more southerly animals spawn only as the temperature decreases after a temperature rise (Belding, 1910; Sastry, 1963, 1968). The maximum amplitude of gonad response is very different in the North Carolina and Massachusetts populations. Maximum amplitude is attained in June in the northern population, but not until near the end of August in the more southern one (Sastry, 1970).

It can be asked whether or not differential responses to temperature between populations are genetically or environmentally induced, but few studies offer answers to this question. One such study is that of Sastry (1966, 1970), who worked with populations of an echinderm, the sea urchin *Arbacia punctulata,*

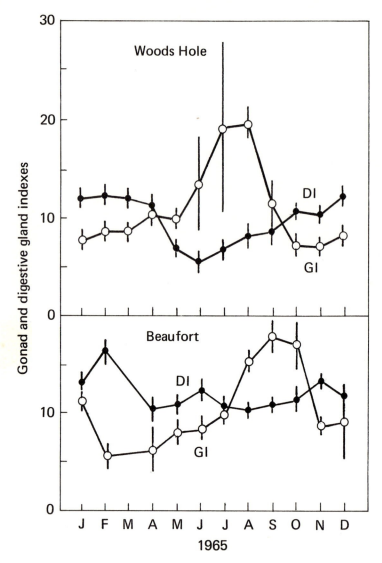

Fig. 4-24. Observed monthly mean gonad and digestive gland index values for the populations of scallops from Massachusetts and North Carolina. Vertical lines show the 95 percent confidence intervals from the mean. (From Sastry, 1968. University of Chicago Press.)

from Massachusetts and from North Carolina. Specimens from Massachusetts typically have two reproductive peaks during the year, while those from North Carolina have only one peak (Fig. 4.24). However, if the Massachusetts animals in the nonreproductive stage are transplanted to North Carolina and maintained under identical conditions with North Carolina *Arbacia*, then the reproductive patterns are identical. These population differences, then, are environmentally controlled and do not reflect genetic differentiation.

Temperature may interact with other factors in the environment, such as photoperiod, to influence phases of the reproductive cycle. In the three-spined stickleback *Gasterostus aculeatus,* for example, gonadal development was found to increase with temperature when the animals were on a 16-hour photoperiod per day at any time between December and May; when the animals were placed on an 8-hour day, minimal gonad development occurred (Baggerman, 1957). The shrimp *Palaemonetes pugio* may be induced to spawn in the winter either by increasing temperature or by combining increased temperature and photoperiod regime. During the experiments, however, only 45% of the females were induced to deposit eggs when temperatures were gradually increased (10° to 25°C), whereas 100% deposited eggs when both temperature and daylength (10.5 to 14.5 hr/day) were increased (Little, 1968).

Temperature may also interact with food supply to influence the reproductive cycle of some animals. When the scallop *Aequipecten irradians* was maintained at 20°C in the absence of food, oocytes did not develop; but if they were fed and kept at 20°C then oocyte development occurred. If, however, the temperature was lowered to 15°C, even when food was present, no oocyte development occurred (Sastry, 1968).

The reproductive potential of the acoelous turbellarian *Archaphanostoma agilis* depends on food supply, for specimens kept without food will first resorb their oocytes, then their sexual organs, and finally the greater parts of their somatic cells. Given an adequate food supply of diatoms of the genus *Nitzschia, A. agilis* constantly produced eggs if maintained at a temperature of 16–18°C (Apelt, 1969). All phases of the reproductive cycle of this turbellarian were greatly influenced by temperature. Although the upper lethal limit of *A. agilis* is 25–26°C, fewer eggs were produced at higher temperatures than at lower ones. For instance, at 22°C egg production was markedly lower than it was at 16–18°C. However, at low temperatures of 5–6°C *A. agilis* survived as well as at 15–18°C, provided that this worm had been adapted to a constant temperature of 5–6°C for a sufficient period of time. But at the slightly lower temperature range of 3–4°C, egg production was lowered considerably and ceased altogether at temperatures ranging from −1° to 1°C.

Concerning reproductive responses of marine fish, Brett (1970) stated: "While in most instances temperature is subordinate to light in its relation to spawning, once the reproductive cell is released its fate is very dependent on temperature within the natural limits of salinity." Indeed, the early developmental stages of most marine organisms are stenothermal; that is, they are incapable of tolerating great fluctuations in the thermal environment. Ushakov (1968) has commented that in multicellular marine animals, the early stages of development including the gametes, zygotes, and developing eggs, have the smallest capacity for nongenetic resistance adaptation of any stage in the life cycle of the individual. Furthermore, temperature tolerance levels of intertidal animals are not necessarily the same throughout the developmental period. Studies on egg developmnt of the teleost fish *Cyprinodon macularius* (Kinne and Kinne, 1962b), for example, showed that during early development (fertilization to gastrulation) there was a period of reduced thermal resistance. This phase was followed by increased resistance and then a second period of low resistance toward the end of embryonic development.

Types of reproduction may also be temperature dependent. The cladoceran *Podon polyphemoides* reproduces both sexually and parthenogenetically, and the sexual reproductive phase is apparently dependent on environmental temperatures. In the Chesapeake Bay, Maryland, parthenogenetic females were found over a wide thermal range, from approximately $5°$ to $30°C$, but males and sexual females occurred only when water temperatures were between $11°$ and $17°C$ (Fig. 4.25) (Bosch and Taylor, 1970).

b. Photoperiod

It has been demonstrated that photoperiod has a profound effect on the reproductive cycles of many animals, but the exact manner in which light initiates these effects is not known. In a discussion on photoperiod effects in marine animals, Segal (1970) has pointed out that it is not known if light is acting directly on the gametes, reproductive structures, neurons serving the oviduct musculature, or an integrative center removed from the reproductive elements.

Because many boreo-Arctic species that have spread into lower latitudes breed with the onset of cold weather, temperature is generally considered to be the major factor in synchronization of these breeding cycles. However, this assumption may not always be true. Consider, for example, populations of the amphipod *Pontoporeia affinis,* which are found in waters ranging from 3 to 35 meters. Despite marked temperature differences between deep and shallow water, the onset of reproduction occurs at approximately the same time in animals found at all depths (Segerstråle, 1970). Previous work had shown that when these animals were brought into the laboratory and put under continuous illumination, the maturation of gonads was inhibited (Segeståle, 1969). In a further series of experiments, two sets of animals were brought into the laboratory and maintained in cool water ($3–6°C$) from the end of May until the end of November; one group was kept in a window aquarium under a natural light-dark cycle, the other group in a darkened container. Although both groups showed reproductive cycling, by the end of November the animals in the darkened aquarium were much further advanced than those maintained in the window aquarium under natural light. On the basis of these studies, Segerstråle (1970) speculated that the decrease in light during late summer triggers the maturation process rather than temperature. This hypothesis seems to be borne out in nature, since *P. affinis* in Baltic waters breeds during both the cold season and warmer periods at depths exceeding 100 m, in spite of low light penetration at these depths.

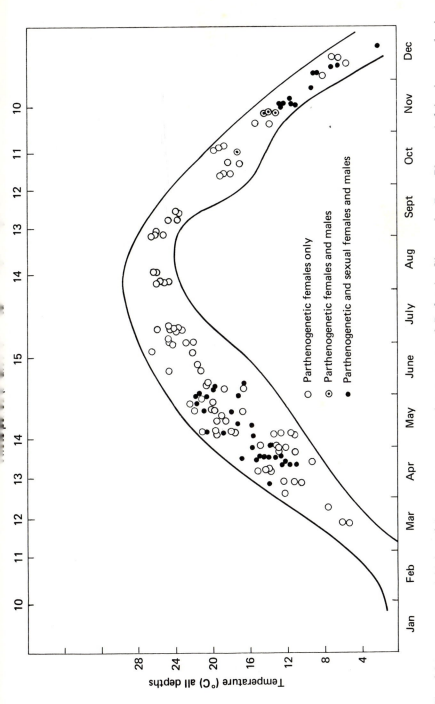

Fig. 4-25. Envelope of high and low temperatures and occurrence of *Podon* in Chesapeake Bay. Photoperiod is given as the interval between sunrise and sunset. (From Bosch and Taylor, 1970.)

Many animals require alternating periods of light and dark to initiate spawning, for when they are kept under constant conditions of light or dark, they will not spawn. For example, it has been known for some time that ascidians spawn in response to light following darkness. In a study on two species, *Ciona intestinalis* and *Molgula manhattensis,* Whittingham (1967) found that a very short exposure to light could initiate spawning. When the animals were exposed to light after a period in darkness, *C. intestinalis* began to shed gametes after only 4.07 minutes' exposure to light, while *M. manhattensis* began spawning 24.16 minutes after they were illuminated. It was found that very high light intensity had no effect at all on inhibition of shedding, but very low light did. A light of only 0.21 foot-candle completely inhibited spawning in *C. intestinalis* and reduced shedding in *M. manhattensis* by three-quarters.

The length of the dark-adaptation period influences the success of spawning. In populations of *C. intestinalis* that had been dark-adapted from 45–55 minutes, then placed in light, only 40% spawned, but if they had been dark-adapted one hour preceding illumination, 78% of the animals spawned (Lambert and Brandt, 1967). The action spectrum for induction of spawning has also been determined for *C. intestinalis*. There were three peaks of effectiveness, with the most effective peak occurring at 416 mμ. The next most effective peak occurred at 550 mμ, and the third peak of 520 mμ was slightly less effective (Fig. 4.26). An interesting comparison may be made between this action

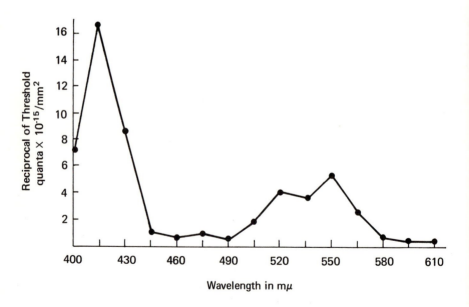

Fig. 4-26. Action spectrum for light-induced spawning of an ascidian, *Ciona intestinalis.* (From Lambert and Brandt, 1967.)

spectrum and that of cytochrome *c*. The close similarity noted between the two suggests that a hemoprotein is a light absorber acting possibly as a chromophore in these animals (Lambert and Brandt, 1967).

c. Salinity

There is much variation in the abilities of estuarine animals to reproduce and grow in the fluctuating salinities characteristic of their environment. Reduced salinity decreases the reproductive potential of many estuarine animals which characteristically migrate to higher salinity waters to breed and spawn. A few species, however, can breed over a very wide range of salinities, and one of the most successful is the brachyuran crab *Halicarcinus australis*, found in the Swan estuary in western Australia (Lucas and Hodgkin, 1970). Salinities in this area range from essentially fresh water in winter to 36 0/00 during the summer, and water temperatures vary from 10–15°C in winter to 25–30°C in summer. Ovigerous females were found throughout the year over the entire salinity range, although the peak breeding season occurred from October to January, when salinities were relatively high.

Some species seek out reduced salinity waters for breeding. Salmon, for example, leave open ocean waters and move inward to fresh water in order to spawn. Panikkar and Aiyar (1939) reported that there are a number of estuarine fish in India which have their peak of breeding activity during the monsoon season when salinities are at the lowest levels. Apparently the lowered salinity serves to stimulate reproductive activity. Even continuous breeders in the estuary spawn at a greater rate after rain, thus confirming the opinion of Panikkar and Aiyar that the fresh water acts as a definite stimulus to spawning.

Salinity in which an animal can breed may also depend somewhat on latitude. For example, the adult polychaete *Mercierella enigmatica* has been found in salinities ranging from 2 to 50 0/00. In the tropics the larvae can settle in salinities that range from 1 to 30 0/00, but their growth is slower in salinities less than 5 0/00 or greater than 33 0/00. The optimal salinity in the tropics appears to be between 6 and 15 0/00 (Hill, 1967). However, a breeding population has been found recently on an exposed Bristol channel shore in water with a salinity of approximately 34 0/00 (Harris, 1970). The discovery of this population led Harris to speculate that in temperate climates the polychaete can breed in high salinities, but in the tropics breeding is limited to low salinities.

Khlebovich (1969) has suggested that the salinity requirements of adult organisms and of their respective larvae and embryos can be grouped into six different categories. (1) Adults can survive salinity reductions as low as 5–8 0/00, but the development of these animals is possible only at a higher salinity. The blue crab *Callinectes sapidus* is an example of such an animal. (2) The larvae can live in salinities as low as 5–8 0/00, but adults require a higher salinity; this type of response is found in marine euryhaline forms, such as the oyster *Crassostrea virginica*. (3) Some adults can live in fresh or almost fresh water, but the larval development is limited to salinities not less than 5 0/00, as illustrated by the polychaete *Nereis diversicolor*. (4) Some organisms cannot tolerate salinities above 5–8 0/00 at any stage of the life cycle. In these animals the salinity resistance of the adults generally is higher than that of developmental stages. These animals are mainly freshwater species. (5) There are larvae that cannot tolerate salinities above 5–8 0/00, whereas the adults can live well at higher salinities. Such species include many freshwater and migrating fish and euryhaline amphibians. (6) Some free-living adults will die at salinities over 5–8 0/00; and endoparasitic larvae require a salinity of not less than 5 0/00 to develop. Khlebovich believes this type of salinity response probably is limited to the family Unionidae.

d. Oxygen tension

The influence of oxygen on the reproductive processes has received little attention in comparison to other environmental parameters. Low oxygen tensions were shown to suppress the reproductive cycle in polychaete worms *Capitella capitata*. When the worms were suspended in water having little or no dissolved oxygen, they did not feed or reproduce, although they did survive for several days. In water with an oxygen content of 2.9 ppm the worms fed, but did not reproduce. Reproduction did not occur until the oxygen content of the water reached 3.5 ppm (Reish and Barnard, 1960).

Oxygen concentration has been shown to influence other phases of reproduction, such as egg fertilization in *Arbacia*. Sperm of *Arbacia*, for example, are irreversibly damaged by oxygen deprivation and become incapable of fertilizing eggs. Although *Arbacia* eggs remain normal for about 8 hours without oxygen, parthenogenesis tends to be stimulated by oxygen lack (Harvey, 1956). Reduced oxygen levels retard the rate of cleavage in a number of species of sea urchins, and cleavage may even be blocked under anaerobic conditions (Harvey, 1956). Eggs of the salmon *Oncorhynchus keta* subjected to subnormal oxygen levels at early incubation stages developed a number of abnormalities (Alderdice et al., 1958).

Too much oxygen can also act to inhibit hatching in some marine animals. In *Fundulus heteroclitus*, Milkman (1954) was able to control time of hatching by bubbling oxygen into a flask containing eggs and sea water. The flask was then stoppered until several days after the expected hatching date. The eggs that were treated in this manner hatched within minutes when they were washed in fresh sea water. This response does not seem to be entirely related to ambient oxygen level, however, but may also be due to the presence of an inhibitory substance in the water.

Since the environment in which estuarine animals live is made up of numerous fluctuating factors, it is not surprising that the combined effects of temperature, salinity, and oxygen can affect embryonic development and egg mortality in some of these euryhaline animals. One example is the euryhaline fish *Cyprinodon macularius* (Kinne and Kinne, 1962a, b). Egg mortality is highest during the first few days after fertilization. A second peak of mortality occurs shortly before and during hatching. Egg mortality is higher in water with 70% air saturation than in fully air-saturated water at temperatures above 20°C. In hypoxial water there is a mortality of 100% at 27.5°C, but in air-saturated water, egg mortality rates reached the 100% level only when temperatures reached 36°C. Embryonic development is also reduced when the levels of dissolved oxygen are reduced. For example, at 27.5°C, 35 0/00 salinity in 100% air-saturated sea water, incubation time is 5.5 days. In 70% air-saturated sea water, incubation time is increased to 6.6 days. If the oxygen saturation of the water is increased artificially by 300%, then incubation time is reduced significantly in high salinities and the ability to withstand high temperatures and high salinities increases.

e. Lunar phases

In his paper on lunar periodicity, Korringa (1957) observed, "The tides and the moonlight itself are probably of a much greater biological significance than we, human beings, are inclined to admit." The spawning migrations of some estuarine animals lend support to Korringa's thesis. Consider, for example, Racek's study (1959) on the breeding migrations of certain species of Australian shrimp. One of these species, the king prawn *Penaeus plebeius* is found in estuaries for 9–11 months during post-larval development. Early in the summer as they approach maturity, the shrimp show a pronounced urge to leave the estuarine environment, as reflected by mass migrations to oceanic waters. Such migrations are essential if they are to reproduce, for shrimp that are prevented from migrating never mature. Shrimp prevented from leaving the estuarine environment for two years were still sexually immature at the end of this period, although they had attained normal adult body length. The migrations from the estuaries are always precisely timed, occurring during the outgoing tide at the dark phase of the moon. Peak migrations occur between the twelfth and sixteenth night after the full moon (Fig. 4.27). After leaving the estuary, the shrimp stay for 2–3 weeks in the shallow part of the inner littoral area at depths of 10–20 fathoms. Then they move suddenly into the outer littoral area where they mate and spawn in depths of water ranging from 50–80 fathoms. During the post-mysis and post-larval stages, the shrimp move back into the estuaries. Racek's studies suggest that these inward movements are also subject to lunar periodicity, since most of the observed post-larvae entered the estuaries during the period between the first quarter and full moon (Racek, 1959).

Some laboratory studies have investigated lunar periodicity in spawning of animals that do not migrate to breed. In the Mediterranean populations of the polychaete *Platynereis dumerilii*, the breeding season extends from March to October and the numbers spawning at the surface of the sea show a monthly periodicity. Maximum spawning occurs during the new moon period, minimum spawning at full moon (Hauenschild, 1960). Under constant illumination, periodicity in spawning ceased; however, if the worms were subjected to a

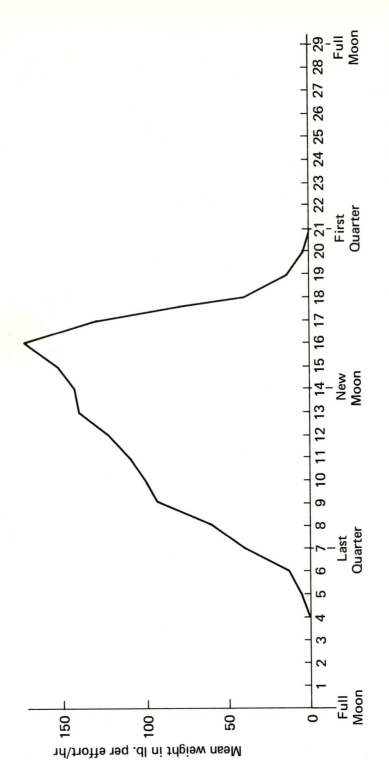

Fig. 4-27. Lunar periodicity in offshore migrations of king prawns as indicated by fluctuations in net catch. (From Racek, 1959. State Fisheries, Chief Secretary's Department, New South Wales.)

12–12 light:dark cycle for 24 days followed by continuous illumination for the remaining 6 days in each month, then periodicity was reestablished, and the worms spawned 16–20 days after the end of the period of continuous illumination. Hauenschild (1966) has suggested that the lunar periodicity observed in the swarming behavior of these worms is due to photoperiodic suppression of cerebral endocrine activity. Other studies on the role of light, endocrines, and swarming periodicity in annelids have been summarized in an extensive review by Clark (1965).

REFERENCES

Ackefors, H. (1969). Seasonal and vertical distribution of the zooplankton in the Askö area (Northern Baltic proper) in relation to hydrographical conditions. *Oikos,* **20:**480–492.

———, and C. G. Rosen (1970). Temperature preference experiments with *Podon polyphemoides* Leuckart in a new type of alternative chamber. *J. exp. mar. Biol. Ecol.,* **4:**221–228.

Alderdice, D. F., W. P. Wickett and J. R. Brett (1958). Some effects of temporary exposure to low dissolved oxygen levels on Pacific salmon eggs. *J. Fish. Res. Bd. Can.,* **15:**229–250.

Allen, J. A. (1966). The rhythms and population dynamics of decapod Crustacea. *Oceanogr. Mar. Biol. Ann. Rev.,* **4:**247–265.

Altman, P. L., and D. S. Dittmer (eds.) (1971). Respiration and Circulation. Federation of American Societies for Experimental Biology, Bethesda, Maryland. 930 pp.

Apelt, G. (1969). Fortpflanzungsbiologie, Entwicklungszyklen und vergleichende Frühentwicklung acoeler Turbellarien. *Mar. Biol.,* **4:**267–325.

Baggerman, B. (1957). An experimental study on the timing of breeding and migration in the three-spined stickleback (*Gasterosteus aculeatus* L.). *Archs néerl. Zool.,* **12:**105–317.

Ballard, B. S., and W. Abbott (1969). Osmotic accommodation in *Callinectes sapidus* Rathbun. *Comp. Biochem. Physiol.,* **29:**671–687.

Baskin, R. J., and K. Allen (1963). Regulation of respiration in the molluscan (*Tivela stultorum*) heart. *Nature.* **198:**448–450.

Basu, S. P. (1959). Active respiration of fish in relation to ambient concentrations of oxygen and carbon dioxide. *J. Fish. Res. Bd. Can.,* **16:**175–212.

Beadle, L. C. (1934). Osmotic regulation in *Gunda ulvae. J. exp Biol.,* **11:**382–396.

Belding, D. L. (1910). A Report upon Scallop Fishery of Massachusetts, Including the Habits, Life History of *Pecten irradians,* Its Rate of Growth and Other Factors of Economic Value. Spec. Rept. Comm. Fish and Game, Mass. Pp. 1–150.

Bishai, H. M. (1962). Reactions of larval and young salmonids to water of low oxygen concentration. *J. Cons. Perm. Int. Explor. Mer.,* **27:**167–180.

Bjornberg, T. K. S., and K. M. Wilbur (1968). Copepod phototaxis and vertical migration influenced by xanthene dyes. *Biol. Bull.* **134:**398–410.

Bosch, H. F., and W. R. Taylor (1970). Ecology of *Podon polyphemoides* (Crustacea, Branchiopoda) in the Chesapeake Bay. Tech. Rep. 66, Chesapeake Bay Institute. 77 pp.

Bouck, G. R., and R. Ball (1965). Influence of a diurnal oxygen pulse on fish serum proteins. *Trans. Am. Fish. Soc.,* **94:**363–370.

Bowden, K. F. (1967). Circulation and diffusion. *In* Estuaries. G. H. Lauff, ed. AAAS Publ. No. 83, Washington, D.C. Pp. 15–36.

Brand, G. W., and I. A. E. Bayly (1971). A comparative study of osmotic regulation in four species of calanoid copepod. *Comp. Biochem. Physiol.,* **38B:**361–371.

Brett, C. E. (1963). Relationships between marine invertebrate infauna distribution and sediment type distribution in Bogue Sound, North Carolina. (Doctoral Thesis) University of North Carolina, Chapel Hill, North Carolina.

Brett, J. R. (1964). The respiratory metabolism and swimming performance of young sockeye salmon. *J. Fish. Res. Bd. Can.,* **21:**1183–1226.

——— (1970). Temperature. *In* Marine Ecology. O. Kinne, ed. Wiley-Interscience, New York. Pp. 515–560.

Bull, H. O. (1931). Resistance of *Eurytemora hirundoides,* a brackish water copepod, to oxygen depletion. *Nature, London,* **127:**406–407.

Bullock, T. H., and C. A. Terzuola (1957). Activity of neurons in crustacean heart ganglion. *J. Physiol.,* **138:**341–364.

Cameron, J. N., and J. J. Cech, Jr. (1970). Notes on the energy cost of gill ventilation in teleosts. *Comp. Biochem. Physiol.,* **34:**447–455.

Carriker, M. R. (1967). Ecology of estuarine benthic invertebrates: A perspective. *In* Estuaries. G. H. Lauff, ed. AAAS Publ. No. 83, Washington. D.C. Pp. 442–487.

Clark, R. B. (1965). Endocrinology and reproduction in polychaetes. *Oceanogr. Mar. Biol. Ann. Rev.,* **3:**211–255.

Collip, J. B. (1921). A further study of the respiratory processes in *Mya arenaria* and other marine mollusca. *J. Bio. Chem.,* **49**(2):297–311.

Copeland, D. E. (1968). Fine structure of salt and water uptake in the land crab, *Gecarcinus lateralis. Am. Zool.,* **8:**417–432.

Costlow, J. D., Jr., C. G. Bookhout, and R. Monroe (1960). The effect of salinity and temperature on larval development of *Sesarma cinereum* (Bosc) reared in the laboratory. *Biol. Bull.,* **118:**183–202.

Coull, B. C., and W. B. Vernberg (1970). Harpacticoid copepod respiration: *Enhydrosoma propinquum* (Brady) and *Longipedia helgolandica* (Klie). *Mar. Biol.,* **5:**341–344.

Creutzberg, F. (1961). On the orientation of migrating elvers (*Anguilla vulgaris* turt.) in a tidal area. *Netherlands Journal of Sea Research,* **1**(3):257–338.

Croghan, P. C. (1961). Competition and mechanisms of osmotic adaptation. *In* Mechanisms in Biological Competition. Symposia of the Society for Experimental Biology. No. 15. Great Britain. Pp. 156–167.

———, and A. P. M. Lockwood (1968). Ionic regulation of the Baltic and freshwater races of the isopod *Mesidotea* (*Saduria*) *entomon* (L.). *J. exp. Biol.,* **48:**141–158.

Cushing, D. H. (1951). The vertical migration of planktonic Crustacea. *Biol. Rev.,* **26:**158–192.

Darnell, R. M. (1961). Trophic spectrum of an estuarine community, based on studies of Lake Pontchartrain, Louisiana. *Ecology,* **42:**553–568.

——— (1967). Organic detritus in relation to the estuarine ecosystem. *In* Estuaries. G. H. Lauff, ed. AAAS Publ. No. 83, Washington, D.C. Pp. 376–382.

Dehnel, P. A. (1962) Aspects of osmoregulation in two species of intertidal crabs. *Biol. Bull.,* **122:**208–227.

Denne, L. B. (1968). Some aspects of osmotic and ionic regulation in the prawns *Macrobrachium australiense* (Holthius) and *M. equidens* (Dana). *Comp. Biochem. Physiol.,* **26:**17–30.

Djangmah, J. S. (1970). The effects of feeding and starvation on copper in the blood and hepatopancreas, and on blood proteins of *Crangon vulgaris* (Fabricius). *Comp. Biochem. Physiol.,* **32:**709–731.

Dodgson R. W. (1928). Report on mussel purification. Fish. Invest. London, **10:**1–498.

Duerr, F. G. (1968). Excretion of ammonia and urea in seven species of marine prosobranch snails. *Comp. Biochem. Physiol.,* **26:**1051–1059.

Dunson, W. A. (1969). Reptilian salt glands. *In* The Exocrine Glands. Stella Y. Botelho, F. P. Brooks, and W. B. Shelley, eds. Univ. of Pennsylvania Press, Philadelphia.

——— (1970). Some aspects of electrolyte and water balance in three estuarine reptiles, the diamondback terrapin, American and "salt water" crocodiles. *Comp. Biochem. Physiol.,* **32:**161–174.

DuPaul, W. D., and K. L. Webb (1970). The effect of temperature on salinity induced changes in the free amino acid pool of *Mya arenaria. Comp. Biochem. Physiol.,* **32:**785–801.

Ekman, S. (1953). Zoogeography of the Seas. Sidgwick and Jackson, Ltd., London.

Enright, J. T., and W. M. Hamner (1967). Vertical diurnal migration and endogenous rhythmicity. *Science,* **157:**937–941.

Fish, J. D., and G. S. Preece (1970). The ecophysiological complex of *Bathyporeia pilosa* and *B. pelagica* (Crustacea: Amphipoda). I. Respiration rates. *Mar. Biol.,* **5:**22–28.

Florkin, M., and E. Schoffeniels (1965). Euryhalinity and the concept of physiological radiation. *In* Studies in Comparative Biochemistry. K. A. Munday, ed. Pergamon Press, London. Pp. 6–40.

Foster, M. A. (1969). Ionic and osmotic regulation in three species of *Cottus* (Cottidae, Teleost). *Comp. Biochem. Physiol.,* **30:**751–759.

Fox, H. M., and A. E. R. Taylor (1955). The tolerance of oxygen by aquatic invertebrates. *Proc. Roy. Soc. London, B,* **143:**214–225.

Fretter, V. (1955). Uptake of radioactive sodium (Na[24]) by *Nereis diversicolor* Mueller and *Perinereis cultrifera* Grube. *J. mar. biol. Ass. U.K.,* **34:**151–160.

Fry, F. E. J. (1958). Temperature compensation. *A. Rev. Physiol.,* **20:**207, 224.

Gray, G. A., and H. E. Winn (1961). Reproductive ecology and sound production of the toadfish, *Opsanus tau. Ecology,* **42:**274–282.

Grindley, J. R. (1964). Effect of low-salinity water on the vertical migration of estuarine plankton. *Nature,* **203:**781–782.

Gross, W. J. (1963). Acclimation to hypersaline water in a crab. *Comp. Biochem. Physiol.,* **9:**181–188.

Gunter, G. (1945). Studies on marine fishes of Texas. *Publ. Inst. Mar. Sci. Univ. Tex.,* **1:**1–190.

Halcrow, K., and C. M. Boyd (1967). The oxygen consumption and swimming activity of the amphipod *Gammarus oceanicus* at different temperatures. *Comp. Biochem. Physiol.,* **23**:233–242.

Haq, S. M. (1965). Development of the copepod *Euterpina acutifrons* with special reference to dimorphism in the male. *Marine Science Laboratories.* Pp. 175–201.

Harris, J. E., and P. Mason (1956). Vertical migration in eyeless *Daphnia. Proc. Roy. Soc. London, B,* **145**:280–290.

Harris, R. R., and H. Micallef (1971). Osmotic and ionic regulation in *Potamon edulis,* a fresh-water crab from Malta. *Comp. Biochem. Physiol.,* **38A**:769–776.

Harris, T. (1970). The occurrence of *Manayunkia aestuarina* (Bourne) and *Mercierella enigmatica* Fauvel (Polychaeta) in non-brackish localities in Britain. *J. exp. mar. Biol. Ecol.,* **5**:105–112.

Harvey, E. B. (1956). The American *Arbacia* and Other Sea Urchins. Princeton University Press.

Hauenschild, C. (1960). Lunar periodicity. *In* Biological Clocks. Cold Spring Harb. Symp. quant. Biol. **25**:491–497.

—— (1966). Der hormonale Einfluss des Gehirns auf die sexuelle Entwicklung bei dem Polychaeten *Platynereis dumerilii. Gen. Comp. Endocr.,* **6**:26–73.

Herreid, C. F., II (1969). Integument permeability of crabs and adaptation to land. *Comp. Biochem. Physiol.,* **29**:423–429.

Hickman, C. P., and B. F. Trump (1969). The kidney. *In* Fish Physiology. W. S. Hoar and D. Randall, eds. Academic Press, New York. **1,** Ch. 2.

Hill, M. B. (1967). The life cycles and salinity tolerance of the serpulids *Mercierella enigmatica* Fauvel and *Hydroides uncinata* (Phillipi) at Lagos, Nigeria. *J. Anim. Ecol.,* **36**:303–321.

Hughes, D. A. (1969). Responses to salinity change as a tidal transport mechanism of pink shrimp, *Penaeus duorarum. Biol. Bull.,* **136**:43–53.

Job, S. V. (1957). The routine active oxygen consumption of the milk fish. *Proc. Indian Acad. Sci.,* **B6**:302–313.

Johannes, R. E., S. J. Coward, and K. L. Webb (1969). Are dissolved amino acids an energy source for marine invertebrates? *Comp. Biochem. Physiol.,* **29**:283–288.

Jürgens, O. (1935). Die Wechselbeziehungen von Blutkreislauf, Atmung und Osmoregulation bei Polychäten (*Nereis diversicolor* O. F. Mull). *Zool. Jb. (Abt. allg. Zool. Physiol. Tiere),* **55**:1–46.

Kähler, H. H. (1970). Über den Einfluss der Adaptationstemperatur und des Salzgehaltes auf die Hitze-und Gefrierresistenz von *Enchytraeus albidus* (Oligochaeta). *Mar. Biol.,* **5**:315–324.

Kalber, F. A., Jr., and J. D. Costlow (1966). The ontogeny of osmoregulation and its neurosecretory control in the decapod crustacean, *Rhithropanopeus harrisii* (Gould). *Am. Zool.,* **6**:221–229.

Keys, A. B. (1931). Chloride and water secretion and absorption by the gills of the eel. *Z. vergl. Physiol.,* **15**:364–388.

Khlebovich, V. V. (1969). Aspects of animal evolution related to critical salinity and internal state. *Mar. Biol.,* **2**:338–345.

Kinne, O. (1952). Zur Biologie and Physiologie von *Gammarus duebeni* Lillj.-V. Untersuchungen über Blutkonzentration an *Heteropenope tridentatus* Maitland (Decapoda). *Kieler Meeresforsch,* **9**:134–150.

—— (1958). Adaptation to salinity variations—some facts and problems. *In*

Physiological Adaptation. C. L. Prosser, ed. Am. Physiol. Soc., Washington, D.C. Pp. 92–105.

———— (1960). Growth, food intake and food conversion in a euryplastic fish exposed to different temperatures and salinities. *Physiol. Zool., 33*:288–317.

———— (1964). The effects of temperature and salinity on marine and brackish water animals. II. Salinity and temperature-salinity combinations. *Oceanogr. Mar. Biol. Ann. Rev., 2*:281–339.

———— (1967). Physiology of estuarine organisms with special reference to salinity and temperature. *In* Estuaries. G. H. Lauff, ed. AAAS Publ. No. 83, Washington, D.C. Pp. 525–540.

————, and E. M. Kinne (1962a). Effects of salinity and oxygen on developmental rates in a cyprinodont fish. *Nature, 193*:1097–1098.

————, ———— (1962b). Rates of development in embryos of a cyprinodont fish exposed to different temperature-salinity-oxygen combinations. *Can. J. Zool., 40*: 231–253.

————, and H. W. Rotthauwe (1952). Biologische Beobachtungen und Untersuchungen über die Blutkonzentration an *Heteropanope tridentatus* Maitland (Decapoda). *Kieler Meeresforsch., 8*:212–217.

Kirschner, L. B. (1970). Refresher course on ionic regulation in organisms. *Am. Zool., 10*:330–436.

Knight-Jones, E. W., and E. Morgan (1966). Responses of marine animals to changes in hydrostatic pressure. *Oceanogr. Mar. Biol. Ann. Rev., 4*:267–299.

Korringa, P. (1957). Lunar periodicity. *In* Treatise on Marine Ecology and Paleoecology. J. Hedgpeth, ed. Geol. Soc. Am. No. 67, Washington, D.C. *1*:917–934.

Krishnamoorthy, R. V., and A. Venkatramiah (1971). Kinetic changes in flexor myosin ATPase of *Scylla serrata* adapted to different salinities. *Mar. Biol., 8*:30–34.

Krogh, A. (1931). Dissolved substances as food of aquatic organisms. *Biol. Rev., 6*:412–442.

Lambert, C. C., and C. L. Brandt (1967). The effect of light on the spawning of *Ciona intestinalis. Biol. Bull., 132*:222–228.

Lance, J. (1962). Effects of water of reduced salinity on the vertical migration of zooplankton. *J. mar. biol. Ass. U.K., 42*:131–154.

———— (1964). The salinity tolerances of some estuarine planktonic crustaceans. *Biol. Bull., 127*:108–118.

Lange, R. (1970). Isosmotic intracellular regulation and euryhalinity in marine bivalves. *J. exp. mar. Biol. Ecol., 5*:170–179.

Lewis, R. M. (1966). Effects of salinity and temperature on survival and development of larval Atlantic menhaden, *Brevoortia tyrannus. Trans. Amer. Fish. Soc., 95*: 423–426.

Lindroth, A. (1938). Studien über die respiratorischen Mechanismen von *Nereis virens* Sars. *Zool. Bidr. Uppsala, 17*:367–497.

Little, G. (1968). Induced winter breeding and larval development in the shrimp, *Palaemonetes pugio* Holthius (Caridea, Palaemonidae). *Crustaceana,* Suppl. *2*:19–26.

Lockwood, A. P. M. (1960). Some effects of temperature and concentration of the medium on the ionic regulation of the isopod, *Asellus aquaticus* (L.). *J. exp. Biol., 37*:614–630.

———— (1962). The osmoregulation of Crustacea. *Biol. Rev., 37*:257–305.

——— (1964). Animal Body Fluids and Their Regulation. Harvard University Press, Cambridge, Mass.

Lucas, J. S., and E. P. Hodgkin (1970). Growth and reproduction of *Halicarcinus australis* (Haswell) (Crustacea, Brachyura) in the Swan Estuary, Western Australia. I. Crab Instars. *Aust. J. Mar. Freshwater Res.,* **21:**149–162.

Mangum, C. (1970). Respiratory physiology in annelids. *Am. Scient.,* **58:**641–647.

Mantel, C. H. (1967). Asymmetry potentials, metabolism and sodium fluxes in gills of the blue crab, *Callinectes sapidus. Comp. Biochem. Physiol.,* **20:**743–753.

Mayer, A. G. (1914). The effects of temperature upon tropical animals. *Pap. Tortugas Lab.,* **6:**3–24. Carnegie Institute, Washington, D.C.

McCutcheon, F. H. (1966). Pressure sensitivity, reflexes, and buoyancy responses in teleosts. *Anim. Behav.,* **14:**204–217.

McLaren, I. A. (1963). Effects of temperature on growth of zooplankton and the adaptive value of vertical migration. *J. Fish. Res. Bd. Can.,* **20:**685–727.

McLeese, D. W. (1956). Effects of temperature, salinity and oxygen on the survival of the American lobster. *J. Fish. Res. Bd. Can.,* **13:**247–272.

Mihursky, J. A., and V. S. Kennedy (1967). Water temperature criteria to protect aquatic life. *Spec. Publ. Am. Fish. Soc.,* **4:**20–32.

Milkman, R. (1954). Controlled observation of hatching in *Fundulus heteroclitus. Biol. Bull.,* **107:**300.

Moreira, G. S., and W. B. Vernberg (1968). Comparative thermal metabolic patterns in *Euterpina acutifrons* dimorphic males. *Mar. Biol.,* **1:**282–284.

Odum, W. E. (1970). Utilization of the direct grazing and plant detritus food chains by the striped mullet *Mugil cephalus. In* Marine Food Chains. J. H. Steele, ed. U. of Calif. Press, Berkeley and Los Angeles.

Oglesby, L. C. (1965). Steady-state parameters of water and chloride regulation in estuarine nereid polychaetes. *Comp. Biochem. Physiol.,* **14:**621–640.

——— (1968a). Responses of an estuarine population of the polychaete *Nereis limnicola* to osmotic stress. *Biol. Bull.,* **134:**118–138.

——— (1968b). Some osmotic responses of the sipunculid worm *Themiste dyscritum. Comp. Biochem. Physiol.,* **26:**155–177.

——— (1969). Salinity-stress and desiccation in intertidal worms. *Am. Zool.,* **9:**319–331.

Pandian, T. J. (1967a). Food intake, absorption and conversion in the fish *Ophiocephalus striatus. Helgoländer wiss. Meeresunters.,* **15:**637–647.

——— (1967b). Intake, digestion, absorption and conversion of food in the fishes *Megalops cyprinoides* and *Ophiocephalus striatus. Mar. Biol.,* **1:**16–32.

——— (1970). Intake and conversion of food in the fish *Limanda limanda* exposed to different temperatures. *Mar. Biol.,* **5:**1–17.

Panikkar, N. K. (1940). Osmotic properties of the common prawn. *Nature,* **145:**108.

———, and R. G. Aiyar (1939). Observations on breeding in brackish-water animals of Madras. *Proc. Indian Acad. Sci., B,* **9:**343–364.

Pierce, S. K., Jr. (1970). The water balance of *Modiolus* (Mollusca:bivalvia:Mytilidae): Osmotic concentrations in changing salinities. *Comp. Biochem. Physiol.,* **36:**521–533.

——— (1971). A source of solute for volume regulation in marine mussels. *Comp. Biochem. Physiol.,* **38A:**619–635.

Potts, W. T. W., and G. Parry (1964). Osmotic and Ionic Regulation in Animals. Pergamon Press, London and Oxford.

Pritchard, D. W. (1967). What is an estuary: Physical viewpoint. *In* Estuaries. G. H. Lauff, ed. AAAS Publ. No. 83, Washington, D.C. Pp. 3–5.

Prosser, C. L., and F. A. Brown, Jr. (1961). Comparative Animal Physiology. Saunders Co., Philadelphia, 2nd ed.

Pütter, A. (1909). Die Ernährung der Wassertiere und der Stoffhaushalt der Gewässer. Fischer, Jena, Germany.

Qasim, S. Z. (1970). Some problems related to the food chain in a tropical estuary. *In* Marine Food Chains. J. Steele, ed. U. of California Press, Berkeley and Los Angeles.

Racek, A. A. (1959). Prawn Investigations in Eastern Australia. State Fisheries, Chief Secretary's Dept., New South Wales. Pp. 5–57.

Read, D. E., Jr. (1968). Salinity Tolerance and Osmoregulation of the Pea Crabs *Pinnotheres ostreum* Say and *Pinnotheres maculatus* Say. Unpubl. M.A. thesis, Duke University.

Reeve, M. R. (1966). Observations of the biology of a chaetognath. *In* Some Contemporary Studies in Marine Science. H. Barnes, ed. Allen and Unwin, Ltd., London. Pp. 613–630.

Reish, D. J. (1955). The relation of polychaetous annelids to harbor pollution. *Publ. Hlth. Rep., Wash.,* **70**:1168–1174.

———, and J. L. Barnard (1960). Field toxicity tests in marine waters utilizing the polychaetous annelid *Capitella capitata* (Fabricius). *Pacif. Nat.,* **1**:1–8.

Remane, A. (1934). Die Brackwasserfauna. *Zool. Anz.,* **7**:34–74.

———, and C. Schlieper (1958). Die Biologie des Brackwassers. E. Schweizerbartische Verlagsbuchhandlung, Stuttgart. 348 pp.

Remmert, H. (1968). Die *Littorina*-Arten: Kein Modell für die Entstehung der Landschnecken. *Oecologia,* **2**:1–6.

Riley, G. A. (1967). The plankton of estuaries. *In* Estuaries. G. H. Lauff, ed. AAAS Publ. No. 83, Washington, D.C. Pp. 316–326.

Rothstein, A. (1968). Membrane phenomena. *Ann. Rev. Physiol.,* **30**:15–72.

Rudy, P. P. (1967). Water permeability in selected decapod crustacea. *Comp. Biochem. Physiol.,* **22**:581–589.

Sastry, A. N. (1963). Reproduction of the bay scallop, *Aequipecten irradians* Lamarck. Influence of temperature on maturation and spawning. *Biol. Bull.,* **125**:146–153.

——— (1966). Variation in reproduction of latitudinally separated populations of two marine invertebrates. *Amer. Zool.,* **5**:374–375.

——— (1968). The relationships among food, temperature and gonad development of the bay scallop, *Aequipecten irradians* Lamarck. *Physiol. Zool.,* **41**:44–53.

——— (1970). Reproductive physiological variation in latitudinally separated populations of the bay scallop, *Aequipecten irradians* Lamarck. *Biol. Bull.,* **138**:56–65.

Schlieper, C. (1966). Genetic and nongenetic cellular resistance adaptation in marine invertebrates. *Helgoländer wiss. Meeresunters,* **14**:482–502.

Schmidt-Nielsen, K. (1960). The salt-secreting gland of marine birds. *Circulation,* **21**:955–967.

Schoffeniels, E. (1960). Origine des acides amines intervenant dans la regulation de la pression osmotique intracellulaire de *Eriocheir sinensis* Milne Edwards. *Archs. int. Physiol. et Biochem.,* **68**:696–697.

Schneider, D. E. (1967). An Evaluation of Temperature Adaptations in Latitudinally Separated Populations of the Xanthid Crab, *Rhithropanopeus harrisii* (Gould) by Laboratory Rearing Experiments. Ph.D. Dissertation, Duke University.

Seelemann, U. (1968). Zur Überwindung der biologischen Grenze Meer-Land durch Mollusken intersuchungen an *Alderia modesta* (Opisth.) und *Ovatella myosotis* (Pulmonat). *Oecologia, 1*:130–154.

Segal, E. (1970). Light. *In* Marine Ecology. O. Kinne, ed. Wiley-Interscience, New York. Pp. 159–211.

Segerstråle, S. G. (1969). Light and gonad development in *Pontoporeia affinis* Lindström (Crustacea, Amphipoda). *Proc. Fourth Europ. Symp. Mar. Biol.*, Bangor, 1969. No pages listed.

———— (1970). Light control of the reproductive cycle of *Pontoporeia affinis* Lindström (Crustacea, Amphipoda). *J. exp. mar. Biol. Ecol., 5*:272–275.

Sharma, M. L. (1969). Trigger mechanism of increased urea production by the crayfish, *Orconectes rusticus* under osmotic stress. *Comp. Biochem. Physiol., 30*: 309–321.

Shaw, J. (1961a). Studies on ionic regulation in *Carcinus maenas* (L). I. Sodium balance. *J. exp. Biol., 38*:135–152.

———— (1961b). Sodium balance in *Eriocheir sinensis* (M. Edw.). The adaptation of crustacea to fresh water. *J. exp. Biol., 38*:153–162.

————, and D. W. Sutcliffe (1961). Studies on sodium balance in *Gammarus dubeni* Lilljeborg and *G. pulex pulex* (L.). *J. exp. Biol., 38*:1–16.

Smith, R. I. (1955). Salinity variation in interstitial water of sand at Kames Bay, Milliport, with reference to the distribution of *Nereis diversicolor. J. mar. biol. Ass. U.K., 34*:33–46.

———— (1963). A comparison of salt loss rate in three species of brackish-water nereid polychaetes. *Biol. Bull., 125*:332–343.

———— (1964). D_2O uptake rate in two brackish-water nereid polychaetes. *Biol. Bull., 126*:142–149.

———— (1967). Osmotic regulation and adaptive reduction of water-permeability in a brackish-water crab, *Rithropanopeus harrisii* (Brachyura, Xanthidae). *Biol. Bull., 133*:643–658.

———— (1970a). Chloride regulation at low salinities by *Nereis diversicolor* (Annelida, Polychaeta). *J. exp. Biol., 43*:75–92.

———— (1970b). The apparent water-permeability of *Carcinus maenas* (Crustacea, Brachyura, Portunidae) as a function of salinity. *Biol. Bull., 139*:351–362.

Stephens, G. C. (1963). Uptake of organic material by aquatic invertebrates. II. Accumulation of amino acids by the bamboo worm, *Clymenella torquata. Comp. Biochem. Physiol., 10*:191–202.

———— (1967). Dissolved organic material as a nutritional source for marine and estuarine invertebrates. *In* Estuaries. G. H. Lauff, ed. AAAS Publ. No. 83, Washington, D.C. Pp. 367–373.

———— (1968). Dissolved organic matter as a potential source of nutrition for marine organisms. *Am. Zool., 8*:95–106.

Subrahmanyam, C. B. (1962). Oxygen consumption in relation to body weight and oxygen tension in the prawn *Penaeus indicus* (Milne Edwards). *Proc. Indian Acad. Sci., 55*:152–161.

Tagatz, M. E. (1961). Reduced oxygen tolerance and toxicity of petroleum products to juvenile American shad. *Chesapeake Sci.,* **2**:65–71.

Theede, H. (1965). Vergleichende experimentelle Untersuchungen über die Zelluläre Gefrierresistenz Mariner Muscheln. *Kiel. Meeresforsch.,* **21**:153–166.

Thompson, R. K., and A. W. Pritchard (1969). Respiratory adaptations of two burrowing Crustaceans, *Callianassa californiensis* and *Upogebia pugettensis* (Decapoda, Thalassinidae). *Biol. Bull.,* **136**(2):274–287.

Thorson, G. (1950). Reproductive and larval ecology of marine bottom invertebrates. *Biol. Rev.,* **25**:1–45.

Tucker, L. E. (1970). Effects of external salinity on *Scutus breviculus* (Gastropoda, Prosobranchia). II. Nerve conduction. *Comp. Biochem. Physiol.,* **37**:467–480.

Tundisi, J., and T. M. Tundisi (1968). Plankton studies in a mangrove environment. V. Salinity tolerances of some planktonic crustaceans. *Bolm. Inst. Oceanogr. S. Paulo,* **17**:57–65.

Umminger, B. L. (1971). Osmoregulatory role of serum glucose in freshwater-adapted killifish (*Fundulus heteroclitus*) at temperatures near freezing. *Comp. Biochem. Physiol.,* **38A**:141–145.

Ushakov, B. P. (1968). Cellular resistance adaptation to temperature and thermostability of somatic cells with special reference to marine animals. *Mar. Biol.,* **1**:153–160.

Van Winkle, W., Jr. (1968). The effects of season, temperature and salinity on the oxygen consumption of bivalve gill tissue. *Comp. Biochem. Physiol.,* **26**:69–80.

Van Weel, P. B., J. E. Randall, and M. Takata (1954). Observations on the oxygen consumption of certain marine Crustacea. *Pacif. Sci.,* **8**:209–218.

Vernberg, F. J. (1969). Acclimation of intertidal crabs. *Am. Zool.,* **9**:333–341.

———, and J. D. Costlow, Jr. (1966). Studies on the physiological variation between tropical and temperate zone fiddler crabs of the genus *Uca*. IV. Oxygen consumption of larvae and young crabs reared in the laboratory. *Physiol. Zool.,* **39**:36–52.

———, C. Schlieper, and D. E. Schneider (1963). The influence of temperature and salinity on ciliary activity of excised gill tissue of molluscs from North Carolina. *Comp. Biochem. Physiol.,* **8**:271–285.

———, and W. B. Vernberg (1970). The Animal and the Environment. Holt, Rinehart and Winston, Inc., New York.

Vernberg, W. B. (1969). Adaptations of host and symbionts in the intertidal zone. *Am. Zool.,* **9**:357–365.

——— (1972). Metabolic-environmental interaction in marine plankton. *Archivo di Oceanografia e Limnologia.* In press.

Virkar, R. A., and K. L. Webb (1970). Free amino acid composition of the soft-shell clam *Mya arenaria* in relation to salinity of the medium. *Comp. Biochem. Physiol.,* **32**:775–783.

Waldichuk, M., and E. L. Bousfield (1962). Amphipods in low oxygen marine waters adjacent to a sulphite pulp mill. *J. Fish. Res. Bd. Can.,* **19**:1163–1165.

Weil, E., and C. F. A. Pantin (1931). The adaptation of *Gunda ulvae* to salinity. The water exchange. *J. exp. Biol.,* **8**:73–81.

Wells, H. W. (1961). The fauna of oyster beds, with special reference to the salinity factor. *Ecol. Monographs,* **31**:239–266.

Whittingham, D. G. (1967). Light-induction of shedding of gametes in *Ciona intesti-nalis* and *Molgula manhattensis*. *Biol. Bull.,* **132:**292–298.

Williams, R. B., M. B. Murdock, and L. K. Thomas (1968). Standing crop and importance of zooplankton in a system of shallow estuaries. *Chesapeake Sci.,* **9:**42–51.

Winn, H. E. (1964). Observations on the reproductive habits of darters (Pisces-Percidae). *The Amer. Midland Naturalist,* **59:**190–212.

Woodhead, P. M. J. (1966). The behavior of fish in relation to light in the sea. *Oceanogr. Mar. Biol. Ann. Rev.,* **4:**337–403.

Sandy beaches and sea stacks on the Oregon coast (Courtesy of Dr. P. J. DeCoursey).

CHAPTER V

COASTAL AND OPEN OCEAN WATERS

The entire water mass covering the bottom, or benthic zone, of the ocean is generally termed the pelagic zone. It is arbitrarily divided into two major subdivisions: the coastal waters or neritic zone, and the open ocean waters. Although the dividing line between these two areas is not well defined, the water over the continental shelves is considered neritic (Hedgpeth, 1957). The range of variation in physical factors typically is greater in coastal water than in oceanic water. These factors vary seasonally as well as geographically, for run-off from land and freshwater systems results in sediment, salinity, and nutrient fluctuations. Since coastal areas are shallow (less than 200 meters), wave action stirs up bottom sediments, allowing mixing and recycling of nutrients. The chemical composition of waters near the shore also varies more than that of oceanic water.

Since the coastal and oceanic water masses are not entirely separated from each other, the edge of one frequently will invade the other. This intrusive water may retain its physical and biotic characteristics for some time. One such area of environmental flux is the Cape Hatteras area off the North Carolina coast. In this area the northerly flowing Gulf Stream, with its relatively warm water, frequently invades coastal waters, where it may alter the thermal characteristics of the region, depending on the season of the year.

Light is one of the most important environmental factors in the upper layers of coastal and open ocean waters. The depth of the photic zone depends on latitude, season, and amount of particulate matter. Even within the photic zone, the quality of light changes because of differential penetration of the wavelengths. Generally, the yellow-green wavelengths penetrate inshore waters, and the blue-green wavelengths reach deep oceanic water.

Plankton dominate the pelagic regions. Phytoplankton are the primary producers, zooplankton the primary herbivores.

A. RESISTANCE ADAPTATIONS

1. Temperature

Unusually prolonged periods of thermal extremes, both high and low temperatures, have resulted in the mass mortality of marine organisms. The following examples will demonstrate the extent of this widespread phenomenon. At the northern limit of their distribution in Florida, many tropical fish have been killed by severe cold spells (Storey and Gudger, 1936), and in Great Britain, the severe winter of 1962–1963 caused great mortality to its marine fauna (Crisp, 1964). Unusually high temperatures also may result in the death of oceanic species, as observed off the coast of Peru where the irregular appearance of a warm countercurrent is lethal to the cold-adapted fauna of Peru (Gunther, 1936). Although most tropical animals normally experience high temperature, mass mortalities of Puerto Rican tropical reef animals were observed after exposure to the unusually high temperature which occurred during the extreme mid-day low tides of the spring and summer seasons. Laboratory studies confirmed that water temperature, which remained at 40°C for prolonged periods, was lethal to many species, especially echinoids (Glynn, P., 1968). In many cases, the thermal environment of tropical marine animals is very near their upper lethal levels.

Irregular thermal variations, as well as "regular" summer high and winter low temperature fluctuations in a given region, may act to limit the distribution of animals. Studies on various coastal animals illustrate how high summer temperatures restrict their distribution into lower latitudes. Resident populations of *Mytilus edulis* (a bivalve) are found in the cooler waters north of Cape Hatteras, and when water currents are favorable, larvae are carried south where they settle and grow. However, as soon as summer temperatures reach about 27°C, all these mussels die (Wells and Gray, 1960). It has also been noted that the southern distribution of three species of North Atlantic actinians can be correlated with their upper lethal thermal limits (Sassaman and Mangum, 1970). Elevated water temperature limits the southward spread of two species of cottid fish on the Pacific coast by the partial inhibition of their osmoregulatory ability (Morris, 1960). In contrast, tropical animals are thought to osmoregulate with greater ease at high temperature (Panikkar, 1940).

At the other end of the thermal spectrum, the normal low winter temperatures found in higher latitudes are known to be lethal to tropical species. For example, the region of Cape Hatteras represents a zone of marked thermal discontinuity, since here the northerly flowing warm water of the Gulf Stream comes into dynamic contact with the southerly flowing, cooler waters of the Virginian coastal current. The offshore fauna south of the Cape could survive high temperature better than cold-adapted fauna found north of the Cape. Moreover, these warm-water species could not survive either the low winter

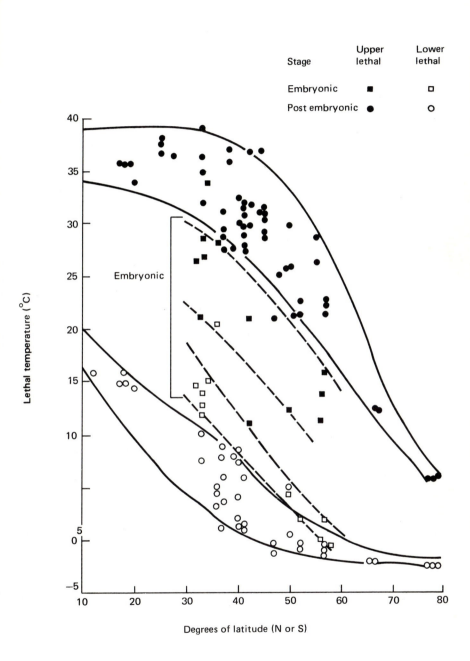

Fig. 5-1. General distribution of upper and lower lethal temperatures for embryonic and post-embryonic fishes in relation to latitude. (From Brett, 1970. John Wiley and Sons, Limited.)

temperature of 4°C, which is characteristic of the water mass north of the Cape, or the low winter temperature of 10°C of the waters between the Gulf Stream and the shore south of the Cape. The few species which have wide geographical limits were able to shift their lethal limits by seasonal acclimation phenomena. Similar general trends were exhibited by larval crustacea from this area of the sea: zoeae of warm-adapted species were less resistant to both low temperature and low salinity than cold-adapted species (Vernberg, F., and W. Vernberg, 1970). Recently, another study demonstrated that tropical animals survived lower temperatures better than previously reported. Generally it was assumed that reef-building coral could not tolerate temperatures below about 16°C, but Macintyre and Pilkey (1969) found these corals on the continental shelf off North Carolina where the winter temperatures are as low as 10.6°C.

Extreme examples of how temperature differentially restricts the distribution of marine animals are found in the comparative studies involving polar and warm-water oceanic animals. The upper lethal temperature of Antarctic animals is lower than the lower lethal temperature of warmer water species. Some fish die of heat death at 6°C (Somero and DeVries, 1967), and an amphipod died at about 12°C (Armitage, 1962). Brett (1970) has summarized much of the existing data on the thermal limits of embryonic and post-embryonic fish from different latitudes (Fig. 5.1).

Temperature extremes also influence the distribution and migratory pattern of oceanic species, such as the squid *Loligo,* which migrates great distances. Apparently temperature is one of the principal factors governing its movements, since this species is restricted to water temperatures of 8°C or higher (Summers, 1969). Utilizing data on tagged lobsters, Cooper and Uzmann (1971) suggested that permanent populations are found on the continental edge and slope where they attain a larger size than inshore lobsters. However, during the summer the temperature of this offshore habitat is too low to permit extrusion and hatching of eggs, molting, and subsequent mating, and thus lobsters migrate shoreward to warmer waters.

To study the influence of temperature on the physiological ecology of oceanic animals, it is not necessary to examine intact organisms. Schlieper and coworkers (1960) have admirably demonstrated that the response to temperature of the cilia of isolated gill pieces from lamellibranchs gives an excellent index of cellular resistance to environmental stress. The upper lethal limits of tissue from species living at various depths are indicated in Fig. 5.2. Tissue from deep-water species living at a depth of 100 meters was less resistant to high temperature than tissue from species living near the shore or at intermediate depths. These responses reflect the thermal characteristics of the organism's habitat, the species from deeper water do not experience the relatively higher temperatures which animals encounter near the shallow coastal regions.

The thermal resistance of some animals may be shifted by acclimation to various temperatures, but this is not true for all species. For example, the upper lethal thermal limits of the American lobster were increased with increasing

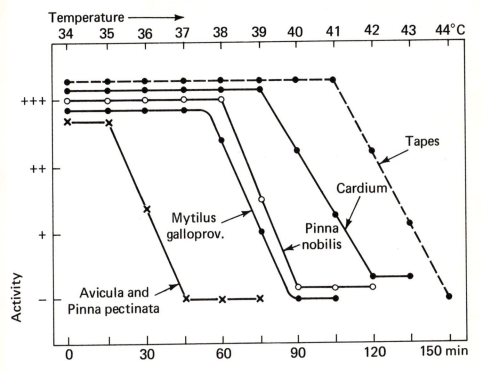

Fig. 5-2. Comparison of cellular thermal resistance in the gill tissue of some bivalves from different depths. *Avicula* and *P. pectinata* are cold, deep water forms; *Mytilus, Tapes,* and *Cardium* are shallow water animals; *P. nobilis* is found at medium depths. (From Schlieper *et al.,* 1960).

temperature (McLeese, 1956), but the upper lethal limits of certain fish were not altered by thermal acclimation (Dean, 1971).

2. Salinity

Despite salinity fluctuations in coastal waters, especially near the mouth of an estuary or areas of freshwater run-off, ocean salinity generally does not vary sufficiently to limit the distribution of oceanic species by acting as a lethal factor. However, this generality does not imply that all oceanic species have the same tolerance to salinity, for some open ocean zooplankton can withstand the low salinities found in estuaries although they are not found there (Hopper, 1960). For the most part, many oceanic species cannot withstand or be acclimated to an appreciable reduction in salinity: most reef corals die within 24 hours in 50% sea water. Many echinoderms do poorly in low salinity waters and are restricted to the open ocean or higher salinity portions of estuaries. For example, the adult sea urchin *Strongylocentrotus purpuratus* cannot tolerate 70% sea water for three hours (Giese and Farmanfarmaian, 1963). However, within some animal groups, diversity in salinity tolerance is noted between species. For example, the shark *Scyliorhinus canicula* died in 60% sea water within a few hours (Alexander et al., 1968), while other species have invaded fresh water.

Adaptation to salinity has been observed at the tissue level, since the gill tissue from deep-water (100 m) bivalves is less resistant to low salinity than species from shallow coastal waters or lagoons (Fig. 5.3) (Schlieper et al., 1960).

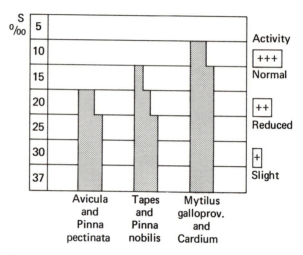

Fig. 5-3. Comparison of cellular osmotic resistance in the gill tissue of some bivalves from different depths. (From Schlieper *et al.*, 1960.)

These responses correlate well with the salinity regimes in each type of habitat. At greater depths the salinity is high and stable, but in lagoons the salinity ranges at least from 37 0/00 to 15 0/00.

On a broad biogeographical basis the salinity tolerances of organisms are of importance, as illustrated by studies involving the Black Sea fauna. Some present-day fauna in the Black Sea have immigrated from the high salinity waters of the Mediterranean Sea, and through time have adjusted to the lower salinities (the high surface salinities are approximately 22 0/00). A number of these immigrant species have low salinity tolerance levels and die if placed in full strength sea water (35 0/00). For example, the barnacle *Balanus improvisus* survives in 3 0/00 salinity, *Palaemon squilla* lives at 8 to 20 0/00, and the optimum salinity for the mysid *Gastrosaccus sanctus* is 18 0/00 (various papers of Pora and coworkers cited by Caspers, 1957).

At any one geographical site the salinity tolerance of an organism probably plays a role in its ability to penetrate estuaries, but the cause and effect relationship is not always clear. For example, 112 species of fish have been reported along the Texas coast in salinities above 30 0/00, 73 species in salinities below 20 0/00, and only 39 species in salinities below 5 0/00 (Gunter, 1945).

3. Oxygen

The lower lethal oxygen concentration for fish varies with habitat. The relatively active scup *Stenotomus chrysops* died when the oxygen tension fell below 16 mm Hg, while another more sluggish fish, *Opsanus tau*, lived for 24 hours in water with an oxygen tension between 0–1 mm Hg (Hall, 1929). Other studies have examined the lower lethal limits for juvenile fish: for the American shad *Alosa sapidissima* and the coho salmon *Oncorhynchus kisutch*, this level is an oxygen concentration of 2 ppm (Tagatz, 1961; Davison et al., 1959). An extensive summary of the lower oxygen limits of salmonids was published by Townsend and Earnest (1940). Although not unexpected, Das (1934) reported that air-breathing fish died when prevented from reaching the water's surface—one species died within 40 minutes, the second survived 15 to 20 hours. The cod *Gadus callarias* died when the oxygen content was approximately 2.7 ml of O_2/liter, a value consistent with the distribution of this species (Sundnes, 1957). The zooplankton of coastal regions have been found in low oxygen waters, but apparently do not survive anaerobic conditions.

Within one sea, interspecific differences in anoxic resistance reflect the different habitats species occupy. For example, in the Clyde Sea harpacticoid copepods, which are restricted to the surface layer of bottom muds, survive anoxia for only 24 hours, while free-living nematodes found deeper in the bottom muds survive anaerobiosis for a month (Moore, 1931).

If a species is to be successful in extending its distributional limits, it must adjust to its newly encountered oxygen regime. An excellent example of this phenomenon, as it relates to a broad zoogeographical problem, was reported by Caspers (1957). In the Black Sea–Azov Basin he found that the present-day fauna, which originated in the Caspian Sea, have not adjusted to the lowered oxygen content found there. Most of these species exist in high oxygen regions while the anoxic habitats are inhabited by species which originated in the Mediterranean Sea.

Temperature influenced the survival of the copepod *Calanus finmarchicus* in reduced oxygen concentrations. For example, at low temperatures a lower oxygen level was tolerated. The hypoxic resistance of this species also changes with different life cycle stages and sex; Stage V is more resistant than the adult, and females are more tolerant than males (Marshall et al., 1934).

Generally, the absence of oxygen in marine habitats is correlated with the presence of sulfides and H_2S. This condition normally occurs in regions with a soft substratum which are poorly aerated, but occasionally it may occur in other habitats, depending on a number of factors, especially water circulation. Frequently these abnormal occurrences cause mass mortality of benthic organisms. Brongersma-Sanders (1957) reported eight cases of massive "kills" apparently caused by oxygen deficency and/or poisonous gases, such as H_2S. As a result of an unusual disappearance of oxygen, much of the benthic fauna from the southern Baltic Sea died (Tulkki, 1965). In addition to "natural"

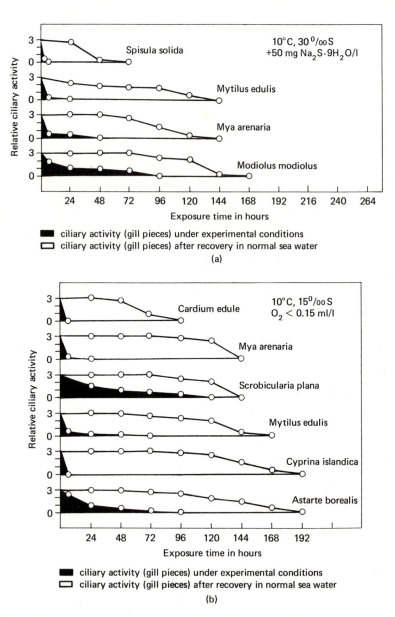

Fig. 5-4. (a) Relative ciliary activity and survival capacity of isolated gill tissues of different bivalve species from the North Sea (⁓ 30 0/00 S) in O₂ deficient water to which 50 mg Na₂S x 9 H₂O/1 had been added. (b) Relative ciliary activity and survival capacity of isolated gill tissues of different bivalves from the Baltic Sea (⁓ 15 0/00 S) in O₂ deficient water. (c) Relative ciliary activity and survival capacity of isolated gill tissues of different bivalves from the Baltic Sea (⁓ 15 0/00 S) in O₂ deficient water to which 50 mg Na₂S x 9 H₂O/1 had been added. (From Theede *et al.*, 1969.)

field disasters, experimental work presents data on the resistance capacity of various marine animals to H_2S. Using animals from the North Sea and Baltic Sea, Theede and coworkers (1969) investigated the resistance of both intact benthic invertebrates and their isolated tissues to oxygen-deficiency and increased H_2S. Results involving whole organisms permit the following generalization: organisms which inhabit mud flats or other soft bottoms are more resistant to anoxia and to increased H_2S levels than inhabitants of sandy bottoms and the phytal zone (Table 5.1). This response can be directly correlated with the levels of oxygen and/or H_2S in these respective habitats. In general, similar interspecific relationships were observed at the cellular level, although the resistance of whole organisms was greater than that of isolated tissues (Fig. 5.4). Apparently, H_2S produces toxic effects by forming insoluble sulfides with heavy metals, and especially influences the cytochrome-oxidase system. Although the resistance capacity of the whole organism is of prime importance in predicting the ability of a species to exist in a sea altered by man, a knowledge of the functional mechanisms influenced is imperative to understanding and controlling the success of organisms in a changing environment.

4. Multiple Factor Interaction

Although environment is the sum of many abiotic and biotic factors inter-

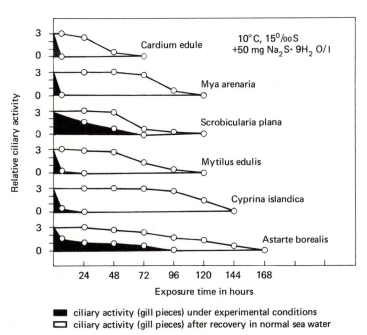

(c)

TABLE 5.1

Resistance of Isolated Gill Tissues of Different Bivalve Species from North Sea and Baltic Sea to Oxygen-deficiency (< 0.15 ml O_2/l) and Hydrogen Sulphide (addition of 50 mg $Na_2S \cdot 9H_2O/l$). Temperautre: $10°C$*

Medium	Species	Salinity ($°/_{oo}$)	Number of Surviving Gill Pieces of 10 Animals after Exposure Times (h) Indicated						
			24	48	72	96	120	144	168
<0.15 ml O_2/l	*Mytilus edulis*	30	10	10	10	10	10	4	0
		15	10	10	10	10	10	2	0
	Mya arenaria	30	10	10	10	10	6	4	0
		15	10	10	10	10	8	0	
	Cardium edule	30	10	10	7	5	0		
		15	10	9	4	0			
$+50$ mg $Na_2S \cdot 9H_2O/l$	*Mytilus edulis*	30	10	10	10	9	5	0	
		15	10	10	9	5	0		
	Mya arenaria	30	10	10	9	7	2	0	
		15	10	10	9	3	0		
	Cardium edule	30	10	6	0				
		15	9	2	0				

* Theede et al., 1969.

acting constantly, frequently each factor acts independently. The organism not only exists in this dynamic, fluctuating complex, but also it is a part of it. To assess the impact of the environment on an organism adequately, data on multiple factor interaction are needed. Most studies cited in this book deal with the influence of one factor on an organism, and although these data are of value, they do not adequately describe the limits of organismic resistance to environmental stress.

One method of depicting the toleration limits of organisms is the tolerance polygon which has been developed by Fry and his coworkers (see Fry, 1964). An illustration of the application of this procedure is found in Fig. 5.5. Although this method is intended only to show the two-dimensional effect of temperature acclimation on thermal lethal limits, other factors, such as oxygen or salinity, also could be used to demonstrate acclimation-lethal limits interaction. Two factors, such as temperature and salinity, may influence survival, as has been graphically shown by Zein-Eldin and Aldrich (1965) (Fig. 5.6). In general, two factors may synergistically interact to produce a lethal effect

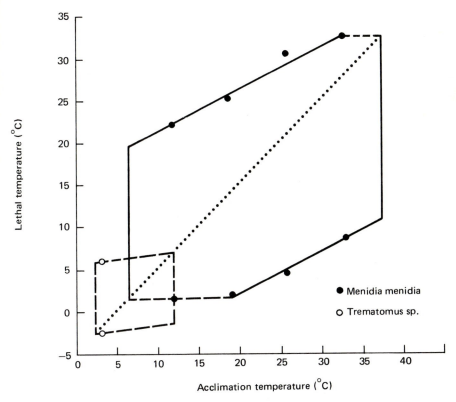

Fig. 5-5. Provisional zones of temperature tolerance illustrating extreme difference between a temperate species *Menidia menidia* from the Atlantic coast and a polar species *Trematomus* sp. from the Antarctic. Broken lines represent likely relation; dotted construction line is where lethal temperature equals acclimation temperature. (From Brett, 1970. John Wiley and Sons, Limited.)

where either acting separately would not result in death. For example, the post-larval shrimp *Penaeus aztecus* can survive 10°C and a salinity of 35 0/00, and it also can withstand exposure to 8 0/00 salinity and 20°C. However, a combination of 8 0/00 and 10°C is lethal.

When the effects of three environmental factors on lethal limits of an organism are considered, the zone of compatibility of an animal may be greatly reduced. One important paper dealing with the influence of three-factor interaction on the zone of lethality is that of McLeese (1956) on lobsters. Fig. 5.7 represents a summary of the combined effects of oxygen, temperature, and salinity on survival. Obviously, simultaneous exposure to sublethal concentrations of these three factors may cause the death of lobsters. These complex interactions are of practical importance in answering environmental problems, such as establishing water quality criteria. For example, fiddler crabs are much

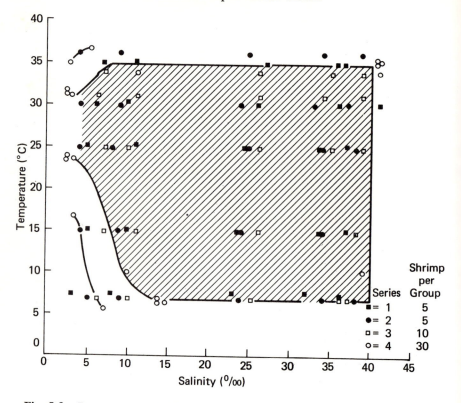

Fig. 5-6. Percent survival of *Penaeus aztecus* postlarvae after 24 hours at indicated levels of salinity and temperature. (From Zein-Elden and Aldrich, 1955.)

more sensitive to mercury when they are stressed by high temperature and low salinity than when temperature and salinity are optimal (Vernberg and Vernberg, 1972). Obviously, the permissible level for mercury in coastal waters determined for animals under optimal conditions of temperature and salinity would not truly reflect the ability of a species to survive this dosage of mercury when stressful temperature and salinity are also encountered.

B. CAPACITY ADAPTATIONS

1. Perception of the Environment

a. Orientation and environment selection

Coastal and open ocean waters are generally much clearer than those of estuaries. Light penetrates to great depths, and within the photic zone light perception is probably the most widely used sensory modality. In the clear waters of the open ocean, the intensity of light due to penetration of sunlight is 10^1 to 10^5 W/m² at 300–700 m; on a bright day, light probably would be

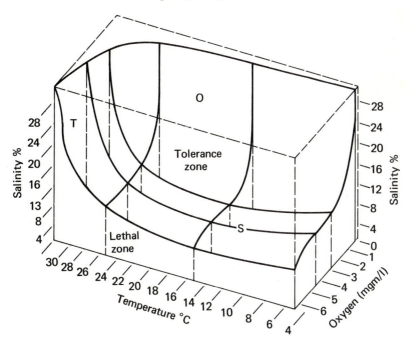

Fig. 5-7. Diagram of the boundary of lethal conditions for lobsters for various combinations of temperature, salinity and oxygen.

T, region in which temperature alone acts as a lethal factor; S, region in which salinity alone acts as a lethal factor; O, region in which oxygen alone acts as a lethal factor. (From McLeese, 1956. Information Canada.)

visible down to about 900 m (Clarke and Denton, 1962). The depth of light penetration can be correlated with certain morphological characteristics of animals, as illustrated in studies on crustacea and fish in the North Atlantic (Foxton, 1970; Badcock, 1970). Above 700 meters these animals characteristically were transparent or semitransparent. Crustaceans often had large dorsal red or orange chromatophores, while the fish tended to be silvery and light reflective. Some of both groups had dermal photophores. The silvering and transparency serve to make the animals relatively invisible, thus lowering their susceptibility to predation. The function of the photophores is less clear, but it has been suggested that these aid in reducing the silhouette of the organisms. The down-dwelling light in the ocean is highly directional, and the visual threshold of mid-water animals very low; thus, a fish or a crab could be quite visible to predators below them. If the prey species has ventral photophores, then the emitted light shining down would act as a countershading mechanism to erase the silhouette (Fraser, 1962; Clarke, 1963; Hastings, 1971).

Below the 700 meter depth, the crustaceans tended to be more or less uniform in pigmentation, varying from deep red to brown reds; some had dermal photophores (Foxton, 1970). Fish were nonreflective, and rather than silvery, they were brown to black (Badcock, 1970). In the western North Atlantic, it has been found that during the day bioluminescence flash rates increase with depth, the maximum occurring at 800–1,000 m (Clarke and Hubbard, 1959; Clarke and Baccus, 1964). Thus, the chance of a fish encountering an illuminating flash increases with depth, and such an increase in bioluminescent activity will nullify any advantage that a silvery reflecting surface may have. Deep-sea fish tend to have greater pupil size and higher pigment density than shallow-water fish, and some have a retina consisting purely of rods, which increases sensitivity to light (Blaxter, 1970).

Many oceanic animals in the pelagic zone undergo diurnal migrations. The vertical migrations appear to be primarily related to sonic deep-scattering layers (Clarke, 1966; Boden and Kampa, 1967). The animals seem to adjust their depth and migrations to some optimum submarine light regime and thus migrate in response to changes in light intensity. Migrating species occupying the upper levels in a community tend to migrate to shallow depths at night, while those living in deeper water communities migrate only up to mid-depth. Nearly all observations on the open ocean organisms are by necessity based on plankton tows or trawl catches at sea, since it has been nearly impossible to keep these animals alive under laboratory conditions.

The sun is used as an orienting or navigational device by some animals, such as rainbow and purple parrot fishes (*Scarus guacamaia* and *S. coelestius*) from Bermuda. These fish live in offshore caves at night and migrate out distances of 400–500 meters during the day to feeding grounds along shores and in bays. When the fish were captured and then released in different and presumably unfamiliar areas, they swam mainly in a southeasterly direction (90° to 180° north), which was the direction from their feeding grounds to their offshore caves. If the fish were released at night or released with opaque eye cups, they circled with no definite directional orientation. If the fish were released during the day and a cloud partly covered the sun, they circled until the cloud had passed by, then swam in a southeasterly direction. When the daily light cycle was slowed by 6 hours, the fish responded to actual rate of change of the azimuth (expected 165° shift clockwise), and not to a constant 15° per hour change (expected 90° shift). On the basis of these results, the authors concluded that the fish were using sun-compass orientation, that is, they were maintaining a constant angle of orientation between their bodies and the sun (Winn et al., 1964).

While some animals undoubtedly orient to the sun, Blaxter (1970) has pointed out that most studies illustrating this behavior have been done on non-migratory animals and do not necessarily show that they are navigating or that they can locate their position on the earth's surface by a true coordinate type of navigation. There has been some speculation that migratory fish utilize sun orientation, but evidence is mainly circumstantial. In fact, one of the most intriguing puzzles in marine biology today is how oceanic animals are able to make long-distance migrations. The migrations of salmon are among the most extensively studied and, although information about their migration patterns and behavior is considerable, controlling mechanisms are still unknown. Recently, Royce and coworkers (1968) summarized some of the information about salmon migrations, and these are given in the following account. The precisely timed migrations are performed by each individual only once in its lifetime, and if the salmon becomes lost or departs from the required time schedule, then the chance of spawning successfully is lost. Therefore, it seems likely that the navigational system is inherited through the kinds of responses elicited by various stimuli. Salmon migrate very long distances, swimming near the surface of the ocean, mainly in the upper 10 m. Pink salmon from southeastern Alaska or British Columbia, for example, cover as much as 5,000 kilometers or more in a 12–15 month period, and even greater distances may be traveled by other species. Because the migration is mainly circular, the salmon are unable to use memorized stimuli that could be followed back in reverse order. Moreover, since salmon do not school as a group in the ocean, migration appears to depend upon individual behavior. The fish are almost continuous travelers, moving at speeds faster than most of the ocean currents that carry them; consequently, migration tends to be a form of active behavior rather than a passive drifting. Termination of the long migration takes place on a remarkably precise time schedule and the arrival of the salmon is less variable than the seasonal change in the weather. Thus it seems unlikely that the timing of salmon migrations is governed by critical temperatures in the water in which they are distributed. If it were, the arrival date could vary as much as two weeks. Migratory routes seem to be unrelated to land or continental shelves, and the normal migratory routes appear to be across open water, yet the directness of the final migration does not appear to be random. When fish that have been tagged and released from different points along a line are recaptured, they tend to proceed directly toward their destination. Since many of the migration routes cross different ocean domains far removed from mixing of home stream waters, olfactory sense would not seem to provide significant guideposts except at the end of the route. Royce and his coworkers also rule out orientation by the sun as a migrating stimulus since the salmon migrate day and night through violent and prolonged ocean storms, often through fog, mist, or haze. Water movement is also dismissed as an orienting mechanism, because rheotropisms in current orientation need some kind of a stationary reference point, such as a shoreline, and none would be available in the open ocean. Although experimental evidence is lacking, Royce and coworkers (1968) suggest instead that electrical receptors in the lateral line allow the fish to depend on cues from ocean currents.

It is possible that orientation to polarized light may be used by some animals in environmental selection, for certain arthropods and fish show distinct reactions to polarized light. Since the pattern of polarization of light from the sky is dependent on the sun's position, directional preferences can be changed by altering polarization patterns. Thus, when the west Pacific half-beak *Zenarchop-*

terus was placed under conditions in which the plane of polarization was at $0°$, $45°$, $90°$, and $135°$ relative to the sun's bearing, the first three polarization planes all evoked significantly different orientation patterns (Waterman and Forward, 1970).

Although as a general rule, oceanic animals use light to accomplish directional orientation in the environment, there is evidence that many marine mammals depend upon sound production interacting with the environment, so that the return of a signal gives these animals orienting information. The generation of sound waves for orientation is termed *echolocation*. A good example of orientation by echolocation can be found in the study by Kellogg (1961) on captive porpoises. A net barrier was stretched across the pool with two openings, one covered with a clear sheet of plexiglass. When forced from one end of the pool to the other, the porpoises selected the unobstructed opening 98% of the time. Porpoises produce high frequency sounds by rapid movements of air from one nasal air sac to another through special valves. They produce intermittent sound bursts even when they are cruising in exceptionally clear water.

Fish are also able to utilize sound in exploration of the environment. Sharks, for example, are known to be able to detect and orientate to sound at least 600 feet away (Wisby and Nelson, 1964), and field observations suggest that they probably utilize sound to locate struggling fish (Limbaugh, 1963).

b. Predator-prey relationships

The sensory modalities utilized in predator-prey relationships depend largely on the animals' habitat and their daily behavior pattern. Predatory animals often depend on vision to locate prey over relatively short distances. It is primarily movement that evokes food-seeking behavior, and, if the object falls within certain size limits, the predator will often ignore color, shape, or pattern of the prey animal. Selection of prey is generally a function of the habitat, or may be based on more specific determination by chemical or other means once the prey is caught. For long distance perception or nocturnal activity, vision is of less importance than other sensory modalities; vision is also of less importance in night feeding animals.

Prey animals have evolved a number of passive mechanisms to conceal themselves from their predators. They include the blending in with the environment which is observed in mid-water fish having silvery surfaces and dermal photophores for countershading, disruptive or cryptic coloration which serves to break up the outlines of the entire body or a particular structure, specific resemblance to objects commonly found in the environment such as occurs in the small Sargasso fish that are nearly indistinguishable from the Sargasso weed in which they live, concealment by burrowing or hiding in a crevice, and possession of a protective exoskeleton or toughened skin. Some prey species have also evolved active defensive adaptations. Adaptations of molluscan prey species have recently been reviewed by Ansell (1969) and offer an illustration of the variety of these responses. A number of species rely on autotomizing certain body parts to distract predators. Some gastropod species are able to autotomize the posterior part of the foot when they are attacked; bivalves autotomize part of the siphon or the mantle edge. The severed organ serves to delay the predator while the animal makes its escape. Behavioral responses of three superfamilies of prosobranch gastropods to potential predators have so much in common that Ansell has suggested that they are expressions of a behavior pattern of great antiquity. These include extension and waving of the cephalic and mantle tentacles, extension of the mantle lobes, elongation of the columellar muscle or "mushrooming," twisting of the shell back and forth through an arc of up to 180°, the turning of the foot to direct subsequent locomotion away from the stimulus, acceleration of the pedal locomotor response or "wave-running," the change of gait or "galloping," inversion and violent movement or "leaping," and movement upward to vertical surfaces or out of the water.

Other potential prey animals rely on chemical protection from predators. In molluscs the glands that produce these repellent secretions are under sensory control and release the secretion by muscular action. False visual-orienting clues are another tactic employed by prey animals in response to a predator's presence. Squid and octopus, for example, eject a black cloud of ink that distracts the predator and gives the prey time to escape. It has been suggested that the ink also may serve as an olfactory anesthetic, destroying chemoreceptor ability in those predators that rely on this modality for prey recognition (MacGinitie and MacGinitie, 1949).

Nocturnal animals may utilize chemoreception in long distance perception of their prey, as illustrated in studies on two species of moray eels, *Gymnothorax moringa* and *G. vicinus* (Bardach et al., 1959). These are nocturnal reef predators that rarely leave their hiding places during the day. In laboratory experiments, blinded morays located food as quickly as fish with sight intact. The behavior patterns of morays with plugged nares, however, were quite different, and these morays took much longer to locate the food. Once the food was approached, receptors other than olfaction were used. The moray touched it with its snout, and taste stimuli apparently elicited a grasping response. Tactile receptors may also play a role in the last response.

Chemoreception can also be used to locate food in animals lacking a well-developed visual sense. A recent study on the starfish *Asterias rubens* illustrates how precise this ability can be (Castilla and Crisp, 1970). When starfish were given a choice between two streams of water, one carrying an effluent and the other not, the starfish was able to navigate toward or away from the effluent-bearing stream, depending on the nature of its odor. When the effluent carried odors of potential prey species, such as barnacles or mussels, the starfish moved toward it. If, on the other hand, the effluent carried odors of its predator, the starfish *Solaster papposus, A. rubens* moved away from that stream.

Porpoises can locate prey animals by echolocation (Kellogg, 1961). When high frequency pulses were emitted every 15–20 seconds and a dead fish was dropped into the tank after one of these pulses the echoes from the next series

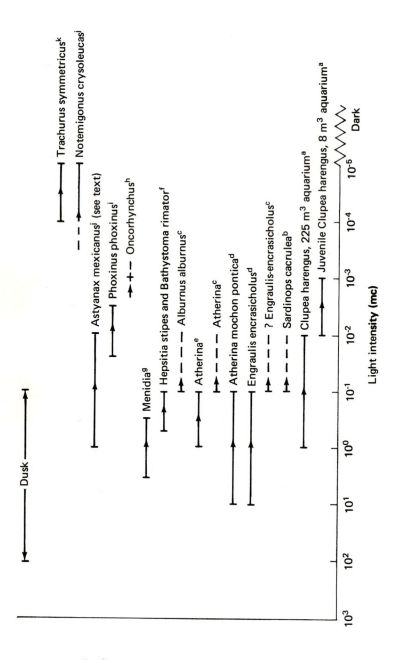

Fig. 5-8. Range of light intensities over which schooling drops. ----, complete range not determined; arrows indicate direction of decrease. (Based on various sources, from Blaxter, 1970. John Wiley and Sons, Limited.)

of pulses seemingly indicated to the porpoise that the environment had changed, for it then emitted an almost continuous sound train until it had located and seized the prey.

c. Communication

Communication between marine organisms is accomplished through visual, olfactory, and auditory cues. In schooling behavior, vision is generally considered the primary, but not necessarily the exclusive, mode of establishing contact between species. Oceanic fish that have a silvery top and bottom often have black spots on the side or on the flanks that could be used to maintain visual contact with each other (Denton and Nicol, 1966). Schooling of these fish usually occurs over a certain range of light intensity (Fig. 5.8), and only infrequently in darkness. Tight schools of *Anchoviella choerostoma* and *Caranx-latus* have been found in the dark; Moulton (1960) also found that the blinded *Anchoviella* have been observed to school as long as they were with moving fish. Olfaction is probably a secondary factor in schooling behavior.

Audition is widely used in communication in fish. Although it has been known for several centuries that fish can also produce sounds, the prevalence and significance of this phenomenon has been appreciated only within the past 20 years. Some of the pioneering work on sound production in fish has been done by Fish and co-workers (1952, 1954), who established a number of facts about the underwater sounds of fishes. There may be a seasonal and daily cycle in sound production. Sound production may be restricted to one sex, usually the male. Sounds are produced within a particular behavioral context, and sounds possess qualitative species specificity. Further work has shown that sound can stimulate courtship activity, attract females to males, stimulate aggressive and escape behavior, and serve as warning signals in escape and investigatory activities (Winn, 1964).

Sounds by fish are not finely organized, and only major moods and responses can be obtained. Sonograms of various types of fish sounds are given in Fig. 5.9. The squirrel fish *Holocentrus rufus* has been found to make two different sounds, grunts and staccatos (Winn et al., 1964). The grunt was usually associated with an aggressive reaction, such as chasing fish. Each individual of the squirrel fish maintains a territory, usually a crevice in a reef. When another similar-sized fish enters this territory, the squirrel fish responds with grunting sounds and attempts to chase the intruder away. When a large strange object such as a large fish comes into view, the squirrel fish utters a staccato sound that is associated with escape patterns, because the squirrel fish usually retreats into its crevice at the same time. An interesting aspect of sound production in these fish is that it follows a daily cycle. Generally, they produce few sounds a night or during the day, but at dawn and dusk they produce large numbers. Both the staccato and grunt sounds follow a similar pattern (Fig. 5.10). It has been hypothesized that there is a relationship between decreased light (with its concomitant decrease in vision of surrounding objects) and the increased activity in the squirrel fish, causing them to move and meet objects

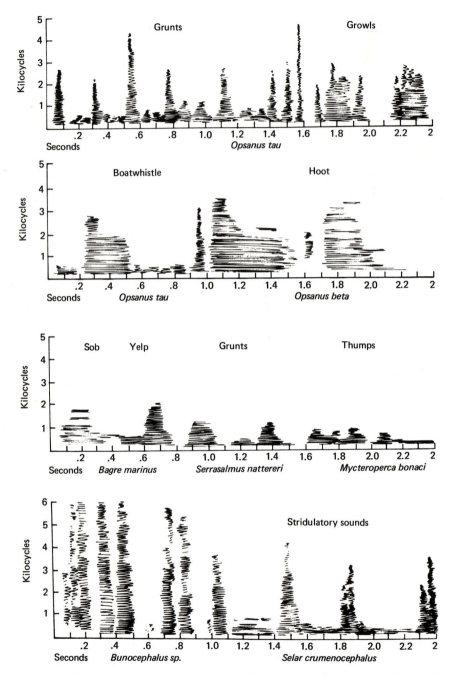

Fig. 5-9. Sonograms of representative samples of the various types of fish calls. (From Winn, 1964. Pergamon Publishing Company.)

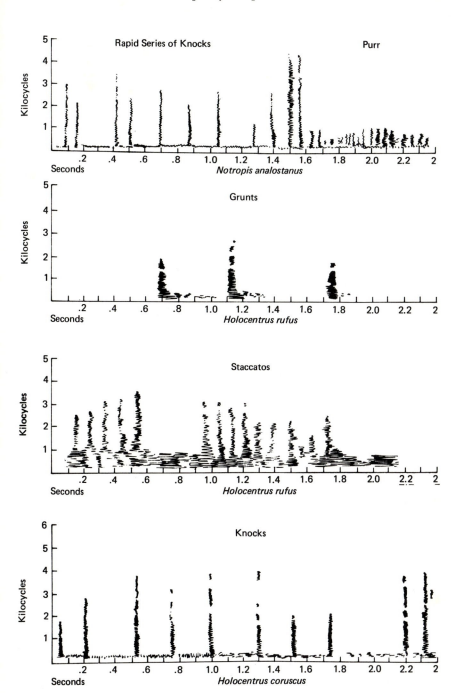

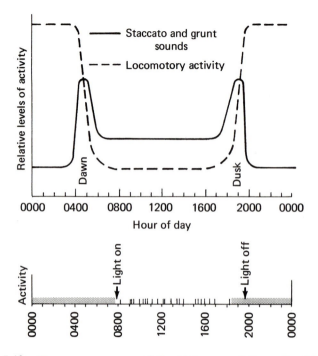

Fig. 5-10. Upper: a summary of the 24 hours activity cycle of locomotion and staccato sounds. Lower: Typical recording of activity. (From
Winn *et al.,* 1964. American Society of Ichthyologists and Herpetologists.)

more often, and in turn to be more vocal. At dawn and dusk both diurnal and
nocturnal species outside the reef move into different environments, and in the
process they pass through the squirrel fish territories.

Reproductive behavior may be mediated through both visual and chemical
cues. Vision is important in courtship and mating of some tropical marine fish
which assume temporary sexual color patterns and undergo elaborate courtship
displays, but many marine fish exhibit very slight sexual dimorphism and
vision is less important to those species. Courtship behavior and display in tropical fish have been reviewed in detail by Reese (1964). Visual stimuli may also
be important in other marine animals, as illustrated by the work of Arnold
(1962) on the mating behavior in the squid *Loligo pealei.* In this animal, sexual
behavior is stimulated by the appearance of an egg mass. Thus, when a naturally laid egg mass was tied to a string and placed in one quarter of an experimental tank, squid immediately investigated the egg mass and pointed toward
it as they swam rapidly with their arms formed into a cone. Since artificial egg
masses elicited the same responses as a real egg mass, the stimulus would seem
to be completely visual. The investigative behavior was followed by the establishment of a social hierarchy and mate selection by the males.

Pheromones, which are chemical signals exchanged between individuals of the same species, have not been studied extensively in marine animals, but there is good evidence that they are produced in the marine environment and influence behavioral responses of the animals living there. Ryan (1966) studied a Pacific species of brachyuran crab, *Portunus sanguinolentus,* which copulates immediately after the molt of the female. The male carries the premolt female for several days and attends molting; copulation occurs within 2–5 minutes after the female extracts herself from the exuvia. The question that has long puzzled marine biologists is how the male crab detects the premolt conditions of the female. Males placed in a holding cage containing premolt females have a distinct display and search pattern, walking around about the cage on the tips of their dactyls with their body elevated as high as possible above the substratum and attempting to pull whatever crabs they come in contact with, male or female, into the precopulatory holding position. This same display and search pattern was observed when water that previously contained a premolt female was siphoned into a tank containing adult males. It was necessary for the premolt female to be in the sea water for at least two hours before it would cause the display-and-search behavior pattern. Thus, the release of the pheromone appeared to be intermittent. When the excretory pore areas of a female were dried with acetone and capped with molten paraffin before placing her in the container of sea water, males failed to show the display-and-search behavior. However, when the paraffin caps of the excretory pores were removed, then the males again gave a positive response that indicated the pheromone was being released into the water.

That sensory basis for complex behavior patterns is not always well understood is well documented in the study on the association between the sea anemone *Calliactus tricolor* and hermit, box, and spider crabs (Cuttress et al., 1970). The relationship of the anemone with these particular crabs seems to be specific, for *C. tricolor* does not display a behavior response to other crabs which would trigger or facilitate its transfer. The behavior pattern of the anemone in transferring itself or in being transferred to one of the crabs depends on the species of crab. When the hermit crab *Dardanus venosus* approaches the anemone and gently taps close to the edge of the base, a general state of inhibition is initiated in the anemone, which releases its pedal disc as the crab continues to stimulate the column, gently at first, then more rapidly; finally, the anemone clings to the shell and attaches. The box crab *Hepatus epheliticus* responds to *C. tricolor* by bringing its carapace up to the tentacular crown of the anemone, a contact that evokes a clinging response in *C. tricolor.* It will subsequently rapidly release its pedal disc and transfer to the crab. But the most elaborate behavior pattern in both commensal partners is found in the relationship between the decorator crab *Stenocionops furcata* and *C. tricolor.* The crab approaches the anemone, pinching and squeezing the entire column with all its anterior appendages, although the chelae are concentrated on the edge of the base trying to get under the pedal disc and lift off the anemone. Once the anemone is detached, it is manipulated for several minutes during which time it becomes completely relaxed. The crab then scrapes the exposed pedal disc and brings it to its mouth as if to taste. The crab passes a cheliped over its head and across the carapace as if to locate an empty spot; this movement is usually carried out twice. During this time the anemone is apparently completely insensitive to external stimulus. Finally, the anemone is deposited on the carapace, whereupon it opens up and becomes responsive. In the crab, one stage triggers the next, and if a single stage is interrupted, the behavior pattern is set back one stage. What sensory modalities are involved in this complex behavior pattern? Since eyes are located on the sides of the head of the crab, it is doubtful whether visual cues are responsible for the pattern. As Cuttress and his coworkers have observed, "There are indications here of special manipulative and representational capacities in the crustacean central and motor nervous systems."

2. Feeding

The pelagic zone of the ocean has been characterized by Conover (1968) as a "nutritionally dilute environment," for although food supply is essentially limitless, it may be so thinly dispersed that the problem of getting enough to eat is formidable. Consequently, animals living here have evolved diverse and efficient methods for feeding.

The zooplankton, which are the primary herbivores in the oceanic eco-system, are estimated to consume 80% of the primary production, the remainder going to the benthic communities (Steele, 1965). Many of the zooplankton are filter feeders, passing water currents through a type of sieve or filter that strains out suitable particulate matter. Filter feeding is often a discontinuous process, and the filtering rate usually diminishes with increasing concentrations of food. Thus, in the laboratory the zooplankton feed best at low concentrations somewhat comparable to those found in nature. Feeding mechanisms may vary, however. For example, three types of feeding have been noted for the copepod *Calanus hyperboreus:* filter feeding, encounter feeding, and predation. During the encounter feeding described by Conover (1966), copepods fed on large diatoms if actual physical contact was made with the diatoms. Encounter feeding and filter feeding may be mutually exclusive, for when an individual is conditioned to small food which it can capture by filtration, the chance encounter with a large cell is ignored. On the other hand, an individual fed only large cells may not filter at all. *C. hyperboreus* can also actively hunt prey and is able to detect presence of a moving prey from several millimeters away (Conover, 1966). A number of copepod species have been found to be "unselective in diet and op-portunistic in what they ingest" (Mullin, 1966). None of these species were obligate herbivores and most ate animal food as readily as phytoplankton. The zooplankton did have one distinct preference, however, for large food particles over small ones.

Some sessile bottom-dwelling animals also may feed on plankton. Three reef-dwelling species of crinoids, *Lamprometra klunzingeri*, *Heterometra savignii*, and *Capillaster multiradiatus*, were found to actively collect food from the water mass moved over them by current or wave action. The animals reacted immediately to any changes in the water current by pointing the dorsal side of their arms with the spread pinnuli opposite the current. When waters were still and there were no water currents over the reef, the animals spread their arms in a regular circle around their body. Potential food seemed to be selected primarily by size, and rarely were animals over 1.0 mm in length selected. The animals consumed a relatively wide diversity of species, but zooplankton always made up approximately 90% of their diet, phytoplankton only 10% (Rutman and Fishelson, 1969).

Although many species can utilize several different sources of food, some clearly prefer one food over another, as illustrated by the herbivorous nudibranch, *Aplysia punctata*. When given a choice of the algae they would normally encounter, small sublittoral *Aplysia* ate the algae in an order of preference relative to growth value: *Plocamium*, *Heterosiphonia*, *Cryptopleura*, and *Delesseria*. *Plocamium* supported nudibranch growth best, and, moreover, the total yield of spawn was about seven times as great for animals eating *Plocamium* as for those eating *Cryptopleura* (Carefoot, 1967).

Food preferences of marine organisms often change during development. The youngest larvae of the sand-eels *Ammodytes marinus* feed mainly on copepod nauplii and some algae, while older larvae feed almost exclusively on appendicularians (Tunicates) (Fig. 5.11). In terms of quantity per larva, copepod nauplii and appendicularians occurred in the sea water in approximately equal numbers, thus this change in diet represented an actual change in food preference. Ganoid fish have two phases in their life cycle, pelagic and demersal, and during the pelagic stage (fish up to 53 mm in length), it has been observed that food consisted almost entirely of zooplankton, particularly copepods. When the fish reach 53–77 mm in length, they become bottom-dwelling forms. The basic diet of young ganoids consists of mysids, amphipods, and isopods; larger fish feed on epifaunal prey species (Nagabhushanam, 1965).

Seasonal differences in diet, presumably correlated with food availability, also have been documented for the ophiuroid *Ophiocomina nigra*, which feeds primarily on phytoplankton during the spring, coincidental with the spring bloom. In late spring and early summer, zooplankton is the predominant food, coinciding with zooplankton bloom. During the winter, benthic invertebrates make up as much as 50% of its food (Taylor, from Fontaine, 1965).

Feeding behavior depends in part on the internal state of an animal, so that a fish which feeds intensively one day may feed lightly the next (Pandian, 1967), and starved fish may eat more than individuals that have not been deprived of food (Rozin and Mayer, 1964). Some animals that normally show a preference for a specific type of food will accept less desirable foods when starved. In correlated studies, the tropical sea urchin *Eucidaris tribuloides* nor-

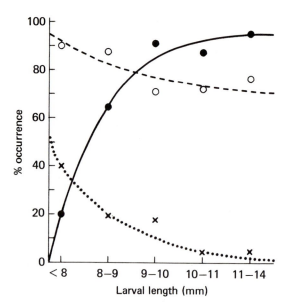

Fig. 5-11. Changes in food eaten by *Ammodytes marinus* larvae with increasing size. •-----•, appendicularians; o-----o, copepod nauplii; x-----x, green food. (From Ryland, 1964. Cambridge University Press.)

mally shows a strong preference for the sponge *Cliona lampa* (McPherson, 1968), but in a starved state *E. tribuloides* consumed the wooden and fiber-glassed sides of holding tanks. Not all species, however, will accept undesirable food species even when they are starving. The sea pen *Astropecten irregularis* is a case in point. These animals are selective feeders, eating primarily bivalves, but if no desirable food species are present the sea pens will nearly stop eating and lose weight (Christensen, 1970).

In some predators food preferences are influenced by previous food supply. Larvae of the herring *Clupea harengus,* for example, seem to become imprinted to particular kinds of food. In a choice situation, some selected nauplii, whereas others took only copepodites, and these choices reflected the food types previously eaten. Although the larvae reacted to other kinds of plankton within the field of vision, they would not snap at them (Rosenthal and Hempel, 1970).

Not all predators respond in this manner, however. The Pacific starfish *Pisaster* are highly selective of mussels, particularly *Mytilus edulis* and *Mytilus californianus,* although they will eat alternative prey. When starfish were fed for three months on turban snails only and then presented with a choice between snails and mussels, conditioning for snails had disappeared by the end of the first week and mussels again were the prey choice (Landenberger, 1968).

Survival of any animal population depends upon total energy available to the individuals. Without enough food, such maintenance is impossible, and a decline in population level inevitably follows. Predators, then, face a dual problem: they must consume enough prey to maintain a population, but at the same time they must not endanger survival by depleting the prey population. The predator-prey relationship between two species, the bivalve *Tellina tenius* and O-group plaice *Pleuronectes platessa,* was recently studied by Trevallion and co-workers (1970). The larval plaice feed on the siphons of the mollusc. The *Tellina* population changes only by growth, recruitment, and mortality, for many of the same individuals live from five to six years. The plaice population, on the other hand, is completely new each year, depending on recruitment of larvae from the breeding stock of adults offshore. Thus, the initial settlement is due to random fluctuation and is not dependent on the density of the food in the bay. The food available to *Tellina* is relatively constant, and it would appear to be another random variable. Laboratory experiments were set up in outside tanks, and the tanks were stocked with local specimens of *Tellina* and newly metamorphosed plaice in varying densities ranging from 300:1 to 30:1. The fish grew rapidly during the first month; the siphons of *Tellina* were cropped intensively and siphon weights were rapidly reduced, roughly in proportion to the predator-prey ratios. This decrease in siphon weights continued during the second month when weights dropped to a very low level. The decrease was faster in the tanks with the lowest ratio. Once the siphon weights were reduced to a level of 0.4 to 0.5 mg, growth of the fish ceased. Gradually the weights of the siphons began to increase, although at a much slower rate than the regeneration rates measured in the laboratory. This suggests that a limited amount of predation continued, enough to maintain the fish but not enough for growth. These experiments indicated that once the number of available siphons dropped below a certain threshold, the plaice fed less intensively on them. This form of density-dependent population regulation is similar to that reported for terrestrial vertebrates.

Feeding rates of some species can be affected by temperature. The optimum feeding rate of sea stars is at a temperature lying a few degrees below the highest mean temperature in the area where they live (Feder and Christensen, 1966). Filtration rate and food utilization were different at different temperatures in two species of offshore lamellibranchs, *Arctica islandica* and *Modiolus modiolus* (Winter, 1970). Habitat temperatures of these two species range during the year from 4° to 16.5°C. Experimental temperatures were 4°, 12°, and 20°C. At 4°C, there was a 50% reduction in filtration rate and in the amount of phagocytized algae. The filtration rate is essentially the same at 12° and 20°C, but the amount of phagocytized algae increases from 108 mg at 12°C to 144 mg algal dry weight per hour at 20°C.

Many organisms show a periodicity of feeding that correlates with light. Feeding in larvae of the sand-eels *Ammodytes marinus* and larvae of the plaice *Pleuronectes platessa* began at first light and continued vigorously for the subsequent 2–3 hours. It continued at a less vigorous rate throughtout the daylight hours and terminated at the onset of darkness (Ryland, 1964) (Fig. 5.12). The sea stars *Astropecten polyacanthus* are crepuscular, feeding early in the morning

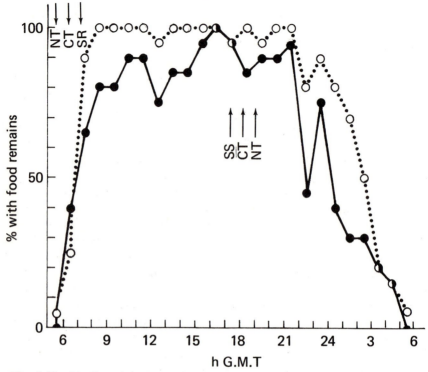

Fig. 5-12. Feeding periodicity of ●-----●, larval sand-eels, *Ammodytes marinus*, and o-----o, plaice larvae, *Pleuronectes platessa*. NT, nautical twilight; CT, civil twilight; SR, sunrise; SS, sunset. (From Ryland, 1964. Cambridge University Press.)

and again around sundown; most remain burrowed in the sand for the remainder of the day. This rhythm depends primarily on light cues, for when sea stars had been under constant darkness, there was no regularity in feeding patterns (Mori and Matutani, 1952). In the open ocean feeding activity of baleen whales is high in early morning, decreasing during the day, then increasing again late in the evening (Nemoto, 1959). Feeding patterns of these animals are probably linked to the diurnal migrations of their food species, for the fin whales do not show such a marked rhythm when feeding on the copepods *Calanus cristatus* and *C. plumchrus* during the summer, when migrations of these copepods are less intense (Nemoto, 1970).

3. Respiration

a. Temperature

(1) *Whole animal adaptations*

A number of years ago three papers appeared that represented important contributions to the problem of latitudinal effects on animals—those of Fox (1936), Sparck (1936), and Thorson (1936).

Thorson compared the metabolic temperature response of molluscs from the Arctic waters of Greenland and the warmer waters of the Mediterranean, and he found that generally species with a northerly distribution have a higher metabolic rate than southerly distributed species at the same temperature. He also reported metabolic diversity within a given geographical area. For example, epifaunal species such as *Pecten* and *Modiolaria* have a higher rate of oxygen consumption, especially at higher temperatures, than burrowing species which have a lower rate of oxygen uptake and also have a lower rate of metabolism at increased temperatures. Thus, the rate of oxygen consumption was correlated with the mode of life of the individual species. He found that deep water forms where the water temperature was below $0°C$ were more temperature sensitive than shallow water forms. Although he did not compute Q_{10} values, calculations from his data show Q_{10} values of approximately 21 for these deep water forms over the temperature range of $-1°$ to $+1°C$.

Working on molluscan species in the marine waters of Denmark and the Mediterranean Sea, Sparck reported results similar to those of Thorson. He further speculated that if northern species moved to warm waters, the increased metabolic rate would cause the animals to starve to death and thus would limit the southern distribution of the species. On the other hand, the northern distribution of southern forms would be limited by the thermal influences on larval development and reproduction. Fox, reporting on what was to be the first of a number of papers comparing metabolism of animals from England with northern cold-water species from Sweden, found the English species had a higher rate of metabolism at any temperature than the corresponding northern species. It was suggested that this difference was due to nonlocomotor (cellular) metabolism. Later, when the rate of oxygen consumption of muscle tissue from the prawn *Pandalus montagui* was determined at different temperatures, Fox and Wingfield (1937) found the metabolic-temperature curve for northern animals to be parallel or to the right of the curve of the southern animals. Although these papers were of fundamental significance, the authors had not always accounted for metabolic fluctuations due to certain variables including season, body size, and especially thermal acclimation. In 1939, when the last of these papers was presented before the beginning of World War II, the fundamental question was raised as to whether these differences in physiological functions were individual modifications, that is phenotypic variation, or whether they reflected genetic differences.

In 1953, Scholander and coworkers measured the oxygen consumption rate at various temperatures of 38 species of tropical and Arctic poikilotherms including fish, crustacea, insects, and spiders. They concluded that there was considerable but incomplete metabolic adaptation in aquatic Arctic forms in comparison to aquatic tropical species, but terrestrial insects showed little if any adaptation. Since these animals occupied different habitats and the influence of laboratory thermal acclimation on metabolism was not determined, it was impossible to make interspecific comparisons.

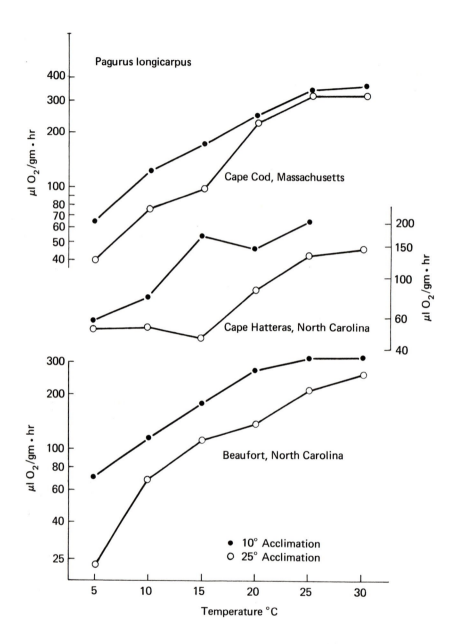

Fig. 5-13. M-T curves for five geographically separated populations of hermit crabs, *Pagurus longicarpus*. (From Vernberg, 1971.)

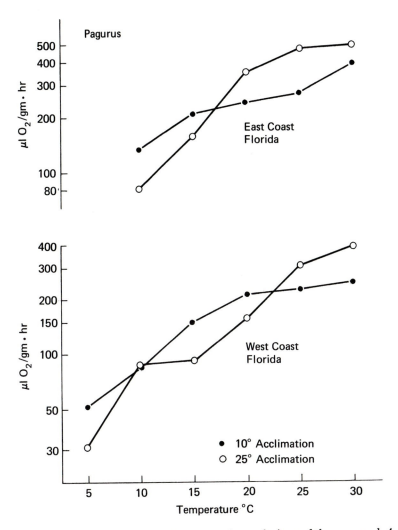

Apparent differences in respiration rates of populations of the copepod *Acartia clausi* from England and Long Island have been reported. The forms from the English waters had a statistically higher metabolic rate during the colder months of the year. Therefore, it was suggested that the American form had the greater capacity for control of metabolism under varying thermal conditions (Conover, 1959).

Diversity in metabolic-temperature response has been reported for latitudinally separated populations of the coastal hermit crab *Pagurus longicarpus* (Vernberg, 1971). These populations were from Massachusetts, from north and south of Cape Hatteras, North Carolina, and from the east and west coasts of Florida. In general the metabolic-temperature curves for the three northern populations demonstrated a common adaptational response (Fig. 5.13) for the cold-acclimated animals consumed oxygen faster than warm-acclimated animals, a response classified as translation to the left (see Chapter I). In contrast, the metabolic-temperature response of the two Florida populations was different and more nearly resembled the rotation pattern.

Comparison of the absolute metabolic rates of these geographically separated populations, rather than the patterns of response, reveals certain generalities. Warm-acclimated hermit crabs from Florida exhibit higher metabolic rates at elevated temperatures than the northern populations, but at lower temperatures the cold-water populations tended to have the higher rates. Since the northern animals normally experience lower temperatures, their relatively higher metabolic rate at lower temperatures would have adaptive significance.

Animals need not be separated by great latitudinal differences to show thermal-metabolic diversity, because thermal barriers may be found within very confined regions of the ocean. The Cape Hatteras area off North Carolina offers a case in point. The environment north of the Cape is a cold-water one, where the waters characteristically flow in a southerly direction, but south of Cape Hatteras are the northerly flowing warm waters of the Gulf Stream. Near Cape Hatteras the two water masses come together and flow out to sea. Inshore from the Gulf Stream south of Cape Hatteras is a zoogeographical area, called the Carolinian Province, which extends from the shores and estuaries over the shelf to the region of the Gulf Stream. Hence, three distinct zoogeographical areas occur within a very short distance, and each has a more or less distinctive faunal assemblage (Cerame-Vivas and Gray, 1966). The metabolic-temperature curves of 17 species of invertebrates from these three areas were determined at different seasons of the year and under different conditions of thermal acclimation. In general, adults having a northern affinity and occurring north of Cape Hatteras are metabolically depressed at high temperatures, whereas species from the Gulf Stream area are metabolically depressed at low temperatures. Those species with wide geographical limits are the most metabolically labile organisms as indicated by thermal acclimation studies. Some representative metabolic-temperature curves are presented in Fig. 5.14. A comparison of the metabolic-temperature pattern of zoeae of animals north of Cape Hatteras and tropical species showed that the metabolic rate of the more northerly distributed species, *Cancer irroratus*, was higher than that of the tropical species over the temperature range used in this study (15–25°C) (Vernberg, W., and F. Vernberg, 1970). Differences in metabolic patterns were found between Gulf Stream animals from North Carolina and tropical species from the Caribbean. The metabolic rate of the reef animals reached a peak at 20°C and leveled off; in the two tropical species, the rate was increased over the range of 15–25°C. The significance of these differences is unknown.

The ability to acclimate metabolically to changing temperatures is influenced by many factors. For example, Conover (1962) reported that stage 5 of *Calanus hyperboreus* almost perfectly acclimates its oxygen uptake over nor-

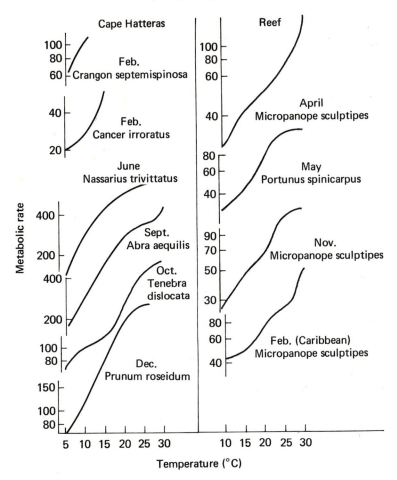

Fig. 5-14. Some representative M-T curves for oceanic animals. (From Vernberg, W. and J. Vernberg, 1970.)

mal temperature ranges (2° to 8°C). Spring populations of *C. finmarchicus* did not acclimate over a temperature range of 4–10°C, but they did at temperatures of 10–20°C. During the summer *Calanus* showed no acclimation over the entire range from 4° to 20°C (Halcrow, 1963). Small and co-workers (1966) reported no acclimation in *Euphausia pacifica* at temperatures between 5° and 10°C, but they did find acclimation at 15°C.

Metabolic-temperature curves may vary with body size. In the Antarctic amphipod *Orchomonella chilensis*, for example, the metabolic rate of 200-mg animals gradually increased with increased temperature up to 10°C, but a sharp decrease in the rate of oxygen uptake took place between 2° and 4°C for animals weighing 20 mg (Armitage, 1962). When considering the population size structure, Armitage presented an acutely determined metabolic-temperature curve which indicated that the metabolic rate for this species was temperature insensitive over the temperature range of about $-1.8°$ to $6°C$. From $8°C$ upward the rate increased sharply, until heat death occurred at about $12°C$.

(2) Tissue adaptations

Metabolic adaptation to the environment has also been demonstrated at the tissue level. Brain and gill tissues from fish exhibiting a high degree of locomotor activity have higher metabolic rates than those of sluggish bottom-dwelling species (Vernberg and Gray, 1953; Vernberg, gill data, unpublished). However, metabolic responses of liver and muscle tissues from different species may not always follow this same pattern (Vernberg, 1954).

Tissue respiration of animals from different latitudes may be different and reflect adaptation to temperature or some other variable. At a common temperature, brain tissue of the toadfish from Massachusetts had a higher metabolic rate than did brain tissue of animals from North Carolina (Vernberg, 1954). On an interspecific basis, the metabolic-temperature curves of liver and brain tissue of a cold-adapted fish, the polar cod, were to the left of those of a warm-adapted fish, the golden orfe (Peiss and Field, 1950). Of particular interest, the metabolic rate of tissues from the warm-adapted species dropped markedly at about $10°C$, whereas the metabolic values of tissues from the polar cod dropped slightly with temperatures below $10°C$. This response pattern suggests that the metabolic machinery of the warm-adapted species broke down at a low temperature which was atypical for this fish. Another example of tissue adaptation to temperature was reported by Somero and co-workers (1968). They found that the metabolic-temperature curves of gill and brain tissue of the Antarctic fish *Trematomus bernacchii* were translated to the left when compared with those of temperate zone species.

Some teleosts, such as the tunas, are able to maintain their body temperature at a higher level than that of their surrounding environment. For example, the temperature of the red muscle from the skipjack tuna *Katsuwonus pelamis* was $9.1°C$ above the ambient water temperature of $25.6°C$, the white muscle was $8.55°C$ above, and the brain was $4.53°C$ above (Stevens and Fry, 1971). The bluefin tuna was able to control its body temperature between $25°$ and $30°C$, and the warmest portion of the muscle mass varied only $5°C$ over the thermal range of $10–30°C$ (Carey and Teal, 1969b). Some sharks are able to conserve metabolic heat and maintain their body temperatures $7–10°$ above ambient temperature (Carey and Teal, 1969a).

(3) Biochemical adaptations

Although metabolic-temperature adaptations have been demonstrated at both the organismic and the tissue levels, in recent years much emphasis has been placed on the influence of temperature on the biochemical machinery of cells. Somero and Hochachka (1971) speculate that biochemical adaptation to the environment results from the production of "the proper molecules in the proper quantities at the proper time to ensure satisfactory biological function in the face of new environmental conditions." They proposed the hypothetical acclimation scheme in Fig. 5.15.

Various studies have described specific aspects of each of these steps, but a critical examination is needed to develop a comprehensive predictive model of the integrated action of these separated steps. However, these types of study are not easily carried out as pointed out by Kunnemann and co-workers (1970). They suggested that the following variables must be taken into account when studying temperature adaptation and enzymes: diurnal fluctuations in enzyme activity, individual differences on successive days, and seasonal changes in

Change in temperature → Endocrine or → Increased protein and →
(or photoperiod?)   neural response   nucleic acid synthesis

→ New ribosomes → New steady
   new isozymes   state

Transitory Changes	New Steady State
1. Altered rates of protein and nucleic acid synthesis 2. Activation of metabolic pathways associated with biosyntheses.	1. New isozymes. 2. New ribosomes. 3. New membrane lipids. 4. Altered ionic compositions of body fluids and tissues. 5. Altered metabolite pools.

Fig. 5-15. Hypothetical acclimation scheme. (Somero and Hochachka, 1971. American Zoologist.)

enzyme activity. In addition, acclimation times and experimental procedures must be standardized.

On the basis of studies on the incorporation of radioactive amino acids into protein, Haschemeyer (1971) reported an increased protein synthetic activity in the liver of the toadfish *Opsanus tau* acclimated to cold temperatures. In addition, one of the critical enzymes of the protein synthetic pathway, an aminoacyl transferase which catalyzes the binding of aminoacyl-transfer ribonucleic acid to ribosomes, was found to have elevated activity in the liver of cold-acclimated fish.

Alternate metabolic pathways can be activated by temperature change. For example, the pentose shunt is activated in cold-acclimated fish (Ekberg, 1958; Hochachka and Hayes, 1962). The following general responses of cold-adapted fish have been reported (Hochachka, 1967). In muscle, Embden-Meyerhof glycolysis was strongly stimulated, but the Krebs cycle activity was unchanged. The rate of lipogenesis was increased, and the activity of the pentose shunt was slightly increased. A greater number of changes were observed in liver as a result of cold acclimation: glycolysis increased; Krebs cycle activity decreased;

lipogensis greatly increased; pentose shunt activity clearly increased; and glycogen synthesis increased. Other studies have indicated that key metabolic control points are sensitive to temperature change (Prosser, 1967).

Different lactic dehydrogenase isozymal patterns have been described for cold- and warm-acclimated fish, but not in all tissues. In muscle, distinct qualitative differences exist after cold acclimation, but only possible quantitative changes in liver lactic dehydrogenase levels were found with no alteration in isozymal pattern (Hochachka, 1966).

It has been suggested that temperature-induced lipid changes affect the physical chemistry of biomembranes, thereby affecting homeostasis in poikilotherms (Roots, 1968; Caldwell and Vernberg, 1970). Data bearing on this problem indicated that significant changes in the lipid composition of fish gill mitochondria resulted in response to temperature acclimation, for lipids were more unsaturated at colder acclimation temperatures. Although the total phospholipid phosphorus levels were unchanged, proportionately more cardiolipin and phosphatidyl ethanolamine and less sphingomyelin were present in cold-acclimated fish (Caldwell and Vernberg, 1970).

Acclimation to low temperature resulted in quantitative changes in the ionic composition of body fluids of various species of poikilotherms (Rao, 1967). Such change could have important effects on cell metabolism. For example, increased potassium levels of fluids surrounding muscle increased its resting metabolism, and potassium modulated by calcium activates glycolysis.

Research on the biochemical basis for environmental adaptation has been very dynamic in recent years and will undoubtedly attract even more concerted interest in the future.

b. Oxygen

Marine animals vary in their metabolic response to ambient oxygen tension irrespective of taxonomic affinity. The oxygen consumption rates of some species are oxygen dependent: *Caudina* and *Stronglyocentrotus* (echinoderms); *Opsanus* and *Clupea* (fish); and *Homarus* and *Limulus* (arthropods). Others are oxygen independent: *Thyone* and *Echinus* (echinoderms); *Blennius* and *Coris* (fish); and *Calanus* and *Pugettia* (arthropods). As is the case with most biological generalizations, all organisms do not fit neatly into man-made categories, and marine organisms are no exception in that some represent an intermediate position between these two groups (consult Altman and Dittmer, 1971, for extensive listing of species).

The oxygen concentration at which an organism switches from oxygen-independent to oxygen-dependent metabolism may be correlated with the animal's habitat. For example, the rate of oxygen consumption of the sea star *Patiria miniata* is dependent on ambient oxygen tension until a high tension is reached; this species lives in rocky surf waters which are always saturated with oxygen (Hyman, 1929). Similarly, in *Holothuria forskali* the critical oxygen tension is high (60–70% air saturation), and this species actively moves toward a source of oxygen when the ambient oxygen concentration drops significantly (Newell and Courtney, 1965). Many bottom-dwelling, sessile, or slow-moving molluscs which might be exposed to low oxygen levels are metabolically independent until a low oxygen tension is reached.

Marked interspecific differences correlated with habitat are seen in teleost fishes. For example, the very active mackerel *Scomber scombrus* fails to withdraw oxygen when the tension drops below 70 mm Hg, while the less active scup *Stenotomus chrysops* is oxygen independent over the range of 40 to 120 mm Hg. In contrast, the sluggish bottom-dwelling toadfish *Opsanus tau* is oxygen dependent and can continue extracting oxygen from the surrounding sea water as long as any oxygen remains (Hall, 1929, 1930).

The oxygen content of sea water not only influences an individual organism, but can alter the schooling behavior of fish, such as menhaden, striped mullet, and northern anchovies. When an individual fish in a school is exposed to reduced oxygen levels and/or increased carbon dioxide, a condition probably encountered in the middle of the school, swimming behavior may be altered, thereby affecting the structure of the entire school (McFarland and Moss, 1967).

Individual animals have different response patterns to reduced oxygen content; some stop respiratory movements entirely while others increase the flow of water. There is no inhalant stream of water in *Branchiostoma lanceolatum* when the ambient oxygen level is low, probably as a result of the cessation of ciliary activity (Courtney and Newell, 1965). In contrast to this response of reduced activity, when barnacles are exposed to lowered oxygen levels their cirral activity is initially increased, but after prolonged exposure to hypoxic conditions activity is reduced (Southward and Crisp, 1965).

The lobster *Homarus* does not increase its ventilation rate in low oxygen tensions. Instead, it apparently increases the amount of oxygen withdrawn from the sea water by a circulatory adjustment (Thomas, 1954). However, fish typically increase ventilation rate and have a fairly constant rate of oxygen removal (Hall, 1929).

The transition point between metabolic dependence and independence varies depending upon a number of factors, such as:

(1) Age and stage of life cycle: Large prawns were more dependent on ambient oxygen tension than small ones (Subrahmanyan, 1962). In copepods the preadult stage V and the adult *Calanus* have different P_c values.

(2) Temperature: In the starfish *Asterias,* the P_c varies inversely with temperature. *Branchiostoma* does not remove oxygen from sea water if the oxygen tension is below 70% air saturation at temperatures from 4° to 10°C, but at 15°C oxygen is removed down to less than 20% air saturation (Courtney and Newell, 1965).

(3) Oxygen acclimation: After acclimation to low ambient oxygen tensions, the P_c values are lower (Prosser et al., 1957; Beamish, 1964).

Hypoxic conditions also influence the behavior of animals. The echinoderm *Holothuria tubulosa* elevates its posterior end out of the water and apparently takes in air through its rectum. However, another species, *H. forskali,* actively seeks a source of better aerated water (Newell and Courtney, 1965). The rate of rhythmic valve closure in lamellibranchs is altered during hypoxia. In *Ostrea* the closure rate increases immediately, there is no effect on *Cardium* and *Ensis,* and a delayed effect is followed by increased closure in *Glycemeris* (Salanki, 1966).

The oxygen-minimum layer, which occurs at intermediate depths in most of the world's oceans, is one of the special habitats in the oceans. Although the oxygen content may be only 0.1 ml/liter, an abundance of animals has been reported there (Marshall, 1954). In addition to resident biota, the deep scattering layer has been recorded at the same depths. Some data on the metabolic capability of animals living in this low oxygen environment are available. Longhurst (1967) stated that resting stocks of zooplankton could reside for long periods of time in this layer at oxygen tensions as low as 0.2 ml/liter, or vertically migrating animals could survive here during their daytime sojourn at greater depths. One resident of the eastern Pacific oxygen-minimum layer, the mysid crustacean *Gnathophausia ingens,* is able to maintain a constant rate of oxygen uptake over a wide range of oxygen tensions until the low level of about 0.26 ml/liter is reached. Then the rate of consumption drops with decreasing tension (Childress, 1968). Another species from this layer, *Euphausia mucronata,* removed oxygen from sea water in a closed test chamber down to an unmeasurable concentration (Teal and Carey, 1967).

c. Salinity

Although animals living in open ocean areas do not encounter decreased salinity waters, those living in coastal waters do. In some coastal species, the metabolic rate is unchanged over a wide salinity range. The sea urchin *Strongylocentrotus purpuratus,* for example, maintained a constant metabolic rate over a salinity range of 25 to 100% sea water (Giese and Farmanfarmaian, 1963). In contrast, the metabolic rate of the copepod *Acartia tonsa* from coastal waters increased in low salinity waters and remained elevated for more than 24 hours (Lance, 1965). This increased rate of oxygen uptake at reduced salinity is accompanied by decreased grazing activity, which would suggest that this species would not do well in brackish water. However, since *A. tonsa* is one of the most common copepods in estuarine systems, Lance argued that acclimation to low salinity waters must occur in the field.

4. Circulation

a. Blood

In general, little is known of the number and types of blood cells and the chemical composition of circulating body fluids in marine animals (see Altman and Dittmer, 1971, for tables of values). With few descriptive data available, it is not surprising that the influence of environmental stress on these parameters has received little attention.

The amount of lipid in the whole blood of the echiurid *Urechis caupo* is high (460–1400 mg/100 ml) compared with other invertebrates (5–100 mg/ 100 ml) (Lawrence et al., 1971). Similarly, the total carbohydrate level is comparatively high. These high values indicate that the blood is a nutritious medium and perhaps may account for the observation that fertilized eggs develop more rapidly in the body cavity than controls maintained under normal conditions. In a more detailed study, DeJorge and co-workers (1969) reported that during prolonged fasting the chemical composition of the coelomic fluid of the echiuran worm *Lissomyema exilii* changed. The concentration of water, calcium, iron, and glucose decreased significantly, sodium, potassium, magnesium, and copper increased noticeably, and nitrogen and phosphorus showed a slight increase.

Antarctic fishes are unable to survive high temperatures (about 6°C) which would represent the low lethal temperature for many fish from other biogeographical zones. The blood plasma proteins of Antarctic fish apparently are one of the adaptive mechanisms that has evolved permitting the fish to prosper in cold water. In general, Antarctic fish sera had less albumin and more lipoproteins than sera of non-Antarctic fishes. Also, freezing point–depressing glycoproteins were present only in the sera of Antarctic species. That the blood of cold-adapted fishes clotted faster at 0°C than did the blood of warm-adapted forms indicates the efficient functioning at low temperature of their blood-clotting mechanism (Komatsu et al., 1970).

Molting may alter the composition of blood constituents. For example, in the lobster *Homarus americanus* the following changes were reported by Telford (1968). The glucose levels were 35% higher in the premolt stage than in intermolt and were 30% lower in postmolt. Fructose and galactose appeared most frequently during premolt, and in postmolt a decrease in oligosaccharides occurred. These changes are closely related to the changing carbohydrate needs of molting.

Working with *Homarus vulgaris*, J. Glynn (1968) found increased concentrations of calcium, magnesium, inorganic and total phosphate, and total protein in the sera of lobsters before molt and decreased levels thereafter. Sodium and chloride concentration in sera varied little with the stage of molting, although serum potassium decreased markedly before molting.

b. Heart rate

The heart rate of marine invertebrates has been shown to be affected by temperature and by body size (Pickens, 1965). The rate was lowered by both decreased temperature and increased body size. The heart rate of larger invertebrates (*Mytilus californianus*, 21 beats per minute at 15°C, and *Homarus*, 100 beats per minute at 16–20°C) is slower than planktonic species (*Calanus,* 200 beats per minute at 5°C).

c. Respiratory pigments

The amount of hemoglobin in the blood of fishes can, in a general way, be correlated with their habitat, degree of locomotor activity, and zoogeography. Active surface-dwelling fish from the temperate zone have higher hemoglobin concentrations than do sluggish bottom-dwelling species, while intermediate values characterize moderately active fish (Hall and Gray, 1929).

The role of hemoglobin in cold-water fish has been of particular interest to environmental physiologists, especially since one species lacks hemoglobin. In cold waters more oxygen is in solution than at higher temperatures, a condition favorable to the evolution of a fish without hemoglobin (Ruud, 1954). How-ever, most cold-water fishes have hemoglobin, although the concentration is rela-tively low in Arctic fish and in deep sea fish of lower latitudes living at low temperatures (Scholander and Van Dam, 1957). Holeton (1970) compared oxygen exchange in the ice-fish *Chaenocephalus aceratus,* the hemoglobin-less Antarctic fish, with that of hemoglobin-bearing fish from the Antarctic and reported the following principal differences: (1) The oxygen consumption of *C. aceratus* is low, but comparable to other Antarctic fish. (2) The blood oxygen–carrying capacity of fish with hemoglobin is about ninefold higher than *C. aceratus.* (3) Even though the cardiac output is about three to four times higher in *C. aceratus,* the fish with hemoglobin have a higher oxygen-transporting capacity by a factor of 2 to 4 times. (4) The high blood-to-tissue oxygen tension gradient (48 mm Hg) which exists in the ice-fish appears to be compensatory for the absence of hemoglobin, since the blood oxygen equilibrium relationship is necessarily linear. The P_{50} of hemoglobin from other Antarctic fish varies from 8.5 to 21.5 mm Hg (Grigg, 1967), and the blood of these fish would be highly saturated at the blood-to-tissue oxygen tension gradient found in the ice-fish. Since the oxygen content is high in cold Antarctic waters and the demand by Antarctic fish for oxygen is low, Holeton felt that the ice-fish were, in general, successful residents of cold water. The major advantage red-blooded fish had over the ice-fish was their increased resistance to hypoxia, for *C. aceratus* died at oxygen tensions below 50 mm Hg, while other fish survived at least to 15 mm Hg.

When comparing the red-blooded Antarctic fish with fish from the tem-perate zone, the oxygen affinity of the blood of Antarctic fish is extremely sensitive to temperature (Fig. 5.16). With an increase in temperature of 6°C, the absolute oxygen capacity of the blood was reduced by 30% (Grigg, 1967). The P_{50} curves for the Antarctic fish were far to the left of the curves for tem-perate zone species. This relationship of the translation of the cold-adapted curve to the left (Prosser, 1958) corresponds to the typical response of the metabolic temperature curves of cold-water versus warm-water species.

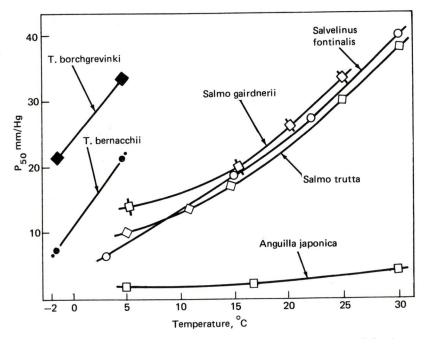

Fig. 5-16. Magnitude of change in P_{50} with temperature in species of the Antarctic fish *Trematomus* as compared with that in trout and eels. (From Grigg, 1967. Pergamon Publishing Company.)

The structure of respiratory pigments can change during the ontogenetic development of an oceanic species. For example, the hemocyanin of the adult cuttlefish (*Sepia officinalis*) shows a constant electrophoretic pattern that is significantly different from young organisms. However, as the young develop, the pattern gradually changes to that of the adult (Decleir and Richard, 1970).

Differences in the affinity of hemoglobins from various fish can be correlated with their ecology. The blood of actively swimming surface-dwelling species, such as the mackerel *Scomber scombrus* and the menhaden *Brevoortia tyrannus*, possesses high oxygen capacities and a hemoglobin with a low affinity for oxygen. In contrast, less active species, *Tautoga onitus* and *Opsanus tau*, have blood with a low oxygen capacity and hemoglobin which has a relatively high oxygen affinity (Hall and McCutcheon, 1938). The dogfish *Squalus suckleyi* has a P_{50} value of 17 mm Hg and no Bohr effect. Both of these responses would be beneficial to this fairly active species living near the bottom where the waters could become deficient in oxygen and high in CO_2 (Lenfant and Johansen, 1966). The oxygen capacity of the blood of sea mammals is high (about 20 vol%) compared with fish (5–14 vol%) or most invertebrates (0.2–10 vol%). In the invertebrates, similar correlations of the oxygen affinity of respiratory pigments and habitats are found. Active species tend to have higher

P_{50} values (the squid *Loligo pealei,* 42 mm Hg) than those of bottom-dwelling sluggish animals (*Limulus,* 11 mm Hg, and the echiuroid *Thalassema,* 1.4 mm Hg).

Various environmental parameters other than oxygen tension, such as temperature, pH, and salinity, influence the ability of respiratory pigments to transport oxygen. After a decrease in temperature, the position of the oxygen equilibrium curve is typically moved to the left (Redmond, 1955; Grigg, 1967). However, an interesting variation on this generality was noted when the responses to temperature of hemocyanins from tropical and temperate zone spiny lobsters were compared. The oxygen equilibrium curve of the temperate zone species determined at its typical habitat temperature (15°C) was the same as the curve for the tropical species *Panulirus argus* determined at its habitat temperature of 25°C (Redmond, 1968) (Fig. 5.17). At one geographical site

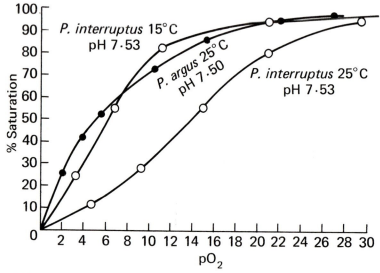

Fig. 5-17. Oxygen equilibrium curves of hemocyanins from two species of spiny lobsters. *Panulirus argus* is a tropical species, *P. interruptus* a temperate one. (From Redmond, 1968. Academic Press, Inc.)

interspecific differences in thermal sensitivity of hemocyanins appeared to have survival significance. The P_{50} values for two species of crabs were similar at 16–17°C, but at 25°C this value was lower in the species that survived best at this elevated temperature (Redmond, 1955).

Increased acidity decreases the affinity of respiratory pigments for oxygen (Bohr effect), a process which facilitates unloading of oxygen in the tissues. In the lobster *Panulirus,* the Bohr effect appears to be of little significance at normal habitat temperatures. However, at elevated temperatures, pH changes may cause a significant shift in the position of the oxygen equilibrium curve (Redmond, 1955).

The hemocyanin from the horseshoe crab *Tachypleus tridentatus* had a relatively high oxygen affinity ($P_{50} = 8.71$ mm Hg, pH 7.2, 25°C). But the oxygen affinity of this hemocyanin was greatly reduced when calcium ions were removed (the P_{50} was now 13.5 mm Hg) (Hwang and Fung, 1970). Similar reduction in the oxygen affinity of hemocyanin was observed in various crustaceans when they were unable to ion-regulate in lowered salinities and the total salt content of their blood was reduced (Djangmah and Grove, 1971).

5. Chemical Regulation and Excretion

a. Ionic and osmoregulation

Except in certain restricted coastal regions or during migrations into estuarine or fresh water, marine organisms do not encounter marked fluctuations in salinity. Hence, marine biologists have assumed oceanic animals have limited osmo- and/or ionic regulation capabilities. This is not necessarily true, however. Pelagic medusae, for example, do have a measure of ionic regulation in that they contain fewer sulfate and magnesium ions than sea water, but have a greater concentration of sodium and chloride. This regulation has a buoyancy effect, since a reduction of 50% in the concentration of sulfate produces a buoyance of 1 mg/ml (Denton and Shaw, 1962). Many pelagic molluscs also have a well-developed ionic regulatory ability which aids in buoyancy, as illustrated in studies involving *Sepia* and *Nautilus*. When salt is transported from the chamber fluid by active transport until the fluid is hypoosmotic with the blood, water is evacuated from the shell chamber, moving by osmosis into the blood against the hydrostatic pressure and hence aiding in flotation (Denton, 1964). It has been estimated that this system would be effective to a depth of over 20 atmospheres (about 600 feet). Not all molluscs rely on this buoyancy technique, for some depend solely on ionic regulation; pelagic gastropods, for example, reduce density by lowering the concentration of magnesium and sulfate ions in the blood in much the same manner as coelenterates. Ionic regulation does not always occur in response to buoyancy. The marine hydroid *Tubularia crocea*, for example, can regulate sodium and chloride in concentrated solutions while the brackish water hydroid *Cordylophora lacustris* can concentrate these ions in waters of reduced salinity (Steinbach, 1963).

A constant exchange of water occurs between an aquatic animal and its environment. In fish, the rate of water exchange can be correlated with habitat characteristics. The euryhaline teleost *Tilapia mossambica* was observed to exchange 115% of its body water per hour in fresh water, 84% in 100% sea water, but only 40% in 200% sea water (Potts et al., 1967). On an interspecific basis, Evans (1969) demonstrated that freshwater species of fish are more permeable to water than marine species. Since the blood of fish is hypoosmotic to sea water, a decreased water permeability in the marine environment would have a functional advantage in reducing the net loss of water from the body to the external environment. As discussed in Chapter III, worms and crustaceans exhibit the reverse response; that is, freshwater species are less permeable to water than marine species. The body fluids of these invertebrate marine species, unlike that of fish, are typically isosmotic with sea water so that water regulation is not a problem. Hagfish, which have body fluids isosmotic with sea water, are extremely permeable to water, a response similar to that of marine crustaceans (Rudy and Wagner, 1970). The marked difference in the rate of water influx in the hagfish and various fresh- and seawater fishes is graphically presented in Fig. 5.18.

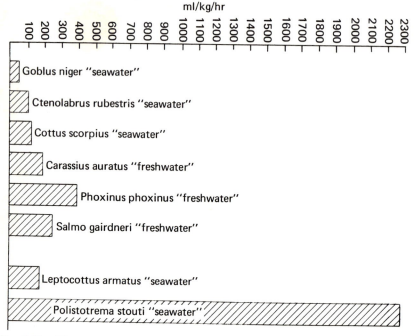

Fig. 5-18. Water influx in the Pacific hagfish, *Polistotrema stouti,* and some marine and fresh-water teleosts. (Rudy and Wagner, 1970. Pergamon Publishing Company.)

Generally, the crustacean hemolymph of oceanic species is isotonic with the ambient sea water, although its ionic composition may differ markedly. Comparative studies on Alaskan subtidal crustaceans showed diversity in responses. Whereas the king crab hemolymph was hyperosmotic (5 to 20 m-osmoles) to sea water, two other species were isosmotic (Mackay and Prosser, 1970). No functional significance in this difference was apparent. Further species differences were observed in respect to ionic regulation. Only chloride was well regulated by the tanner crab, and the level was less than that of sea water. In the king crab magnesium was lower, but chloride was higher, in the hemolymph than in sea water. Of the three species, the coon-striped shrimp was the most efficient ionic regulator: magnesium and chloride levels were lower and calcium higher in the hemolymph than in sea water. The values for these three species (Table

TABLE 5.2

*Ionic and Osmotic Concentration of Hemolymph and Sea Water from Which the Animals were Sampled**

Species	Fluid	Osmolarity (m-osmoles)	Ion Concentration (m-moles/L.)				
			Na	K	Ca	Mg	Cl
Chionoecetes tanneri	Hemolymph	998 ± 9 (4)	403 ± 11 (5)	11.1 ± 0.5 (5)	8.1 ± 0.5 (5)		495 ± 3 (5)
	Urine	997 ± 2 (5)	420 ± 23 (4)	11.4 ± 1.4 (4)	10.4 ± 1.4 (4)		482 ± 23 (4)
	Hemolymph	862 ± 2 (5)	409 ± 7 (5)	10.5 ± 0.2 (5)	9.3 ± 0.1 (5)	44.8 ± 1.2 (5)	462 ± 2 (5)
	Urine		393 ± 19 (5)	9.9 ± 0.7 (5)	8.7 ± 0.6 (5)	45.2 ± 3.5 (5)	456 ± 6 (5)
	Sea water	860	417	9.0	8.9	52.5	490
Pandalus hipsinotus	Hemolymph	896 ± 16 (6)	395 ± 10 (6)	7.4 ± 0.7 (6)	12.3 ± 0.5 (6)	5.8 ± 1.0 (6)	466 ± 9 (6)
	Sea water	907	380	7.8	9.3	52.0	493
Paralithodes camtschatica	Hemolymph	933 ± 5 (35)	461 ± 9 (17)	13.5 ± 0.4 (16)	13 (2)	36.9 ± 1.2 (13)	577 ± 12 (13)
	Sea water	902	452 ± 3 (4)	12.1 ± 0.2 (4)		52.0	494 ± 4 (5)

The first number in each column is the mean of the samples followed by the standard error of the mean (S.E.) and in brackets the number of individuals in the sample.

* Mackay and Prosser, 1970.
Pergamon Publishing Company.

5.2) differ in certain respects from those of other marine crustacea, chiefly in that chloride concentration is greater in the king crab than all other crabs, and that the coon-striped shrimp regulates magnesium best.

Copepods from coastal and oceanic waters generally are thought to be osmo-conformers, although few detailed studies have documented this hypothesis. One inshore marine euryhaline species, *Centropages hamatus,* is an osmocon-former in diluted and slightly concentrated sea water (Bayly, 1969). Of evolu-tionary and ecological significance is the adaptive capacity to osmotic stress demonstrated by copepods from different habitats. Brand and Bayly (1971) have suggested the following evolutionary sequence for centropagid copepods from Australia. Oceanic and coastal water species are euryhaline osmocon-formers, while truly estuarine and brackish water species are hypo- and hyper-osmotic regulators. Freshwater species are hyperosmotic regulators, and species from inland saline waters are conformers with great cellular tolerance and/or cellular osmoregulation.

The hemolymph of the common whelk *Buccinum undatum,* which occurs subtidally, is slightly hyperosmotic to sea water. This species has limited ability to survive wide salinity changes, for a salinity of 11 0/00 proved fatal and usually *B. undatum* is found at salinities above 18 0/00. The limited ability of this species to invade low saline waters may be correlated with its inability to completely regulate cell volume at lowered salinities (Staaland, 1970). In this species the concentration of intracellular ninhydrin-positive substances, chiefly alanine, glycine, and taurine, decreased with decreasing salinity.

The Sipuncula are considered to be exclusively marine, and Oglesby (1968) found that *Themiste dyscritum* survived in the laboratory in 30% sea water, although in nature they are typically found in high salinity waters. This species is a passive osmoconformer and has an apparent ion deficit in the coelomic fluid which may be made up by small molecular weight organic mole-cules. Its water content is not controlled by volume regulation.

Echinoderms are also considered to be restricted to marine waters, although populations of *Asterias rubens* have been reported living in salinities of about 8 0/00 (Schlieper, 1957). This group is poikilosmotic, but ionic regulation has been demonstrated in some species. The perivisceral fluid of the sea urchin *Stronglocentrotus purpuratus* has approximately the same osmoconcentration as that of the surrounding sea water over the range of 50 to 125% sea water (Giese and Farmanfarmaian, 1963). When placed in 50% sea water, *Asterias forbesi* initially gained weight and then lost it as a result of salt loss. When comparing populations of *Asterias rubens* from the North Sea, where the salinity is about 30 0/00, and the Baltic Sea, with a salinity of about 15 0/00, Schlieper (1957) found that populations from low salinity water tended to be smaller, have a softer integument, and exhibit a higher water content. Other echino-derms have some degree of ionic regulation in that potassium and magnesium are regulated. Moreover, marked ionic differences between body fluids and muscles exist.

Most marine vertebrates maintain body fluids at an osmoconcentration much below sea water. However, the blood of hagfish is isosmotic with sea water, and apparently this group is poikilosmotic. In contrast, the elasmobranchs are homoiosmotic, but their blood is nearly isosmotic with sea water. Unlike the hagfish, whose osmotic pressure is due almost entirely to inorganic ions, the elasmobranchs employ urea and trimethylamine oxide in maintaining a high osmotic pressure. However, the freshwater sharks, genus *Potamotrygon*, which are permanent residents of the Amazon basin, have abandoned the retention of urea. They have only 2–3 mg of urea nitrogen/100 ml of body fluid compared with 300 to 1,300 mg of urea nitrogen/1,100 ml in other marine and freshwater sharks (Thorson et al., 1967).

The blood of marine teleosts is hypoosmotic to sea water. Freezing point depressions of 0.66 to 1.0 (400 mOsm/liter) are characteristic of fish in sea water, and approximately 90% of the osmocentration is due to inorganic salts, chiefly NaCl. Marine species face a problem of osmotic dehydration, and to conserve water a low rate of urine production is typical.

Generally, basic osmotic and solute regulation is the function of the kidneys, but extrarenal organs are importantly involved (Hickman and Trump, 1969). In marine animals, the principal extrarenal organs are the gills, the gut, and specialized organs. Although insufficient quantitative data are available to assess the relative importance of these various organs in all marine organisms, some studies have been published on certain species. Over 40 years ago Smith (1930) proposed that marine teleost fish drank sea water to replace water osmotically lost across the body surface. Sodium, chloride, and potassium swallowed with sea water are absorbed by the gut and excreted by an extrarenal route, probably the gills. Since this original hypothesis was suggested, various workers have estimated the amount of sea water swallowed by teleosts. Hickman's (1968b) excellent study on the southern flounder presents a balance sheet on ingestion, absorption, and elimination of salts from sea water. His data are presented in Table 5.3. About 24% of the electrolytes in swallowed sea water is excreted along with rectal fluid, while of the 76% of the electrolytes absorbed, most (95%) is excreted by extrarenal organs. Electrolytes are also differentially treated. Much of the ingested magnesium and sulfate ions and about 32% of the calcium are eliminated along with the rectal fluid, and almost all of the monovalent ions are absorbed and subsequently excreted by extrarenal routes. Of the calcium absorbed (69% of the total ingested), nearly all was also excreted by extrarenal mechanisms, such as the gills. Water drinking by fish may serve the function of eliminating substances, such as calcium being excreted into the gut (Dall and Milward, 1969). The rate of water drinking has been shown to decrease with salinity in some fish (Evans, 1968) but not all (Dall and Milward, 1969).

Among the elasmobranchs the rectal gland of the dogfish shark *Squalus acanthias* secretes a fluid isosmotic with the blood, but containing twice the concentration of sodium and chloride and little urea or trimethylamine oxide

TABLE 5.3

*Major Electrolyte Composition of Rectal Fluid, Urine, and Plasma of Southern Flounder and Flow Rates of Rectal Fluid and Urine**[**]*

	Osmo-lality, mosmoles per liter	mmoles/liter of:							H_2O, ml/l	pH	Flow, ml/h × kg	No. samples averaged	Colln. period, h	Exptl. salinity	Body wt., kg
		Na	K	Mg	Ca	Cl	SO_4	PO_4							
Rectal fluid Fish No.															
1	344	30.0	0.60	174.5	13.6	112.5	102.5	—	949	8.45	1.65	2	6	33.1	0.840
2	327	14.6	0.40	157.5	10.5	150	94.8	0.13	—	7.90	0.34	2	52	32.3	0.730
3	370	23.8	0.58	182	11.9	150.5	105.8	0.17	—	8.00	2.78	2	4	31.8	1.416
4	301	18.6	0.53	187.5	4.0	82.3	108	0.08	—	8.55	0.33	2	16	31.7	2.091
5	341	18.0	1.60	198.5	20.5	136.8	116.7	0.29	966	8.30	0.462	1	6	29.7	0.897
Average	336.6	21.0	0.74	180.0	12.1	126.4	105.6	0.16	957.5	8.24	1.11			31.72	
Urine Fish No.															
1	333	8.0	1.07	130.5	24.6	148	62.8	—	943	7.85	0.306	5	20		
2	337	49.1	3.29	115	15.1	130.3	43.5	27.4	—	7.78	0.056	1	28		
3	386	9.2	1.02	207.5	12.8	128	102.2	8.09	954	7.07	0.105	2	4		
4	293	1.8	1.10	92.5	22.5	96.3	27.7	1.87	970	7.37	0.11	4	2		
5	318	7.0	1.60	166	17.8	130.9	65.2	9.02	959	6.05	0.319	1	6		
Average	333.4	15.0	1.61	142.3	18.6	126.7	60.3	11.6	956.5	7.22	0.179				
Plasma Fish No.															
1	344	182	3.42	1.25	2.67	167.5	0.19	—	946.6	—					
2	334	178	4.54	1.3	2.65	161.1		—	—	7.60					
3	380	189	4.40	1.85	2.50	168.3		4.83	940.0	8.08					
4	300	158	3.00	1.0	2.60	138.0		1.86	926.0	7.93					
5	329	192	3.85	1.7	2.5	165.5		—	935.0	7.64					
Average	337.4	179.8	3.84	1.42	2.58	160.1	0.19	3.34	936.9	7.81					
Seawater average	912	432.2	9.16	49.26	9.42	504.1	25.96	0.00025*	906.5	8.10					

* Seawater phosphate is highly variable. Its concentration on the Carolina Shelf (Beaufort vicinity) ranges from virtually zero to 0.0005 mmoles/l (U. Stefansson, personal communication).

** Hickman, 1968b.

National Research Council of Canada from the Canadian Journal of Zoology.

(Burger and Hess, 1959). However, even after this gland has been removed, salt balance can be maintained although the salt content of the urine is increased. Further, Burger and Tosteson (1966) measured the influx and efflux of sodium and chloride and suggested that the efflux must take place elsewhere than from the kidneys and the rectal gland.

Marine birds and reptiles have salt glands which produce a solution more concentrated than sea water. Thus, these animals can eliminate any salt load resulting from either drinking sea water or eating marine organisms. The comparative data on rates of electrolyte secretion by reptilian salt glands in Table 5.4 (Dunson, 1970) indicate that marine-dwelling species secrete much more

TABLE 5.4

Maximum Rates of Electrolyte Secretion of Reptilian Salt Glands*

Species	Habitat	Secretion rates μmoles/100 g hr		
		Na	Cl	K
False iguana (*Ctenosaura*)	Terrestrial	1.0	—	9.4
Chuckwalla (*Sauromalus*)	Terrestrial	3.3	—	31.0
Land iguana (*Conolophus*)	Terrestrial	25.5	18.4	5.0
Diamondback terrapin (*Malaclemys*)	Estuarine	26.6	19.2	1.0
Banded sea snake (*Laticauda*)	Marine	73	74	3.3
Green sea turtle (*Chelonia*)	Marine–pelagic	134	—	4.9
Yellow-bellied sea snake (*Pelamis*)	Marine–pelagic	218	169	9.2
Marine iguana (*Amblyrhynchus*)	Marine–coastal	255	237	50.7

* Dunson, 1970.
Pergamon Publishing Company.

sodium and chloride than either estuarine or terrestrial species. The kidneys also play a role in eliminating electrolytes, at least in some birds (Table 5.5).

Marine mammals appear to depend on the kidneys to eliminate excess

TABLE 5.5

*Distribution of Electrolyte Elimination Between Cloacal and Nasal Secretion in Cormorants**

	Cl mEq	Na mEq	K mEq	H_2O
Total ingested	54	54	4	50
Cloacal elimination	27.5	25.6	2.66	108.9
Nasal elimination	26.1	23.8	0.31	51.4
Total elimination	53.5	49.4	2.97	160.3

* Schmidt-Nielsen et al., 1958.
American Physiological Society.

salt loads, since they produce urine which may be hyperosmotic to blood or to sea water (Krogh, 1939). However, salivary glands or other extrarenal glands may play a role in some animals. As is the case with other groups of marine organisms, marine water may be ingested by drinking or along with food intake. Not only is salt elimination important to marine organisms, but also a water balance must be maintained to sustain life. Several alternatives for effective water utilization have been described. If the mammals eat marine vertebrates which are hypoosmotic to sea water, relatively more water is available to them than if they depended on invertebrates which are isosmotic with sea water. Krogh (1939) estimated that the kidneys of large mammals like walruses and whales could excrete salt and urea with the water available from their food.

Some marine animals migrate between oceanic waters and estuaries, and some species continue on into fresh water. The spawning migrations of salmon into estuaries or fresh water have been studied extensively. One example is that of the coho salmon *Oncorhynchus kisutch*. Although a number of physiological processes are operative in successful migration, this organism is faced with the problem of osmo- and ionic regulation as a result of leaving the ocean for a life in fresh water. When adult fish enter fresh water, the concentration of plasma ions decreases. In one study it was found that the freezing point depression of blood from salmon living in sea water ($-0.762°C$) was greater than that of fish taken from fresh water ($-0.613°C$). Similar results have been demonstrated in other fish. The time taken to reach a new stabilized osmotic concentration in various fish species appears to be correlated with their type of migratory pattern and their habitat. Estuarine fish such as blennies and flounders, which are subjected to fluctuating salinities, acclimate within hours (Hickman, 1959). In contrast, animals from more stable environments require longer periods of time; steelhead trout, for example, took 80–170 hours (Houston, 1959), and young salmon fry needed 200 hours (Parry, 1960). In addition, the loss of specific ions, particularly sodium and chloride, has been reported in various species. Miles (1971) reported a definite decrease in sodium, magnesium, and chloride concentrations in the plasma of the adult coho salmon, but potassium and calcium were unchanged. Similar changes occurred when juveniles were acclimated to sea water. However, the potassium, sodium, magnesium, and calcium concentrations of various tissues of the chinook salmon (*Oncorhynchus tshawytscha*) were remarkably constant throughout four stages of the life cycle (juveniles in fresh water, juveniles in sea water for 2 weeks, mature adults in sea water, and after 3–4 months of fasting on the spawning grounds), except for juveniles in sea water. This difference might reflect that the acclimation period to a new environment takes longer than 2 weeks (Snodgrass and Halver, 1971).

That the kidney is of paramount importance in the spawning migration of the coho salmon was demonstrated in the detailed paper of Miles (1971). When kidney function of adults captured in sea water was measured both in sea and fresh water, three principal changes in kidney function were observed: the rate of urine flow increased ten-fold in fresh water; the glomerular filtration rate increased about sevenfold in fresh water; and the urine ion concentration of sodium potassium, calcium, magnesium, and chloride decreased in fresh water although the excretion rates of sodium, potassium and calcium increased. Miles concluded that in salt water the salmon kidney conserves water and excretes magnesium, whereas in fresh water it functions to excrete water while conserving salts.

When eels were transferred from fresh water to sea water, the number of chloride cells of gills increased by five- to sixfold within one week and the number decreased to the freshwater level within four weeks after being returned to fresh water (Utida et al., 1971). In addition, the $Na^+ - K^+$ ATPase activity showed similar changes with salinity, a response pattern similar to that of other euryhalinic fish. This enzyme appears to be localized chiefly in the chloride cells. When eels were acclimated to sea water, other adaptive changes were observed. Within the muscles, a significant increase in the free amino acids resulted chiefly from changes in alanine, glutamine, glutamic acid, glycine, proline, tyrosine, cystathionine (Huggins and Colley, 1971).

The southern flounder *Paralichthys lethostigma* shows a seasonal migration during the winter from estuaries and low salinity coastal waters to deeper offshore waters of the continental shelf. Hickman (1968a) reported seasonal differences in the renal performance of this species. During the winter the glomerular filtration rate, which is important in volume regulation, is relatively low, but it returns to a high rate during the warmer months when the fish lives in dilute sea water.

Williams (1960), working on two species of shrimp *Penaeus duorarum* and *P. aztecus,* demonstrated the influence of the environment on physiological response and distribution of animals. These species can hyperosmoregulate in reduced salinity at temperatures normally encountered during the summer. At low temperatures this ability to osmoregulate is impaired in both species, but not to the same extent and they survived low temperatures better at high salinities. *P. duorarum,* which is less affected by reduced temperatures, overwinters in the colder inshore waters of the Carolina coast.

b. Excretion

In marine invertebrates, ammonia is the principal nitrogenous waste product excreted. However, other nitrogenous compounds are lost by these organisms, such as amino acids, urea, and uric acid. The widespread occurrence of ammonotelism in invertebrates is consistent with the general correlation of the type of excretory product and the availability of water. Marine invertebrates do not face an acute water balance problem; therefore, they do not experience any difference in excreting the more toxic ammonia. Various sites of excretion have been described, including the body surface, "kidneys" and extrarenal organs such as gills. The gills appear to be the principal route of ammonia excretion in the crustaceans.

Elasmobranch fish excrete urea as their principal waste of nitrogenous metabolism. Apparently all tissues synthesize urea except the brain and the blood. As indicated elsewhere, a significant amount of urea is retained in the blood of fish, where it plays a key role in maintaining osmoconcentration. Excess urea is excreted mainly through the gills. Some studies have emphasized the influence of environmental change on excretion.

When the shark *Scyliorhinus canicula* is exposed to dilute sea water, urea synthesis was increased. However, studies have shown the total amount of urea within the shark tends to be somewhat constant under different salinity conditions, suggesting that urea in the tissues may play an important role in providing chemical homeostasis (Alexander et al., 1968). Conceivably, proteins of this species have evolved to depend metabolically on urea.

Muscles of the spurdog *Squalus acanthias* were analyzed to determine the effect of certain environmental and biotic factors on concentrations of important nitrogenous compounds (Vyncke, 1970). Feeding habits and the sex of the animal had no influence on the concentration of these compounds, but amino nitrogen, alanine, glutamic acid, and glycine decreased after the fish became sexually mature. Urea, ammonia, and trimethylamine oxide concentration did not significantly vary with season. However, it should be noted that the population studied lives in the North Sea under fairly constant environmental conditions. Little seasonal variation in salinity occurs, and behavioral responses keep the animals in waters with a temperature of 7° to 12°C. In contrast and of unknown functional significance, peptides were in highest concentration during the spring/summer period, but the stage of sexual maturity had no significant influence on concentrations.

Trimethylamine oxide (TMO) is another excreted nitrogenous product. Although earlier papers reported as much as one-third of the excreted nitrogen to be TMO, Wood (1958) maintained that typically relatively small amounts were excreted by teleosts. Food is the principal source of TMO. In sharks the relatively high amounts of blood TMO play an osmoregulatory role.

Lampreys spend most of their adult life in the sea but migrate to fresh water to spawn. The lamprey *Entosphenus tridentatus* captured in fresh water apparently does not have a complete ornithine-urea cycle with only carbomethyl-phosphate synthetase and arginase activities detected in the liver. Whether sea water-adapted individuals have a complete complement of ornithine-urea cycle enzymes is not known (Read, 1968). Ammonia is the chief excretory product of eels, and it is excreted mainly via the gills.

Teleost fish have been considered ammonotelic since their major end product of nitrogen metabolism is ammonia; however, significant quantities of urea have been found in teleost blood. Studies on the biosynthesis of urea in liver slices from teleost fish demonstrated that the rate of conversion of uric acid to urea varied from a low of 5 μmoles of urea per g-hr in the goosefish to 23 μmoles per g-hr in the winter flounder. Allantoin and allantoic acid were also converted to urea at approximately the same rate as uric acid. The enzyme allantoicase was found in the livers of 18 species of teleosts, indicating its widespread occurrence (Goldstein and Forster, 1965).

Marine birds excrete uric acid, as apparently sea snakes and lizards also do. These animals face a water scarcity, and uricotelism is one method of conserving water. On the other hand, the green turtle excretes chiefly urea and ammonia. Marine mammals appear to excrete mostly urea in their urine, but up to 15% of the total nonprotein nitrogen is excreted as ammonia.

6. Reproduction and Development

Animals that live and reproduce in coastal and open waters live in a somewhat more stable environment than do intertidal and estuarine organisms. Salinities are relatively constant and high, and, although temperature is an important factor throughout the reproductive cycle, particularly in animals living in coastal waters, temperature fluctuations in temperate zone areas tend to be less than in intertidal zones or estuaries.

a. Temperature

In tropical coastal and open ocean waters, where both temperature and salinity are relatively constant, there has been speculation for many years about whether or not reproduction in tropical species is cyclical. The idea that tropical species do have continuous breeding cycles was first introduced by Semper in 1881, when he reported on animals found around the Philippine Islands. In his classic paper on breeding and distribution in marine animals, Orton (1920) hypothesized that in areas where the sea is always warm, animals should breed continuously. Other workers (Thorson, 1950; Giese, 1959; Boolootian, 1966; Gunter, 1957) have repudiated this viewpoint, concluding instead that tropical marine species do have discrete reproductive periodicities. However, Pearse (1968) has observed that although many of the more recent studies were based on tropical species, a number of these were made in subtropical environments characterized by definite seasonal changes in temperature. In studying patterns of reproductive periodicity in species of endo-Pacific echinoderms, Pearse (1968) found that four species of echinoderms—the echinoids *Diadema setosum* and *Echinometra mathaei,* the asteroid *Linckia laevigaten,* and the holothuroid *Holothuria astra*—spawned more or less continuously throughout the year when they occurred near the equator. However, populations of these species living at higher or lower latitudes showed a more restricted reproductive activity. For example, the reproductive cycles of *D. setosum* become more cyclical at latitudes 10–15° from the equator, and in populations occurring at approximately latitude 30°, spawning is limited to the summer months. Thus, whether or not these animals have definite reproductive cycles seems to be a function of temperature fluctuations associated with higher latitudes.

In some tropical species living in subtropical areas where seasonal fluctuations are pronounced, however, reproductive cycling relates only indirectly to fluctuations in temperature. Populations of the echinoid *Lovenia elongata* in the Gulf of Suez are subject to pronounced seasonal fluctuations in sea temperature, yet the temperature fluctuations do not coincide with reproductive periodicity (Pearse, 1969). Gametogenesis begins in mid-winter when temperatures are below those experienced by this species in other parts of its distributional range. Spawning begins in late April when the maximum sea temperatures are about 25°C and is completed in September when the minimum sea temperatures drop below 25°C. Salinities are also relatively constant, averaging about 42.5 0/00. Since the mid-winter period of gametogenesis in the Gulf of Suez corresponds to a time when the photoperiod is shorter than in other habitats of this species, photoperiod would not seem to be an initiating factor in the reproductive cycle. Pearse suggests that the reproductive cycle in this population is correlated with the ability of the animals to accumulate nutrient material. When temperatures are high, respiratory and activity rates increase, and energy supplies are channeled into these activities. Gametogenesis then occurs during the cooler months of the year, when metabolic demands are not great and nutrient material can be accumulated; conversely, gametogenesis is terminated in mid-summer when temperatures are high, because the food supply is inadequate to maintain gametogenesis in the face of heightened respiratory and activity rates.

Relatively little is known about gamete release in animals living in deeper waters of the open ocean. Ophiuroids *Gorgonocephalus caryi,* collected off the coast of Washington in water 60–210 meters deep, showed an annual reproductive cycle in which they spawned six months of the year, from June through November (Patent, 1969); temperature fluctuations within the collecting area were not given. Another deep sea echinoderm, *Allocentrotus fragilis,* which lives at depths of 90 to 840 meters, also has a distinct breeding cycle although ambient temperatures vary only 2°C (Giese, 1959).

Temperature also may determine whether the animal reproduces sexually or asexually, as illustrated by Werner (1963) with the anthomedusa *Rathkua octopunctata.* At temperatures below 6° to 7°C, it reproduces sexually (Fig. 5.19).

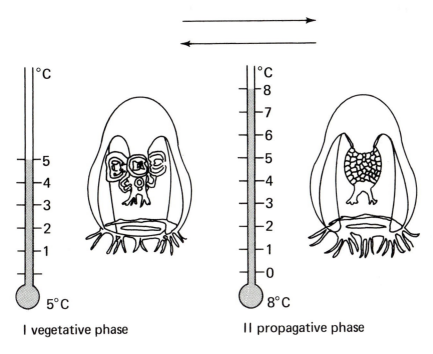

I vegetative phase   II propagative phase

Rathkea octopunctata

Fig. 5-19. The influence of temperature on the type of reproduction in the anthomedusa, *Rathkea octopunctata*. (From Werner, 1963.)

Adaptations of Arctic and Antarctic animals to extreme cold commonly include brooding of the young and vivipary, thereby protecting the young from the harsh surface waters. In a study on lamellibranchs from the waters around East Greenland, Ockelmann (1958) reported three types of larval development. (1) Pelagic plankton-feeding larvae; eggs of these species are small and have little nutritive material. Only 15% of the 47 species studied had this type of larval development. (2) Species with lecithotrophic larvae (large and yolky eggs) with a short or entirely suppressed pelagic stage. (3) Species with internal brood-protection and with no pelagic stage. Forty species, 85%, had one of the last two types of larval development.

Recently it was suggested that the demersal behavior pattern of some cold-water animals also serves the same purpose. In the laboratory, embryos and larvae of the Antarctic sea star *Odontaster validus* were found to be primarily bottom swimmers, indicating that they were not pelagic (Pearse, 1969).

b. Photoperiod

Few studies have investigated photoperiodic effects on reproduction in animals found in the open ocean. The detailed study of Wiebe (1968) on the reproductive physiology of the viviparous seaperch *Cymatogaster aggregata* shows that photoperiod markedly affects spermatogenesis and development of secondary sex characteristics at certain times of the year. During winter, temperature was more effective in stimulating sexual development than photoperiod. Fish maintained at 10°C on either a short- or long-day photoperiod showed neither spermatogenesis nor secondary sex structure, but a temperature of 20°C, regardless of photoperiod regime, caused spermatogenesis and development of secondary sex structures on the anal fins. In late winter and spring, however, photoperiod had more influence than temperature. Warm temperatures and short days were far less effective than a regime of warm temperature and a long day. Growth also occurred when the fish were maintained on a long day at 10°C, although the rate of development was slower (Fig. 5.20). Wiebe has speculated that these responses to temperature and light can be directly correlated with environmental conditions normally encountered by the fish. In the spring they may be found both near the shore, where water temperatures may reach 20°C, and at depths of approximately 60 feet, where water temperatures are about 11°C. Such a wide range of temperature would not offer a reliable cue for development, but the photoperiod in the ocean environment is relatively constant and predictable from year to year. Thus, spermatogenesis in *Cymatogaster* begins early in the spring as daylength increases, and is further enhanced by the warmer temperatures encountered by the males as they move shoreward.

Circumstantial evidence from field studies indicates that photoperiod may also influence reproductive events of nonmigrating animals found in relatively deep water. Studies have shown that a Neapolitan population of the cidaroid sea urchin *Stylocidaris affinis,* taken from a depth of 70–75 meters, had an annual reproductive cycle (Holland, 1967), although the animals were subject to an environment where there were scarcely any environmental fluctuations throughout the year. Temperatures in their habitat varied from 13.7° to 14.8°C, salinities from 37.58 to 38.06 0/00, and oxygen concentrations from 7.4 to 8.4 mg/liter. Thus, it is unlikely that any of these nonfluctuating exogenous factors could control or synchronize gametogenesis and spawning. There are, however, annual fluctuations in photoperiod, which Holland has suggested influence synchronization of reproductive events. But the effect is not altogether clear, since the long periods of oocyte growth, the short periods of spawning, the initiation of oocyte growth, and the initiation of spermatocyte accumulation were not related to photoperiod. Two possible photoperiodic effects have been postulated. Photoperiod could be used as a periodic reference point for synchronizing an endogenous reproductive rhythm, or photoperiod could indirectly influence reproduction by influencing the quantity and quality of available algal food.

c. Lunar phases

One of the best known examples of lunar periodicity in reproductive be-

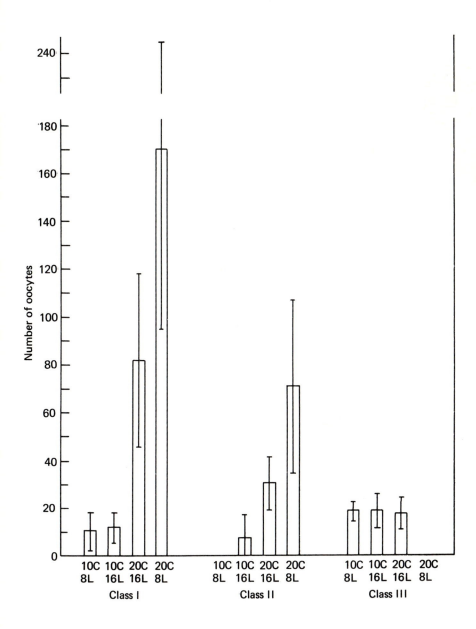

Fig. 5-20. Effect of temperature and photoperiod on oocyte formation and maturation in the viviparous sea perch, *Cymatogaster aggregata*. (From Wiebe, 1968. National Research Council of Canada.)

havior is that of the grunion *Leuresthes tenius*. This small, edible fish, common off southern California, leaves the sea to deposit its eggs on open sand beaches. The spawning cycle is semilunar, extending from March through August, with the heaviest spawning occurring in late spring and summer. During a spawning run the females, usually accompanied by several males, swim ashore on an ingoing wave just after the tide begins to fall. The fish quickly dig pits 2–3 inches deep in the sand, release eggs and sperm in the wet sand, and return to the ocean with the next wave. Spawning must be timed precisely so that the runs coincide not only with the highest waters of each tide, but also with the highest tides each month. Since the eggs take about two weeks to develop in the wet sand, their emergence must be concurrent with the next high tide that occurs within 12–14 days, for if the eggs are not laid at the proper time, they cannot complete development before being washed out to sea. Korringa (1957) has suggested that the semilunar rhythm is linked to a tidal rhythm rather than to light intensity. The exact mechanism remains unknown, however, and Korringa asks the really intriguing question about this rhythm: "How do the animals know which is the maximum tide of the spring series, and when the particular tide has passed its peak?"

The Pacific palolo worm *Eunice viridis* offers another example of lunar spawning. These animals, which live in crevices and burrows of corals and rocks, leave their burrows during the spawning season and swim freely on the surface of the water. Their bodies break into two parts, and the egg- and sperm-filled posterior segments are left behind in great masses on the surface of the water; the anterior half returns to the burrows in the rock or coral and apparently regenerates the reproductive parts. The bulk of these worms swarm on a single day sometime between November 7 and December 22, seven to nine days after a full moon, and usually at dawn. Stimulus for spawning is probably not linked to tidal fluctuations, since tidal amplitude in this area is small and new- and full-moon spring tides alternate cyclically in magnitude over several years. Moonlight itself is not essential for swarming to occur, for the worms spawn regardless of moon and cloud conditions. Temperature and salinity are also relatively unchanging. It is believed instead that the cycle is directly related to change in light. Korringa (1957) has suggested that in these worms, which are nocturnal in habit, the rhythmical sequence of dark and moonlit nights and not the light intensity are of great importance, and that moonlit nights evoke a simultaneous ripening of the gonads. Thus, by the last quarter of the moon, all of the worms reach maturity on approximately the same day.

Some evidence has indicated that deep-living animals also experience semilunar spawning. Sudden sperm release has been observed in tropical demosponges living at depths as great as 49 meters during new and full moon (Reiswig, 1970).

A general review of lunar rhythms in marine animals can be found in a recent review by McDowall (1969).

light produced by their photophores. Some species of fish with poorly developed eyes also may have photophores, but their function is even less obvious.

2. Feeding

Since light is absent in the abyssal region, there is no primary production of food by photosynthetic plants. Most of the energy in this ecological system probably is introduced from the overlying water mass, although deep sea and heterotrophic algae and fungi as well as bacteria may be an important energy source. Various mechanisms for transport of food to deep sea communities have been proposed, but as is true in other phases of deep sea studies, few data are available to support definitive generalizations. One possible mechanism of food transport is the gradual settling to the deep sea of the remains of excreta and dead bodies of surface and mid-water organisms in the form of an endless rain. Observable amounts of terrestrial and intertidal debris, chiefly grasses and large algae which have food value, have been found at abyssal depths. For example, considerable amounts of turtle grass (a tropical grass) have been recovered from the Hatteras Submarine Canyon, off the shore of North Carolina (Menzies, et al., 1967).

A "ladder of migration" theory has also been proposed as a food source. This theory postulates a series of partly overlapping vertical migration ranges of living animals, extending from the surface to the ocean depths. Thus, food would be passed from one depth to another in a manner resembling an overlapping chain or ladder (Vinogradov, 1962). This mechanism would permit a relatively rapid method of food transport as opposed to the generally slower sinking velocities of inanimate objects.

It is known that organically rich sediments from shallow waters can be carried to deeper waters by means of swift submarine turbidity currents (Heezen et al., 1955). However, as Sanders (1968) has pointed out, the organic material moved by these currents may not necessarily be available to the deep sea fauna as a food source. Another source of food may be organic aggregates which adsorb dissolved organics from the sea. Bacteria or filter-feeding animals could feed directly on these aggregates (Riley et al., 1965).

The majority of deep sea animals that have been studied are detritus feeders. The relative abundance of animals collected at one oceanographic station will serve to demonstrate this point: 60% of polychaetes, more than 90% of the Tanaidacea, 90% of the Isopoda, in excess of 50% of the Amphipoda, and 45% of the Pelecypoda, were detritus feeders (Sanders and Hessler, 1969). Of these groups the isopods occur in both deep sea and shallow water environments, but the deep sea forms do not have the wide diversity in feeding types reported for shallow water species. Among the deep sea species there are no known scrapers, parasites, or commensals. Using gut contents as an index, Menzies (1962) found that the abyssal isopods were primarily deposit feeders and carnivores and less than 1% were filter feeders. He also found a high incidence of sediment particles in their digestive tract and suggested that these abyssal animals could survive and grow on adsorbed organic matter and its associated microflora.

Many deep sea fish have jaws and stomachs which are extensible, permitting the fish to swallow relatively large quantities of food at one time. In this manner the fish have adapted to meet the stress of scarcity of food in the environment.

3. Chemoregulation and Excretion

Since salinity varies only slightly in the deep sea, it might be expected that most of the invertebrates would be stenohalinic and poikilosmotic, and that deep sea fish probably osmoregulate as do their surface-dwelling cousins. The low temperatures experienced at these great depths may be close to the freezing point of the body fluid of the animals, especially the fish. The blood of most marine fish freezes between −0.7° and −0.9°C. Based on work with Arctic fish, the suggestion has been made that animals living at great depths and never exposed to ice live in a supercooled condition, with no special antifreeze substance in their blood (Gordon et al., 1962).

4. Reproduction and Development

In most intertidal zone and shallow water animals, exogenous and endogenous factors interact to initiate breeding cycles. But what about breeding cycles of the deep sea organisms living in their uniform, unvarying environment? The limited data available on this subject indicate that both cyclic and continuous reproduction occurs among these animals. Evidence from studies on two species of brittle stars, *Ophiura ljungami* and *Ophiomusium lymani,* supports the theory of cyclic reproduction (Schoener, 1968). These two species of ophiuroids were sampled seasonally over a three-year period, and their size was recorded and gonadal activity indexed. A definite periodicity was suggested by the well-developed gonads and eggs found during the winter. Also the greatest number of small brittle stars were collected during the summer. Schoener postulated that this cyclic breeding season might be correlated with a periodic change in nutrient content either as a result of past plankton blooms or the more rapid "ladder of migration" mentioned earlier. Certain deep sea isopods also apparently exhibit cyclic reproduction (George and Menzies, 1967). In contrast, however, the deep sea isopod *Ilyarchna* sp. and the bivalve *Nucula cancellata* were found to breed continuously throughout the year (Sanders and Hessler, 1969).

Unlike many related shallow water species, planktonic larval stages tend to be suppressed in abyssal organisms. Generally, deep sea animals produce a few large eggs, and the developing organism is in an advanced stage of development by the time of hatching. Viviparity is not uncommon. These modifications to the reproductive scheme appear to be correlated with the shortage of suspended food in the deep sea environment (Thorson, 1950). There are, however, some deep sea fish, especially eels, that live at the surface of the water as larvae before metamorphosing into small eels. After metamorphosing, the eels migrate to great depths, later returning to the surface as sexually mature adults to spawn.

In shallow water forms, the eggs of certain molluscs, tunicates, and annelids would not cleave at pressures in excess of 270 atmospheres, and in six species of echinoderms cleavage was inhibited at 400 atmospheres. Obviously, the deep sea animals have evolved adaptations to overcome the effects of pressure during early development, but it is not known exactly what these adaptations include (Pease and Marsland, 1939).

5. Adaptations to Hydrostatic Pressure

Hydrostatic pressure has long been recognized as an important environmental factor in the life of abyssal organisms. Johnson and coworkers (1954) reviewed earlier studies on pressure and its effects on animals. Hydrostatic pressure, the weight of the water, is entirely separate from the high pressure of air, oxygen, nitrogen, or some other gas. The concentration of solutes changes only slightly under greater compressibility, changing only a few percent even at the greatest depths. Hydrostatic pressure is uniformly exerted in all directions at any one level, but the effects of pressure may be different between different pressure ranges. For example, between 1 and 1,000 atm, protein denaturation is retarded or reversed; between 1,000 and 5,000 atm, protein denaturation is accelerated. Finally, the effects on biological systems during compression are quite different from those effects after the release of pressure.

Two main physiochemical effects of high pressure have biological consequences: (1) the slight compression of sea water with increased depth, going from 1.8% at 4,000 meters in 35 0/00 sea water at 2°C to 4.14% at 10,000 meters; and (2) the rise in degree of dissociation with increasing pressure. The degree of dissociation, which is essential for the buffering capacity of the carbonate system of sea water, changes from a surface pH of 8 at 5°C to 7.88 at 1,000 atm (Schlieper, 1968).

It is probable that the physiochemical effects are largely responsible for the radical changes in decreases in population densities with increasing depth and the radical changes in faunal composition. Zenkevitch (1954) reported that holothurians and starfish were the dominant organisms at 5,000 meters; at 8–10,000 meters, there were no starfish, but large increases in numbers of polychaetes, pogonophorans, isopods, and amphipods. Although some abyssal forms tend to be larger than their shallow water counterparts, there are no clear-cut anatomical differences between related species found in shallow and deep water. The complexity of animals has been said to decrease as depth increases, as illustrated by the work on the crab *Pachygrapsus crassipes* and the bivalve *Mytilus edulis* (Menzies and Wilson, 1961). These two shallow water tidal species were gradually lowered to different depths in the ocean. The lethal point for the more physiologically complex crustaceans was 100 atm; the less complex mollusc survived to a depth of 200 atm.

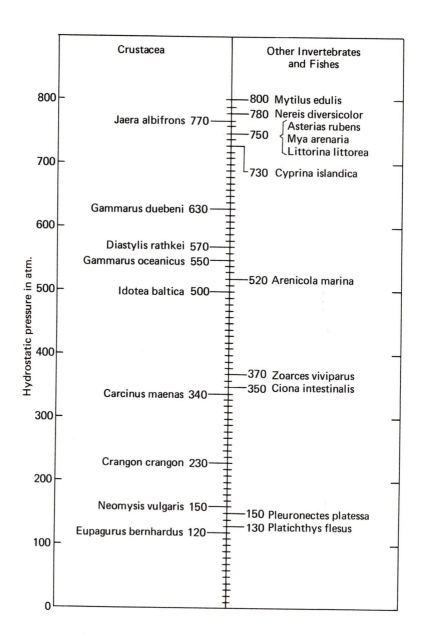

Fig. 6-3. Experimental pressure limits (LD 50—atm) of some marine invertebrates and fishes from shallow waters of the western Baltic Sea. Duration of pressure exposure: 1 hour 10°C; 15 0/00 S. (After Naroska, 1968.)

Laboratory experiments on pressure effects have been conducted primarily on shallow water species, since living deep sea organisms are difficult to obtain. Based on these studies, three different types of pressure resistance in marine species are recognized. Species that are limited to the upper sea layers are termed *stenobathic-barophobic* animals. Organisms that can live only under great hydrostatic pressure are said to be *stenobathic-barophilic,* and animals that can tolerate great ranges of pressure are called *eurybathic.* A surprising number of shallow water species are relatively independent of pressure (Fig. 6.3). Interestingly, the most pressure tolerant organisms living in shallow water have closely related taxonomic representatives in the deep sea (Schlieper, 1968).

The stimulating effect commonly caused by sublethal pressures is illustrated in the medusa of *Cyanea capillata* and in fish and shrimp up to 200 atm. Muscle contraction increased with pressure until at 500 atm the animals died, their muscles totally contracted (Ebbecke, 1935). The oxygen consumption of some animals is also stimulated by increased sublethal pressures. Naroska (1968), working with the flounder *Platichthys flesus,* the green crab *Carcinus maenas,* and the starfish *Asterias rubens,* found that an increase in pressure to 100 atm greatly stimulated respiratory rates in the *P. flesus* and *C. maenas.* However, in *A. rubens,* which is a pressure-resistant species, increased pressure had no effect on oxygen uptake (Fig. 6.4).

There seems to be some correlation between resistance to pressure and eurythermy, as illustrated by the study of Schlieper and coworkers (1967) on excised gill tissue from tropical and temperate zone molluscs. Cellular activity in the gills of the tropical zone species *Chama cornucopia* and *Modiolus auriculatus* decreased at 300 atm, reaching zero activity at 350–400 atm. Cellular activity in tissue of the temperate zone species *Mytilus edulis* and *Ostrea edulis* decreased only when pressures reached 400 atm and ceased completely between 700 and 800 atm (Fig. 6.5). Similar results were obtained when the cold stenothermal tissues of the shallow water bivalve *Cyprina islandica* were compared with *M. edulis,* with the latter being more resistant to pressure than the stenothermal species. Tissue resistance to pressure of the boreal species *Modiolus modiolus,* which occurs at the 15–50 meter level, was found to be much more resistant than the tropical shallow water species *M. auriculatus.* Thus, Schlieper and his coworkers (1967) concluded that cellular resistance to pressure increases with depth distribution and eurythermy. They speculated that the greater stability of proteins and protein complexes that accompany eurythermy also provide these animals with greater pressure resistance.

To test the hypothesis that pressure resistance is somehow linked to protein complexes, Schlieper and coworkers (1967) placed pressure-sensitive hermit crabs, *Eupagurus zebra,* in sea water of double calcium concentration. Previous work (Schlieper and Kowalski, 1956) had shown that temperature and osmotic resistance of tissues of mussels could be shifted by preadapting the animals in calcium-fortified water. Without preadaptation to sea water with added calcium, *E. zebra* died at 250 atm. After adaptation for at least three days in sea water

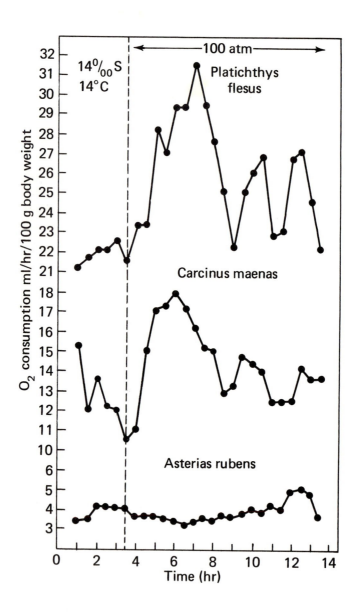

Fig. 6-4. Influence of increased hydrostatic pressure (100 atm) on the oxygen consumption of the sea star *Asterias rubens,* the brachyuran crustacean *Carcinus maenas,* and the flat fish *Platichthys flesus.* (After Naroska, 1968.)

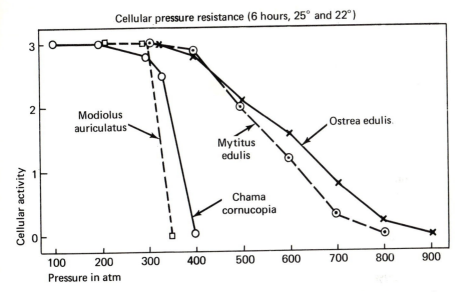

Fig. 6-5. Cellular reaction of excised gill tissues after tolerating different hydraulic pressures for six hours. Cellular activity of 3 is normal; 0 is no ciliary activity. (From Schlieper *et al.*, 1967. University of Chicago Press.)

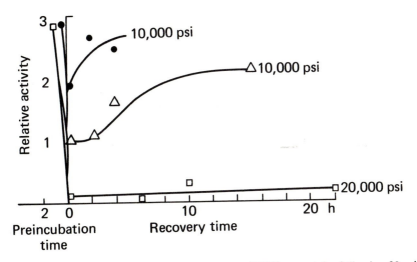

Fig. 6-6. *Coryphaenoides acrolepsis.* Recovery of FDPase activity following 30 min exposure to 10,000 psi (●); 60 min exposure to 10,000 psi (△); and 60 min exposure to 20,000 psi (□). Partially purified enzyme was pre-incubated in the absence of substrate. Recovery of activity was measured at 3°C and 1 atm pressure. (From Hochachka *et al.*, 1970.)

plus calcium, the crabs survived without damage at 250 and 270 atm, and their survival suggested a further link between protein complexes and resistance to pressure.

Evidence from sea urchin eggs indicates that the enzyme systems of cells are a part of pressure effects on marine animals. Metabolic energy is needed to maintain protoplasmic gel structures, and this energy is supplied by splitting of ATP. In fertiized sea urchin eggs, pressure resistance is increased when ATP is added to the medium and decreased if ATP hydrolysis is inhibited (Marsland, 1958). The same tendency is shown in *Escherichia coli* and *Allomyces macrogynus* (Morita, 1967a, b; Hill and Morita, 1965). With increased pressure there was a reduction in dehydrogenase enzyme activity. Gillen (1971) also reported that the muscle lactate dehydrogenase activity of both bathypelagic and shallow water fish was reduced when the hydrostatic pressure was increased.

Another enzyme study on diphosphatase (FDPase) from the liver of fish taken from different pressure levels demonstrated that enzymatic activity was inhibited at great pressure even in an abyssal fish, *Coryphaenoides acrolepsis,* which lives at depths of 1–3,000 meters (Hochachka et al., 1970). FDPase, an important control in the pathway of glucose synthesis from amino acids and other precursors, catalyzes the conversion of fructose diphosphate (FDP) to fructose-6-phosphate (F6P) and inorganic phosphate in the presence of Mg^{2+} or Mn^{2+}. After a 30-minute exposure to a pressure of 10,000 psi, the enzyme showed a 50% initial inactivation, although the enzyme recovered full activity within the first two hours after pressure release (Fig. 6.6). After exposure to this pressure for 14 hours, however, part of the inactivation was irreversible, and only 70% of enzymatic activity was recovered after the pressure was released. A pressure of 10,000 psi is much greater than would be encountered in nature; usually abyssal depth pressure averages about 5,000 psi. Prolonged ex-

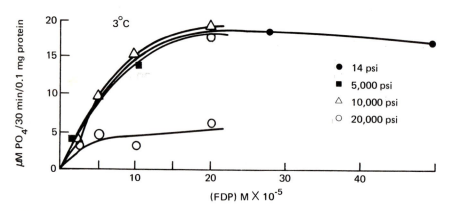

Fig. 6-7. *Coryphaenoides acrolepsis.* Catalytic activity of partially purified liver FDPase at different FDP concentrations and different pressures. Assayed at 3°C, Mg^{2+} concentration was 1 mM. (From Hochachka *et al.,* 1970.)

posure to pressures within the biological range that the fish would encounter in nature also produced enzymatic inhibition at a temperature of 3°C. This low temperature is also within the range the fish would encounter in nature; and, in fact, the enzyme was less sensitive to pressure effects at the higher temperature of 9.5°C. What mechanisms have evolved, then, which protect the enzymes of deep sea organisms against high pressure and low temperatures? Hochachka and his coworkers found that FDPase was protected from pressure sensitivity when both substrate (FDP) and cofactor (Mg^{2+}) were added (Fig. 6.7). The homologous enzyme from a surface fish, *Pimelometopon pulchrum,* showed much

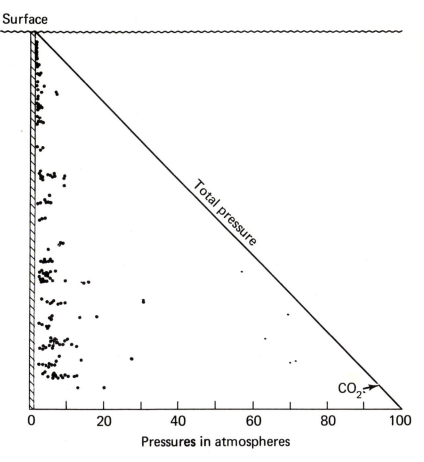

Fig. 6-8. The partial pressures of nitrogen, oxygen, and CO_2 calculated from the percentages and the depths. The shaded area is equal to the partial pressure of nitrogen in the sea water. The partial pressure of nitrogen in the swimbladder is to the left of each point; the partial pressure of oxygen is to the right. The partial pressure of the CO_2 is usually less than what is represented by the thickness of the drawn diagonal line. The overall picture suggests a linear increase in the partial pressure of the nitrogen with depth. (From Scholander and Van Dam, 1953.)

greater sensitivity to high pressures and low temperatures than did the abyssal one (Hochachka et al., 1970).

Deep sea animals are often exposed to reduced oxygen levels in their ambient environment, and in invertebrates it has been found that either very high or low oxygen tensions exert a profound influence on resistance to increased hydrostatic pressure. Pressure resistance of both the whole animal and isolated gill tissue decreased with increasing oxygen tension up to about air saturation or higher. Increased pressure resistance at low oxygen levels occurred not only during exposure, but also pressure resistance increased in isolated tissues kept under oxygen deficiency before the experiment (Theede and Ponat, 1970). It has been suggested that the primary reasons for this decreased resistance at higher oxygen levels are oxidation and inhibition of sensitive SH-enzymes. Pressure resistance is also decreased when dissolved oxygen is almost absent, and Theede and Ponat (1970) have hypothesized that insufficient rates of ATP synthesis may explain the decreased resistance at these extremely low oxygen concentrations.

The air bladder of fish represents another adaptation to hydrostatic pressure. This organ, which functions principally as a hydrostatic organ enabling fish to move from one level to another, has long been of interest to biologists. When a fish is rapidly brought from a great depth to the surface, the gas in the air bladder expands and the bladder may rupture. The presence or absence of a swim bladder can be broadly correlated with the habitat and vertical distribution. Many bathypelagic and bathybenthic species do not have swim bladders, whereas mid-water pelagic species which undergo vertical migration have both swim bladders and well-developed gas glands. The swim bladder also may function in sound production, reception of underwater sounds, and as a source of oxygen during hypoxic conditions. Scholander and Van Dam (1953) analyzed the composition of the swim bladder gas in deep sea fish (known depths between 0 and 2,660 meters) and reported that both oxygen and nitrogen are deposited into the swim bladder against a concentration gradient. The total pressure of gas is approximately equal to 1 atm per 10 meters in depth (Fig. 6.8). The mechanisms of gas secretion are not completely known. Apparently oxygen does not come from oxyhemoglobin, but the glandular epithelium of the gas gland is important in release and uptake of gases (Scholander and Van Dam, 1954).

REFERENCES

Bruun, A. (1957). Deep sea and abyssal depths. *In* Treatise on Marine Ecology and Paleoecology. J. W. Hedgpeth, ed. Mem. Geol. Soc. Am. **67**:641–672.

Denton, E. J., and F. J. Warren (1957). The photosensitive pigment in the retinae of deep-sea fish. *J. mar. biol. Ass. U.K.*, **36**:651–662.

Ebbeke, U. (1935). Über die Wirkung hoher Drucke auf marine Lebewesen. *Pflügers Arch. ges Physiol.*, **236**:648–657.

George, R. Y., and R. J. Menzies (1967). Indication of cyclic reproductive activity in abyssal organisms. *Nature, London,* 215:878.

Gillen, R. G. (1971). The effect of pressure on muscle lactate dehydrogenase activity of some deep-sea and shallow-water fishes. *Mar. Biol.,* 8:7–11.

Gordon, M. S., B. H. Amdur, and P. F. Scholander (1962). Freezing resistance in some northern fishes. *Biol. Bull.,* 122:52–56.

Heezen, B. C., M. Ewing, and R. J. Menzies (1955). The influence of submarine turbidity currents on abyssal productivity. *Oikos,* 6:170–182.

Hill, E. P., and R. Y. Morita (1965). Dehydrogenase activity under hydrostatic pressure by isolated mitochondria obtained from *Allomyces macrogynus. Limnol. Oceanogr.,* 9:243–248.

Hochachka, P. W., D. E. Schneider, and A. Kuznetson (1970). Interacting pressure and temperature effects on enzymes of marine poikilotherms: Catalytic and regulatory properties of FDPase from deep and shallow-water fishes. *Mar. Biol.,* 7:285–293.

Johnson, F., H. H. Eyring, and M. J. Pollisar (1954). The Kinetic Basis of Molecular Biology. John Wiley & Sons, Inc., New York. Chapters 9 and 10.

Koczy, F. F. (1954). A survey on deep-sea features taken during the Swedish deep-sea expedition. *Deep-Sea Res.,* 1:176–184.

Kuenen, H. P. (1950). Marine Geology. John Wiley & Sons, Inc., New York.

Laverack, M. S. (1968). On the receptors of marine invertebrates. *Oceanogr. Mar. Biol. Ann. Rev.,* 6:249–324.

Marsland, D. A. (1958). Cells at high pressure. *Sci. Amer.,* 199:36–43.

Menzies, R. J. (1962). On the food and feeding habits of abyssal organisms as exemplified by the Isopoda. *Int. Revue ges Hydrobiol.,* 47:339–358.

———, R. Y. George, and R. Gilbert (1968). Vision index for isopod Crustacea related to latitude and depth. *Nature, London,* 217:93–95.

———, and J. B. Wilson (1961). Preliminary field experiments on the relative importance of pressure and temperature on the penetration of marine invertebrates into the deep sea. *Oikos,* 12:302–309.

———, J. S. Zaneveld, and R. M. Pratt (1967). Transported turtle grass as a source of organic enrichment of abyssal sediments off North Carolina. *Deep-Sea Res.,* 14:111–112.

Morita, R. J. (1967a). Effect of hydrostatic pressure on succinic, formic, and malic dehydrogenases in *Escherichia coli. J. Bact.,* 74:251–255.

——— (1967b). Effects of hydrostatic pressure on marine microorganisms. *Ann. Rev. Oceanogr. Mar. Biol.,* 5:187–203.

Naroska, V. (1968). Vergleichende Untersuchungen über die Wirkung des hydrostatischen Druckes auf die Überlebensfähigkeit und die Stoffwechselintensität mariner Evertebraten und Teleostier. *Kieler Meeresforsch.,* 24:95–123.

Pease, D. C., and D. A. Marsland (1939). The cleavage of *Ascaris* eggs under exceptionally high pressure. *J. Cell. Comp. Physiol.,* 14:407–408.

Riley, G. A., D. Van Hemert, and P. A. Wangersky (1965). Organic aggregates in tropical and subtropical surface waters of the North Atlantic Ocean. *Limnol. Oceanogr.,* 9:546–550.

Sanders, H. L. (1968). Marine benthic diversity: A comparative study. *The American Naturalist,* 102:243–282.

——, and R. R. Hessler (1969). Ecology of the deep-sea benthos. *Science,* **163:** 1419–1424.

Schlieper, C. (1968). High pressure effects on marine invertebrates and fishes. *Mar. Biol.,* **2:**5–12.

——, H. Flügel, and H. Theede (1967). Experimental investigations of the cellular resistance ranges of marine temperate and tropical bivalves: Results of the Indian Ocean expedition of the German Research Association. *Physiol. Zool.,* **40:**345–360.

——, and R. Kowalski (1956). Über den Einfluss des Mediums auf die thermische und osmotische Resistenz des Kiemengewebes der Miesmuschel *Mytilus edulis* L. *Kieler Meeresforsch.,* **12:**37–45.

Schoener, A. (1968). Evidence for reproductive periodicity in the deep sea. *Ecology,* **49:**81–87.

Scholander, P. F., and L. Van Dam (1953). Composition of the swimbladder gas in deep sea fishes. *Biol. Bull.,* **104:**75–86.

——, —— (1954). Secretion of gases against high pressures in the swimbladder of deep sea fishes. I. Oxygen dissociation in blood. *Biol. Bull.,* **197:**247–259.

Sverdrup, H. V., M. W. Johnson, and R. H. Fleming (1942). The Oceans, Their Physics, Chemistry and General Biology. Prentice-Hall, New York.

Theede, H., and A. Ponat (1970). Die Wirkung der Sauerstoffspannung auf die Druckresistenz einiger mariner Wirbelloser. *Mar. Biol.,* **6:**66–73.

Thorson, G. (1950). Reproductive and larval ecology of marine bottom invertebrates. *Biol. Rev.,* **25:**1–45.

Vinogradov, M. E. (1962). Feeding of the deep-sea zooplankton. *Rapp. Proc. Cons. Int. Explor. Mer.,* **153:**114–120.

Welsh, J. D., and F. A. Chase, Jr. (1938). Eyes of deep-sea crustaceans. 2. Sergestide. *Biol. Bull.,* **74:**364–375.

Zenkevitch, L. A. (1954). Erforschung der Tiefseefauna im nordwestlichen Teil des Stillen Ozeans. *Publ. Un. int. Sci. biol. (Ser. B),* **16:**72–85.

Debidue Banks at the Belle W. Baruch Coastal Research Institute field station near Georgetown, South Carolina (Courtesy of Dr. P. J. DeCoursey).

CHAPTER VII

PERSPECTIVES

Vast regions of the sea are as yet little known and unexploited. In fact, these areas represent the last unknown frontier on the Earth and also one of the few relatively unpolluted regions, although patches of oil and detectable quantities of DDT have been reported in some of the most remote parts of the ocean. Coastal and estuarine areas, however, are quite another story because they have been exploited for so many years and to such an extent that they are almost biologically dead, and man is entirely responsible. A quick glance at a world map shows that coastal areas generally are population centers, and consequently, these regions are heavily used for industrial development, shipping, fishing, and recreation. Competition for this limited resource is becoming ever keener. Since all parts of the ocean are interconnected, destruction of one area of the marine environment is not a limited event, and the problems resulting from such destruction are international in scope.

A. TYPES OF POLLUTION

1. Thermal

The causes of pollution and destruction are legion. One new problem facing the industrialized world is that of thermal pollution. Whereas many industrial processes produce heat as a waste product, the demand of the steam electric–generating utilities for water as a heat transfer medium is greater than all other industrial processes combined. Estimates of the projected growth of electrical generating capacity by the year 2000 indicate that it will be necessary to build the equivalent of about 300, then 600 and 900 power plants of about 1,000-megawatt capacity each during the next successive three decades (Seborg, 1970). This demand is being met by nuclear installations which are now being built in addition to the conventional fuel plants, and, since they have a larger heat-generating capacity than these conventional plants, the demand for water as a coolant will be even greater. It is expected that nuclear electric–generating capacity will grow from the present 1% to about 50% by the year 2000 (Thompson, 1970). There are now 22 nuclear power plants completed in the United States and 100 others are either under construction or in the planning stages. Of these 122 plants, at least 41 will be located along the sea coast, on estuaries, or on major rivers that are within 100 miles of the sea coast (AEC, 1971). Obviously, the uncontrolled dumping of waste heat into the environment could have deleterious side effects for the animals living there. Although most of the nuclear reactor plants have cooling towers or enclosed ponds to receive heated water before it is released into the environment, one of the problems has been to determine what is an acceptable thermal level for release. There can be no common level for all areas because the thermal regime and the lethal thermal limits of an animal living in the cold waters of Maine are very different from those of an animal living in an estuary in South Carolina. The prospects of solving these problems are not, however, entirely bleak since some imaginative approaches have been suggested, such as the constructive use of heat in climate control, desalting of water, or heating buildings (Singer, 1968). Another suggested approach is to redirect the temperature loads into controlled ecosystems. In this manner "waste" heat could be used to stimulate the breakdown of sewage, stimulate the growth of shellfish, or to introduce new species of desirable organisms into areas that were previously too cold to support them (Mihursky, 1967).

2. Radionuclides

The use of nuclear reactors for the production of electrical energy also introduces another potential hazard into the marine environment, that of radioactivity. The ecological aspects of radioactivity in the marine environment have recently been reviewed by Rice and Baptist (1970), who point out that until recently the principal source of man-made radioactivity in the environment has been fallout from the atomic test explosion, but unless extensive testing is resumed or if there is a nuclear war, the contribution from this activity will continue to diminish. On the other hand, the amount of radioactive wastes will increase, and most of this increase will come from the use of nuclear reactors in electric power plants and ships.

Studies have shown that the greatest concentration of radioisotopes is in the top 5 cm of the ocean in the marine environment; marine animals absorb these isotopes primarily from the water or through food webs. Cycling of radionuclides through the environment is given in Fig. 7.1. The fate of radionuclides depends on the interaction of a number of factors in the environment, some acting to disperse the material, others to concentrate them (Fig. 7.2). Unlike many chemical compounds, the threshold of radiation damage is unknown, but in a number of instances a very small amount has been demonstrated to affect living systems. The establishment of thresholds is complicated by the

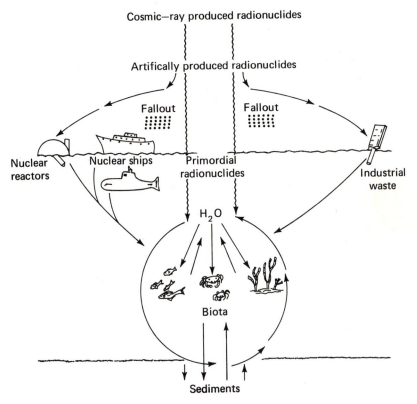

Fig. 7-1. Naturally occurring and artificially produced radionuclides continuously being cycled through the three components of the marine environment—water, biota, and sediments. (Alter Rice and Duke, 1969. The Johns Hopkins University.)

fact that some stages in the life cycle of an animal are more sensitive to radiation damage than others, and it often takes several generations for pathological genetic changes to manifest themselves.

Radioactivity may be introduced into the biota by adsorption, absorption, and digestion and lost by exchange, excretion, and decomposition. Algae, which are among the most resistant of the marine organisms to increased radioactive levels, are able to concentrate ionic radionuclides by adsorption and particulate radionuclides by surface adsorption. Since they have a tremendous surface area per unit volume, rapidly dividing phytoplankton cells can accumulate radionuclides from the water within a few hours (Boroughs et al., 1957). Organisms on a higher trophic level may accumulate radionuclides either by consuming living algae or seaweed or by eating organic detritus, for when the plants die the accumulated radionuclides become part of the organic detritus. Radionuclides that can pass through membranes or that have the same chemical characteristics as essential elements are accumulated in carnivores, and their concentra-

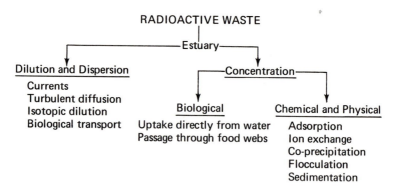

Fig. 7-2. A schematic diagram showing the processes tending to dilute or concentrate radioactive materials added to a marine environment. (From Rice and Baptist, 1970. The Johns Hopkins University.)

tion depends on the turnover rate or the amount available in the food supply, as well as on certain factors in the environment, such as temperature and salinity. Although carnivores can accumulate some from the digestive tracts of food organisms and a little from their tissues, generally they are able to rapidly excrete the major portion of the radionuclides so that no significant concentration is reached (Rice and Baptist, 1970).

Radionuclide concentrations, however, do not ordinarily increase as they move through the food chain in the environment either in the assimilated (in tissues of the prey) or unassimilated form (in gut of prey). In a study on the transfer of ^{65}Zn (assimilated form) and ^{51}Cr (unassimilated form) through a four-step food chain to the fourth trophic level, the levels of concentration generally declined at each higher level in the food chain (Fig. 7.3). Thus, radioactive concentrations decreased rather than increased at the higher trophic levels. (Baptist and Lewis, 1969).

It is impossible to generalize about the sensitivity of marine organisms to ionizing radiation. The vast differences in radiation sensitivities of marine species can be illustrated by comparison of two decapod crustaceans, the grass shrimp *Palaemonetes pugio* and the blue crab *Callinectes sapidus*. The shrimp is much more sensitive than the blue crab; the 40-day LD_{50} for *P. pugio* was 215 rads, for *C. sapidus,* 420,000 rads (Engel, 1967). Some animals are able to take up large amounts of radioactivity and lose it at a slow rate. Consequently they build up high levels in comparison with levels in the water. Such organisms are referred to as "biological indicators," and are of value in detecting low-level radioactive pollution.

Generally, gametes and eggs through the one-cell stage are very sensitive to radiation, and this sensitivity decreases with increasing age. Sensitivity and metabolic rate have also been correlated. Dormant eggs of *Artemia,* for example, are quite resistant to radiation, but once they are hydrated and embryonic development begins, the eggs are twice as sensitive to radiation as dry

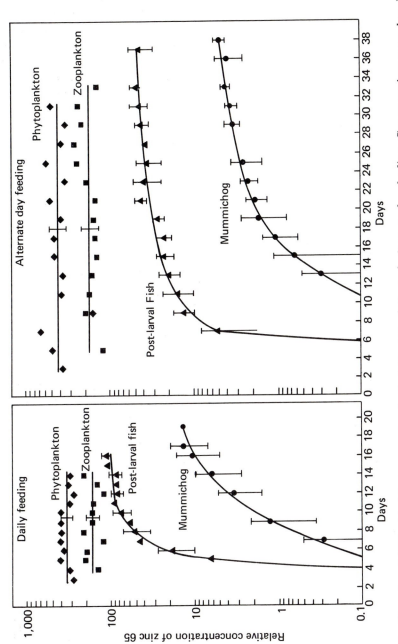

Fig. 7-3. Transfer of ^{65}Zn through an estuarine food chain with daily and alternate-day feeding. Concentrations were based on the initial amount of ^{65}Zn in the phytoplankton culture medium, which was assigned a relative value of 1. Vertical line represents one standard deviation above and below the mean. (After Baptist and Lewis, 1969.)

dormant eggs (Engel and Fluke, 1962). There has been some controversy over the question of possible radiation damage to developing marine animals exposed to low level concentrations of radionuclides. Polikarpov (1966) reported a significant increase in abnormalities in larvae of marine fish exposed to the same concentrations of radioactive strontium and yttrium as is found now in the Irish Sea. Levels of ^{90}Sr and ^{90}Yr are higher in this area than in many other regions since radioactive wastes of British nuclear plants are discharged into the Irish Sea. Other workers using similar concentrations of ^{90}Sr and ^{90}Yr, however, have reported no observable increases in mortality or in the production of abnormal fish larvae (Brown, 1962; Brown and Templeton, 1964; Templeton, 1966). A more complete discussion of this controversy is available in a recent review by Templeton et al. (1970).

The effect of radionuclides on animals in the environment is not necessarily the same under optimum conditions as under environmental stress. Consider the work on the euryhaline fish, *Fundulus heteroclitus,* exposed to the combined effects of ionizing radiation, temperature, and salinity (Angelovic et al., 1969). The LD_{50} of this fish was markedly affected with ^{22}Na efflux in combination with temperature and salinity stress. An increase in salinity at temperatures above $20°C$ increased mortality, but in lowered salinities mortality was reduced at temperatures below $20°C$ (Fig. 7.4). It is possible that the higher temperature enhanced radiation sensitivity by limiting osmoregulatory capabilities, for an irradiated fish generally lost ^{22}Na more rapidly than unirradiated fish (White et al., 1967).

Each radioactive isotope has a characteristic rate of disintegration which is indicated by its half-life. The half-life is constant and unaffected by environmental factors. One of the problems involved in the production and release of radioactive waste is that it cannot be made biologically harmless before it is released into the environment, unless stored indefinitely—for there is no known method for destroying radioactivity. Thus, an increase in environmental radioactivity would seem to be inevitable. To insure the protection of man and aquatic organisms against such an increase, man must be able to provide knowledge of the cycling of radionuclides and of the organismic effects of radiation in association with other environmental factors. Long-term effects of low-level radiation need particular attention. Only when the radiation biology of different species in the marine environment is understood will it be possible to determine permissible levels for the release of radioactivity to the environment.

3. Industrial Wastes

Radioactive pollution is still at a very low level, but industries for years have been releasing large amounts of industrial wastes into the rivers and the estuaries, and the practice continues. Although these wastes may be greatly diluted in some areas, even very low concentrations have marked effects on the animals living there, particularly on their developmental stages. For example,

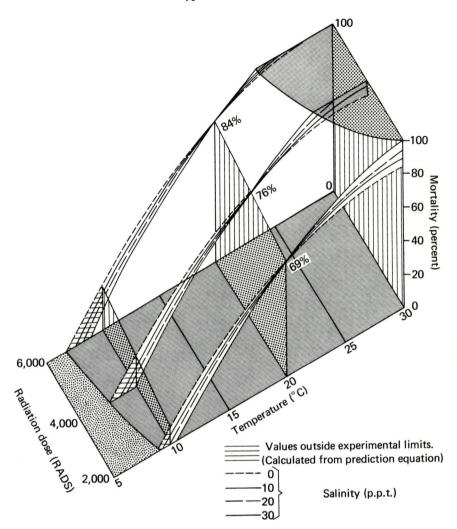

Fig. 7-4. Predicted effects of temperature, salinity, and radiation dose on the percentage mortality of mummichog 20 days after irradiation. (After Angelovic *et al.,* 1969.)

a concentration of 1:32,000 parts of sulfuric acid has only limited effects on fertilization and anatomical development of the herring *Clupea harengus,* but it does affect larval behavior by preventing young larvae from performing prey-catching maneuvers. A concentration of 1:8,000 resulted in death within a few days. Sublethal concentrations had notable harmful effects, such as a lowered percentage of successful fertilization and egg survival, retardation of embryo growth, a shortened period of incubation, and decreased percentage of hatching (Kinne and Rosenthal, 1967).

Many heavy metals are present in trace amounts in the sea and are normal constituents of marine organisms; some are essential for normal growth and development. Rather large quantities of heavy metals are now being emptied into the estuaries and coastal regions from industrial plants, from sewage, and from atmospheric pollution. Many of the heavy metals may be quite toxic in high concentrations. The effects of heavy metals on marine organisms have recently been reviewed by Bryan (1971), who noted that mercury and silver are the most toxic, followed by cadmium, zinc, lead, chromium, nickel, and cobalt. When these are released into the sea water they are, of course, greatly diluted. Some may precipitate out, others are absorbed on clay particles or phytoplankton. Although some animals absorb heavy metals from the water across certain body surfaces, such as the gills, food is probably a more important source than sea water for most. The flounder *Paralychthys*, for example, took up more ^{65}Zn from food than from the sea water (Hoss, 1964), and the oyster *Ostrea edulis* probably absorbs more zinc and cobalt from ingested particles than from solution (Preston and Jeffries, 1969). In some animals the metals tend to accumulate in certain tissues. For example, when *Carcinus maenas* was exposed for 32 days to different levels of zinc, the highest concentration was found in the hepatopancreas and gills (Bryan, 1966). In contrast, concentrations in the blood only doubled even when there was a 500-fold change in the zinc concentration in the water (Fig. 7.5). Fiddler crabs, *Uca pugilator*, maintained in sea water containing 0.18 ppm mercury for 28 days, concentrated mercury primarily in the gills, green glands, and hepatopancreas (Vernberg and Vernberg, 1972). Concentrations of mercury were highest in gill tissue, reaching 14 ppm. The mercury caused extensive alteration of the ultrastructure of the gill filaments. Normally the filaments are characterized by tightly packed, interdigitating epithelial cells. The basal plasma membranes are thrown into folds that penetrate the cell almost to their apical surfaces. The cells contain numerous mitochondria localized within the folds (Fig. 7.6a). In gill tissue from crabs maintained in the sublethal concentration of mercury, the filaments showed loss of cytoplasmic protein, disappearance of membrane folds, and decreased number of mitochondria (Fig. 7.6b). Swelling and loss of cristae of mitochondria were also observed (Watabe and Vernberg, 1971). In some animals, however, the accumulation of heavy metals is nearly proportional to the concentration in sea water. For example, the accumulation of lead in the tissue of *Ostrea virginica* was shown to be nearly proportional to the concentration in the sea water (Pringle et al., 1968). Concentrations of zinc, copper, and lead in seaweed also reflect the concentrations in the sea water.

One of the difficulties in setting standards for the release of heavy metals is that the effects differ from species to species and also for different stages in the life cycle of any one particular species. Thus, larvae of the barnacle *Eliminus modestus* can withstand concentrations of copper as high as 10 ppm whereas the copepod *Acartia clausi* dies at concentrations as low as 3 ppm. Adult fiddler crabs, *U. pugilator*, can survive in a concentration of 0.18 ppm Hg for at least

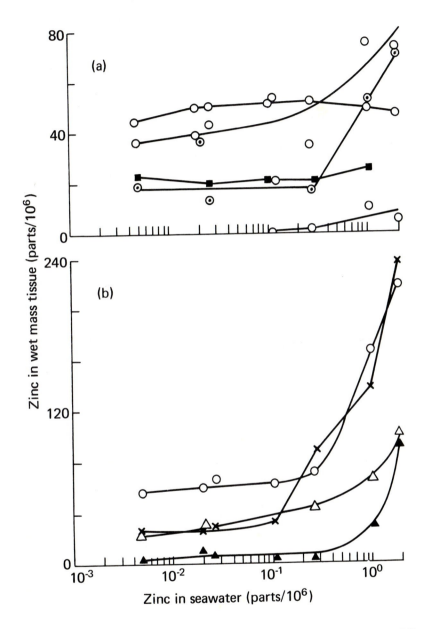

Fig. 7-5. Zinc in the tissues of the crab, *Carcinus maenas*, after exposure to different concentrations in seawater for 32 days at about 13°C. Points are mean values for at least 2 crabs. (a) Concentrations in: ◑, blood; ⊙, excretory organs; ●, muscle; ■, male gonad; ◓, urine. (b) Concentrations in: ▲, shell; x, gills; o, hepatopancreas; △, whole crab. (From Bryan, 1971. Council of the Royal Society.)

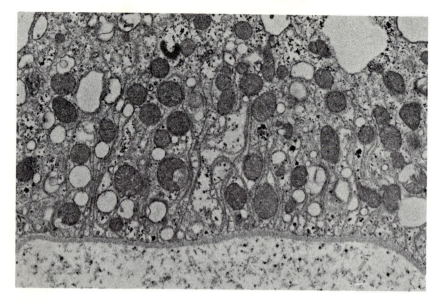

(a)

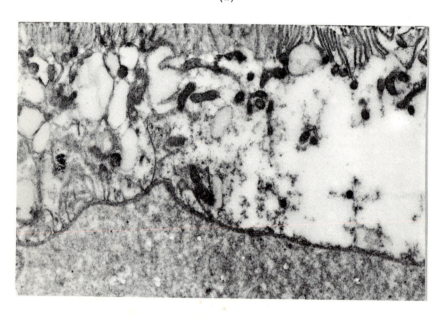

(b)

Fig. 7-6. Ultrastructure of gill tissue of fiddler crabs, *Uca pugilator*, maintained in 0.18 ppm Hg for six weeks. Analysis of the tissues showed a concentration of approximately 17 ppm. (a) Normal gill filament showing interdigitating epithelial cells situated on the basal lamina. Mitochondria are localized within the folds of plasma membrane (x 6,500). (b) Gill filament of experimental animals; note the loss of cytoplasmic protein. Membrane folds are not evident and mitochondria are scarce (x 7,000). (Watabe and Vernberg, 1972.)

six weeks, but first stage zoeae of this same species die within 48 hours in sea water of the same mercury concentration (DeCoursey and Vernberg, 1972). Environmental factors can also influence the lethality of heavy metals. At optimum conditions of salinity and temperature (30 0/00, 25°C) fiddler crabs live almost indefinitely in sea water containing sublethal concentrations of mercury (0.18 ppm). They can also survive prolonged periods of time in low salinity water and high temperature (5 0/00, 35°C), but under the latter conditions, the addition of sublethal concentrations of mercury resulted in an LD_{50} at 26 days for the females and 17 days for the males (Fig. 7.7) (Vernberg and Vernberg, 1972). Another example of larval sensitivity can be seen in studies of the development of the fertilized eggs of *Lytechinus pictus*. Timourian (1968) found that if embryos were exposed to 0.06 ppm zinc from the time of fertilization to formation of the larvae, 25% of the larvae were structurally abnormal and their development was slowed. Even at concentrations that might be normally encountered in coastal waters, more abnormal larvae were found than in controls from untreated sea water. Larvae of the fiddler crab *U. pugilator* showed behavioral responses in the form of decreased swimming rates when exposed to extremely low concentrations of mercury (DeCoursey and Vernberg, 1972). The modification of the swimming pattern of a first stage zoea in 0.18 ppm mercury is shown in Fig. 7.8. Metabolic rates of these larvae were also found to be decreased by exposure to very low concentrations of mercury.

4. Pesticides

Pesticides are another potential source of pollution to marine and brackish water species since diffused pesticides from drainage systems accumulate in estuaries (Eisler, 1968). The chlorinated hydrocarbons are a particularly serious threat because they persist for very long times and are characteristically nonbiodegradable. DDT is probably the most well-known representative of this pesticide group, although a host of others are equally, if not more, toxic. One basic pesticide problem is the accumulation of persistent varieties in the food chain. This has been perhaps best documented by Woodwell et al. (1967) who analyzed DDT residues in animals in an east coast estuary, and found an increased concentration of DDT residues in each higher trophic level of a food web. For example, concentrations of DDT residues were found to be higher in a carnivore than in a planktonic form. In water having 0.0005 ppm of DDT residues, the plankton had 0.04 ppm, the hard clam *Mercenaria mercenaria* had 0.42 ppm, but the herring gull *Larus argentatus* at the top of the trophic chain had 18.5 ppm. This process is termed *biological magnification*.

Another problem is created by the increased toxicity resulting from interaction of temperature with pesticides. Studies on the effects of chlorinated hydrocarbons in combination with thermal stress in the fish *Fundulus* showed that mortality was consistently less at the highest temperature tested, 30°C (Eisler, 1970a). Similar results have been recorded by Ogilvie and Anderson (1965),

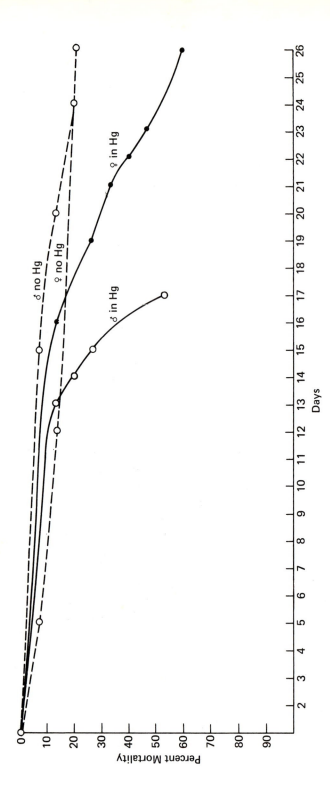

Fig. 7-7. Survival of male and female *Uca pugilator* maintained at 35°C in 5 0/00 sea water with and without 0.18 ppm Hg. (From Vernberg and Vernberg, 1972.)

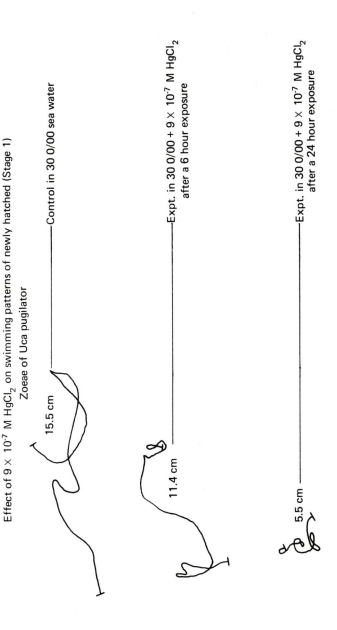

Fig. 7-8. Effect of 9×10^{-7} m $HgCl_2$ (0.18 ppm Hg) on swimming patterns of newly hatched (Stage 1) zoeae of *Uca pugilator*. Adapted from DeCoursey and Vernberg, 1972.)

who observed that survival of fishes subjected to concentrations of insecticides and high temperature survive better at high temperatures than low. They also found that salmon immersed in high concentrations of DDT for 24 hours at 17°C selected the higher temperature in a choice situation, and these results suggested that DDT may interfere with the normal thermal acclimation mechanisms.

The effects of organophosphorous compounds in combination with thermal stress are just the opposite to those of the chlorinated hydrocarbons, for survival is increased with decreasing temperature. There also seems to be a wide range of relative toxicity of the two types of pesticides in marine organisms: teleosts are less resistant to chlorinated hydrocarbons than molluscs, and about equal in sensitivity to decapod crustacea. Crustaceans, however, are highly susceptible to organophosphorus compounds; molluscs relatively resistant; and teleosts are intermediate between these two groups (Eisler, 1970b).

B. MANIPULATION OF THE ENVIRONMENT

The dumping of pollutants is not the only activity that endangers the marine environment. It is well known that erosion of a stream bank or changes in the shoreline can eventually alter the surface of the earth, but this knowledge has largely been ignored. The filling of marshes, cutting of new river channels, and building of new canals often seem to be done with complete disregard of possible consequences. That is not to say that all such changes per se are adverse, but it cannot be denied their ecological consequences may be far reaching. The ecological changes that have occurred since the Suez Canal was opened in 1869 offer an excellent case in point (Thorson, 1967). This canal connects the tropical Red Sea with the subtropical eastern Mediterranean, and when it was opened in 1869, the water level at the Red Sea entrance was 26–40 cm higher than the water level at the entrance to the Mediterranean. The current flowed toward the Mediterranean except during August, September, and October when the Nile, at its highest overflow, caused the current to run in the opposite direction. During this period of time the canal water mass was stagnant.

The biological and chemical composition of the Red Sea and the Mediterranean was very different when the canal was first opened. At that time, the fauna in the tropical Red Sea was very diverse and the salinity was 42–43 0/00 at the entrance of the canal, while the diversity of the fauna in the subtropical Mediterranean was limited and the salinity of the canal and the Mediterranean was 32–39 0/00 except at the highest Nile flood, when the salinity was 26 0/00. In addition, the Great Bitter Lake, which lies in the area between the Mediterranean and the Red Sea, had salinities of 76–80 0/00. The salt layer in the Great Bitter Lake consisted of 970 million tons and was 13.4 meters thick. With such great salinity differences, it is not surprising that there was very little faunal mixing. In the first 60 years after the opening up of the canal, Steinitz (as quoted in Thorson, 1967) recorded only 15 valid species that had passed from the Red Sea to the Mediterranean and none had passed from the Mediterranean to the Red Sea. During the next 40 years, however, conditions changed radically. The salinity at the entrance to the Red Sea is still 43–46 0/00, but the salt layer of the Great Bitter Lake has dissolved and the salinity is now 43–44 0/00, nearly half the level when the canal first opened. The Aswan Dam has reduced the Nile flood so that the salinity of the Mediterranean entrance is now about 40 0/00, thus effectively removing the salinity barrier to the migration of animals through the canal. The canal has also been deepened over the years so that the speed of the current has increased. Under all these conditions, during the past 40 years 8 times more species of animals have gone through the canal and become established in the Mediterranean than during the first 60 years, and it is anticipated that there will be a steadily increasing number of Red Sea animals migrating into the Mediterranean in the coming years. Eventually, the whole fauna and the food webs in the Mediterranean will change, and in fact, a number of migratory species of fish, shrimp, and pearl oysters are already being commercially exploited in the east Mediterranean.

These changes resulting from the construction of the Suez Canal suggest that there should be questions raised about the proposed sea-level canal in Central America, and how it will alter the ecology of the areas involved. The waters of the Pacific are much colder than those of the Atlantic, salinity differs, and marine populations are quite dissimilar. Water level differences between the Caribbean and Pacific side also will be great. The Pacific side will average about 6 feet higher and could be as much as 18 feet higher since the tides are out of phase. Therefore, it is unknown what effects a new canal will have on the final composition of oceanic fauna or on the climate in this area of the world (Cole, 1968).

The dredging and filling of marshes along coastal areas is another example of man's manipulation of the environment. More often than not, it has been done without any thought of possible ecological consequences, and thousands of acres of these marshes have been destroyed. Since these areas serve as nursery grounds for many commercially important marine species, continued destruction could eventually affect significant sections of the fishing industry. These estuarine marshes are high in productivity and could be used for the efficient expansion of potential food resources. Aquiculture techniques are only now beginning to emerge, but as Longwell and Stiles (1970) have pointed out, it was only when modern resource management and selective breeding techniques were combined that highly useful results were achieved in terrestrial agriculture. Development of successful aquiculture techniques will depend in large part on understanding the basic physiological adaptations of potential food species.

C. SUMMARY

All of these environmental problems involve principles of adaptive physiology. The fact that fragmentary information about the total picture of ecological-physiological interaction is available has been painfully highlighted. We are at present unable to answer many vital questions, such as why does an organism live where it does, will it survive thermal addition to the sea, what levels of pollutants are permissible, and how can aquiculture be best developed.

Now that many of the questions of physiological adaptation have been formulated, at least at a new level of ignorance, it is up to the students of today to answer these and raise more significant ones for their successors.

REFERENCES

Atomic Energy News Releases (1971). **2**:1–3. U.S.A.E.C., Washington, D.C.

Angelovic, J. W., J. C. White, Jr., and E. M. Davis (1969). Interactions of ionizing radiation, salinity, and temperature on the estuarine fish, *Fundulus heteroclitus. In* Symposium on Radioecology. D. J. Nelson and F. C. Evans, eds. USAEC Conf-670503. Oak Ridge, Tenn. Pp. 131–141.

Baptist, J. P., and C. W. Lewis (1969). Transfer of [65]Zn and ^{51}Cr through an estuarine food chain. *In* Symp. on Radioecology. D. J. Nelson and F. C. Evans, eds. USAEC Conf-670503, Oak Ridge, Tenn. Pp. 420–430.

Boroughs, H., W. A. Chipman, and T. R. Rice (1957). Laboratory experiments on the uptake, accumulation, and loss of radionuclides by marine organisms. *In* The Effects of Atomic Radiation on Oceanography and Fisheries. National Research Council, Publ. No. 551. National Academy of Sciences, Washington, D.C. Pp. 80–87.

Brown, V. M. (1962). The accumulation of strontium-90 and yttrium-90 from a continuously flowing natural water by eggs of alevins of the Atlantic salmon and sea trout. Rep. U.K. Atom. Energy Auth. (PG–288): 16p.

——— and W. L. Templeton (1964). Resistance of fish embryos to chronic irradiation. *Nature,* **203**:1257–1259.

Bryan, G. W. (1966). Concentrations of zinc and copper in the tissues of decapod crustaceans. *J. mar. biol. Ass. U.K.,* **48**:303–321.

——— (1971). The effects of heavy metals (other than mercury) on marine and estuarine organisms. *Proc. Roy. Soc. London, B,* **177**:389–410.

Cole, L. C. (1968). Can the world be saved? *BioScience,* **18**:679–684.

DeCoursey, P. J., and W. B. Vernberg (1972). Effects of mercury on behavior and metabolism of larval *Uca pugilator*. In press.

Eisler, R. (1968). Pesticides in the marine environment. *Underwater Naturalist,* **5**:11–13.

——— (1970a). Factors Affecting Pesticide-Induced Toxicity in an Estuarine Fish. U.S. Bur. Sport. Fish. Wildl., Tech. Paper, 45. 20 pp.

——— (1970b). Acute Toxicities of Organochlorine and Organophosphorous Insecticides to Estuarine Fish. U.S. Bur. Sport. Fish. Wildl., Tech. Paper, 46. 11 pp.

Engel, D. W. (1967). Effect of single and continuous exposures of gamma radiation on the survival and growth of the blue crab, *Callinectes sapidus. Radiat. Res.* **32**:685–691.

——— and D. J. Fluke (1962). The effect of water content and post-irradiation storage on radiation sensitivity of brine shrimp cysts (eggs). *Radiat. Res.,* **16**: 173–181.

Hoss, D. E. (1964). Accumulation of zinc-65 by flounder of the genus *Paralichthys*. *Trans. Am. Fish. Soc.,* **93**:364–368.

Kinne, O., and H. Rosenthal (1967). Effects of sulfuric water pollutants on fertilization, embryonic development and larvae of the herring, *Clupea harengus. Mar. Biol.,* **1**:65–83.

Longwell, A. C., and S. S. Stiles (1970). The genetic system and breeding potential of the commercial American oyster. *Endeavour.,* **29**:94–99.

Mihursky, J. A. (1967). On possible constructive uses of thermal additions to estuaries. *BioScience,* **17**:698–702.

Ogilvie, D. M., and J. M. Anderson (1965). Effect of DDT on temperature selection of young Atlantic salmon, *Salmo salar. J. Fish. Res. Bd. Can.,* **22**:503–512.

Polikarpov, G. G. (1966). Radioecology of Aquatic Organisms. Reinhold, New York. 314 pp.

Preston, A., and D. Jeffries (1969). Aquatic aspects in chronic and acute contamination situations. *In* Environmental Contamination by Radioactive Materials. International Atomic Energy Agency, Vienna. Pp. 183–211.

Pringle, B. H., D. E. Hissong, E. L. Katz, and S. T. Mulawka (1968). Trace metal accumulation of estuarine mollusks. *J. sanit. Engng. Div. Am. Soc. civ. Engrs.,* **94**:455–475.

Rice, T. R., and T. W. Duke (1969). Radioactivity in the sea: general classification of radionuclides. *In* Encyclopedia of marine resources. F. E. Firth, ed. Van Nostrand Reinhold Co., New York. Pp. 566–569.

———, and J. P. Baptist (1970). Ecological aspects of radioactivity in the marine environment. *In* Environmental Radioactivity Symposium. J. C. Clopton, ed. Dept. Geog. and Envir. Engineer., The Johns Hopkins University, Baltimore. Pp. 107–180.

Seaborg, G. T. (1970). The environment: A global problem—an international challenge. *In* Environmental Aspects of Nuclear Power Stations. Preprint from International Atomic Energy Agency Symposium, New York. IAEA-SM-146/ opening address. 8 pp.

Singer, S. F. (1968). Waste heat management. *Science,* **159**:1184.

Templeton, W. L. (1966). Resistance of fish eggs to acute and chronic irradiation. *In* Disposal of radioactive wastes into seas, oceans, and surface waters. Vienna, I.A.E.A., pp. 847–859.

———, R. E. Nakatoni, and E. Held (1970). Effects of radiation in the marine ecosystem. FAO Tech. Conf. Mar. Poll. and its effects on living resources and fishing. Rome, Italy, 19 p.

Thompson, T. J. (1970). Role of nuclear power in the United States. *In* Environmental Aspects of Nuclear Power Stations. Preprint from International Atomic Energy Agency Symposium, IAEA-SM-146. 42 pp.

Thorson, G. (1967). Recent migration and larval transport via the Suez Canal (Preprint Abstr.). Conference on Marine Invertebrate Larvae with Emphasis on Rearing. Duke Univeresity Marine Lab., Beaufort, N.C.

Timourian, H. (1968). The effect of zinc on sea-urchin morphogenesis. *J. exp. Zool.,* **169**:121–132.

Vernberg, W. B., and F. J. Vernberg (1972). Effects of mercury on survival and metabolism of adult fiddler crabs, *Uca pugilator,* under optimum and stressful environmental conditions. Fishery Bulletin (In Press).

Watabe, N., and W. B. Vernberg (1972). Electron microscopic studies on tissues of fiddler crabs, *Uca pugilator,* maintained in sublethal concentrations of mercury. (In manuscript.)

White, J. C., Jr., J. W. Angelovic, D. W. Engel, and E. M. Davis (1967). Interactions of radiation, salinity, and temperature on estuarine organisms. *In* Annual Report of the Bureau of Commercial Fisheries, Radiobiological Laboratory, Beaufort, N.C., for fiscal year ending June 30, 1966. U.S. Fish, Wildl. Serv., Circ. 270. Pp. 29–35.

Woodwell, G. M., C. F. Nurster, Jr., and P. A. Isaaccson (1967). DDT residues in an east coast estuary: A case of biological concentration of a persistent insecticide. *Science,* **156:**821–824.

Subject Index

Species Index

343